INTERNATIONAL SERIES OF MONOGRAPHS IN

NATURAL PHILOSOPHY

GENERAL EDITOR: D. TER HAAR

VOLUME 22

FOUNDATIONS OF STATISTICAL MECHANICS

FOUNDATIONS OF STATISTICAL MECHANICS

A deductive treatment

BY

O. PENROSE

PROFESSOR OF MATHEMATICS

THE OPEN UNIVERSITY, LONDON

THE QUEEN'S AWARD TO INDUSTRY 1966

PERGAMON PRESS

OXFORD · LONDON · EDINBURGH · NEW YORK

TORONTO · SYDNEY · PARIS · BRAUNSCHWEIG

Pergamon Press Ltd., Headington Hill Hall, Oxford
4 & 5 Fitzroy Square, London W.1

Pergamon Press (Scotland) Ltd., 2 & 3 Teviot Place, Edinburgh 1

Pergamon Press Inc., Maxwell House, Fairview Park, Elmsford, New York 10523

Pergamon of Canada Ltd., 207 Queen's Quay West, Toronto 1

Pergamon Press (Aust.) Pty. Ltd., 19a Boundary Street, Rushcutters Bay,
N. S. W. 2011, Australia

Pergamon Press S.A.R.L., 24 rue des Écoles, Paris 5e

Vieweg & Sohn GmbH, Burgplatz 1, Braunschweig

First edition 1970

Library of Congress Catalog Card No. 70-89513

Printed in Germany

08 013314 2

Contents

Contents

Preface

THE thesis of this book is that statistical mechanics can be built up deductively from a small number of well-defined physical assumptions. To provide a firm basis for the deductive structure, these assumptions have been converted into a system of postulates describing an idealized model of real physical systems. These postulates, which are listed immediately after this preface, thus play a role in the theory similar to the role of the first and second laws in thermodynamics.

Of these five postulates, the crucial one is the fourth, expressing the assumption that the successive observational states of a physical system form a Markov chain. This is a strong assumption, whose influence is felt throughout the book. It is possible, indeed, that this postulate is too strong to be satisfied exactly by any real physical system; but even so, it has been adopted here because it provides the simplest precise formulation of a hypothesis that appears to underlie all applications of probability theory in physics. Our treatment may thus be regarded as a first approximation to the more elaborate theory that would be obtained if this postulate were replaced by a less idealized statement of the same basic hypothesis. Our main concern is not so much to find out whether a real system can exactly obey all the postulates—although we do discuss this difficult question in Chap. IV, § 7—but to show that all the fundamental results of statistical mechanics, both for equilibrium and non-equilibrium situations, can be derived from the postulates in a logical and unified way, avoiding the paradoxes and *ad hoc* assumptions that tend to appear in more informal treatments.

Traditionally, there have been two main approaches to the fundamental problems of statistical mechanics. One of these approaches attempts to base the theory on purely dynamical arguments, using ergodic theorems of general dynamics; the other is based on an assumption about the *a priori* probabilities of dynamical states. These two approaches are discussed in more detail at the end of Chapter I. The theory to be described here differs from the ergodic approach in that it gives explicit recognition to the limitations on one's powers of observation, and from the *a priori* probability approach in that its basic probability assumption (the Markovian postulate) refers only to observable events and is therefore, in principle, experimentally testable.

The first chapter deals with the main physical assumptions and their idealization in the form of postulates. In the next three chapters the consequences of these postulates are explored, culminating in a derivation, in Chapter IV, of the fundamental formulae for calculating probabilities in terms of dynamical quantities. Finally, two chapters are devoted to a careful analysis of the important notion of entropy, showing the links it provides between statistical mechanics and thermodynamics and also between statistical mechanics and communication theory. Since the book is concerned mainly with general principles rather than with particular cases, the only applications considered in detail are to the system with the simplest possible dynamics: the ideal classical gas, which is considered both in its equilibrium and its non-equilibrium aspects. More complicated systems can be treated by methods based on the same principles, but a considerable amount of further mathematical apparatus, starting with the theory of the canonical ensembles and the partition function, is needed. Such material can be found in many other places and has therefore been omitted here in order to keep down the length of the book. At the end of each section (with a few exceptions) some exercises are given to help when the book is used for teaching.

The book is intended for readers with a knowledge of basic physics and an interest in fundamental questions. It could be used for the initial stages of a course in statistical mechanics for students of theoretical physics in their final undergraduate or first postgraduate year. Parts of the book may also be of interest to probability theorists, statisticians, communication theorists, and philosophers. A knowledge of quantum mechanics is useful, but not essential; for the theory applies to both classical and quantum systems so that paragraphs and sections dealing with quantum systems can be skipped without impairing the reader's understanding of the rest. The last two chapters presume that the reader has made the acquaintance of the thermodynamic concept of entropy. No previous knowledge of probability theory or of statistical mechanics is assumed.

I am indebted to numerous colleagues, including D. Baldwin, J. S. R. Chisholm, D. R. Cox, J. S. N. Elvey, R. B. Griffiths, P. Johannesmaa, B. Kumar, J. Lebowitz, R. Penrose, I. C. Percival, E. Praestgaard, R. Purves, G. E. H. Reuter, D. W. Sciama, and G. Sewell for discussions which helped me to formulate the point of view of this book, and for helpful criticisms of various parts of the typescript. I am also indebted to the Physics Department at Ohio State University for the invitation to give a course of lectures (in 1957) which set in motion the project of writing this book, and to the U.S. Air Force Office of Scientific Research for financial support at Yeshiva University, New York, during a part of the time when I was working on it.

The Main Postulates of this Theory

1. *The dynamical description of matter* (Chap. I, § 2)

Macroscopic physical systems are composed of molecules obeying the laws of classical or quantum mechanics with a suitably chosen Hamiltonian.

2. *The observational description of matter* (Chap. I, § 3)

An observation of a macroscopic physical system may be idealized as an instantaneous simultaneous measurement of a particular set of dynamical variables called *indicators*, each of which takes the values 1 and 0 only. The instants at which these observations are possible are discrete and equally spaced.

3. *Postulate of compatibility* (Chap. I, § 3)

The disturbance to the system caused by observing it is negligible, in the sense that it has no effect on later observations.

4. *Markovian postulate* (Chap. I, § 5)

The successive observational states (sets of values for the indicators) of a macroscopic physical system constitute a Markov chain.

5. *Accessibility postulate* (Chap. IV, §§ 3 and 4)

There are no artificial restrictions on the dynamical states available to a system, apart from those implied in quantum mechanics by the Bose and Fermi symmetry conditions.

CHAPTER I

Basic Assumptions

1. Introduction

Statistical mechanics is the physical theory which connects the observable behaviour of large material objects with the dynamics of the invisibly small molecules constituting these objects. The foundations of this theory derive their fascination from the interplay of two apparently incompatible theoretical schemes for describing a physical object. One of these descriptions is the *observational*, *coarse-grained*, or *macroscopic* description, which confines itself to observable properties of the physical object, such as its shape, size, chemical composition, temperature, and density. The other is the *dynamical*, *fine-grained*, or *microscopic* description, which treats the physical object as a dynamical system of molecules, and therefore must include a complete description of the dynamical state of every molecule in the system. Both descriptions may be regarded as simplified models of a reality that is more complex than either. It is the task of statistical mechanics to find and exploit the relationship between the two schemes of description.

In the dynamical description a physical object is regarded as a dynamical system† made up of a large number of simple units which we shall call *molecules*, using the word to include not only the polyatomic molecules of chemistry, but also single atoms, ions, and even electrons. Each molecule moves under the influence of conservative forces exerted on it by the other molecules of the system and by bodies outside the system, particularly the container holding it. These forces are here assumed to propagate instantaneously; thus the theory is non-relativistic. Strictly, one should always use quantum mechanics in studying the motion of molecules; but since classical mechanics often provides a good approximation to the quantum results and is both conceptually and mathematically simpler, much practical statistical mechanics is done classically. In studying fundamentals, too, it is useful to consider the classical treatment alongside the quantum one. In this book, wherever there is a divergence between the classical and quantum treatments, the classical treatment will be given first and the quantum treatment immediately afterwards (unless it is omitted). In this way we can take full advantage of the analogies between classical and quantum mechanics.

One of the simplest systems considered in statistical mechanics is a system of N identical molecules. If each molecule has f degrees of freedom the system as a whole has fN, so that in classical mechanics its *dynamical state*

† Some authors write "assembly" for what is here called a system, and "system" for what is here called a molecule.

(or *microstate*) at any instant may be specified by giving the values of $2fN$ variables: for example, fN position coordinates $q_1, \ldots, q_{fN}$ and their time derivatives, the fN velocity coordinates $\dot{q}_1, \ldots, \dot{q}_{fN}$. The dynamical state can be usefully visualized as a point in a $2Nf$-dimensional space, in which the variables $q_1, \ldots, q_{fN}, \dot{q}_1, \ldots, \dot{q}_{fN}$ may be used as a coordinate system. This space may be called the *dynamical space* of the system. For example, if the molecules are monatomic, as they are in inert elements such as argon, f is 3, so that there are $3N$ position coordinates. These can be defined by making (q_1, q_2, q_3) the Cartesian components of the position of the first particle, (q_4, q_5, q_6) those of the second, and so on; then the velocity coordinates $\dot{q}_1, \ldots, \dot{q}_{3N}$ are related in the same way to the Cartesian components of velocity.

The dynamical state of any classical system evolves with time. It may be visualized as tracing out a curve in dynamical space, called a *trajectory*. This evolution is governed by the Newtonian equations of motion; if each molecule is a particle of mass m these equations are

$$m\frac{d^2q_i}{dt^2} = -\frac{\partial}{\partial q_i}U(q_1, \ldots, q_{3N}) \quad (i = 1, 2, \ldots, 3N), \tag{1.1}$$

where $U(q_1, \ldots, q_{3N})$ is the potential energy function. Since these differential equations are of the second order in the time, their joint solution is in principle fully determined by the values of the q_is and their first time derivatives at any chosen time. That is, if we could solve the differential equations and knew the dynamical state at any one time, we could calculate the dynamical state for all times. It follows that if two dynamical systems have the same laws of motion and are in the same dynamical state at some particular time t_0, then they must be in the same dynamical states at all times. This property is called the *determinism* (or *causality*) of classical dynamics. It is reflected in the geometry of phase space by the fact that just one trajectory passes through each point in dynamical space. The idea of representing our unpredictable world by a deterministic mechanical model goes back to Descartes and Laplace.

A macroscopic physical object contains so many molecules that no one can hope to find its dynamical state by observation. There is no insult to the skill of experimental physicists in the assertion that they could never observe the dynamical state of every molecule in, say, a glass of water containing over 10^{24} molecules. This limitation on our powers of observation is an essential part of statistical mechanics; without it the theory would be no more than a branch of ordinary mechanics. The simplest way to describe this limitation is to use an idealized model of observation based on the assumption that an elementary observation is an instantaneous act by means of which the observer can only distinguish between a limited number of possible *observational states* (also called *macrostates*) of the system he observes.

It will also be assumed that, at least in classical mechanics, the dynamical state of a system completely determines its observational state; that is, if two systems are in the same dynamical state they must be in the same observational state. On the other hand, because of our limited powers of observation, the observational state does not completely determine the dynamical state; that is, if two systems are in the same observational state they can be in different dynamical states. The set of all dynamical states compatible with an observational state A may be called the *dynamical image* of A.

To specify the details of the model of observation it is necessary to specify a dissection of the entire dynamical space into a set of such dynamical images. Physically, the choice of this dissection depends on what physical properties of the system are regarded as measurable, and with what accuracy. As an illustration, suppose that the length of a rod is measured to the nearest millimetre. If we define the dynamical variable $\lambda(q_1, \ldots, q_{fN})$ to be the true length of the rod—that is, the distance between the two most widely separated molecules in the rod—then the dynamical image regions will be the regions of dynamical space corresponding to the tolerance intervals $0 \leqq \lambda \leqq \frac{1}{2}$ mm, $\frac{1}{2}$ mm $< \lambda < 1\frac{1}{2}$ mm, $1\frac{1}{2}$ mm $\leqq \lambda \leqq 2\frac{1}{2}$ mm, etc. (exact half-integral true lengths being rounded off to the nearest even number). If instead we had measured lengths to the nearest centimetre, the dynamical image regions would have been $0 \leqq \lambda \leqq 5$ mm, 5 mm $< \lambda < 15$ mm, etc. A more complicated example is Boltzmann's description of a gas in terms of the occupation numbers of cells in the dynamical space of a single particle: this is considered in some detail in Chap. V, §§ 6 and 7. Fortunately, however, there is no need for us to discuss at length the physical considerations affecting the choice of observational states and their associated dynamical images since they have no effect whatever on the deductions to be made from it. All that matters is that whatever choice has been made must be used consistently.

The observational state of a system, like the dynamical state, changes with time, but unlike the dynamical state it need not change in a deterministic way. The observational state of a system at one time does not in general determine the observational state at any other time: two systems in the same observational state at some particular time t_0 can be in different observational states at times other than t_0. This is because the two systems can be in different dynamical states at time t_0, and if so their dynamical states at other times will also be different and may be observably different. For example, two women in the same observational state, both expecting babies, may be in different dynamical states, one having a boy foetus and one a girl; the difference between these two dynamical states is not observable when they go into the maternity hospital but it does lead later on to observable differences when one woman has a boy baby and the other a girl. In physical systems, a similar lack of determinism becomes important whenever molecular fluctuations become important; an example is the unpredictability of the Brownian motion of small colloid particles suspended in a liquid and observed with a microscope.

Although this lack of determinism makes it impossible, in general, to predict reliably the future observational states of a system given its present observational state, it may still be possible to make reliable *statistical* predictions about a large population or ensemble of systems in the same observational state. To continue the obstetrical example, we cannot predict the sex of any particular expected baby, but we can predict fairly accurately the fraction of boy babies to be born during the next year in a large maternity hospital. Likewise, in the Brownian motion example, we cannot predict the position of any particular colloid particle 10 min ahead, but we can predict fairly accurately what fraction of the colloid particles will be in a particular region of the fluid (say the northern half) in 10 min time. The theory used to study statistical regularities like these is the theory of probability, and we shall find that probability theory performs for observational states the same service that analytical dynamics performs for dynamical states: it provides a mathematical framework for describing their laws of change.

This distinction between the determinism of dynamical states and the indeterminism of observational states for a classical system can be carried over without major changes to quantum systems, despite the well-known indeterminacy of the quantum description of matter. The dynamical state of a quantum system with fN degrees of freedom may be defined to be its Schrödinger wave function $\psi(q_1, \ldots, q_{fN})$ at that instant. As in classical mechanics, the dynamical state may be visualized as a point in a *dynamical space* whose points are all the possible dynamical states of the system; this dynamical space is called the *Hilbert space* of the system and, unlike classical dynamical space, has an infinite number of dimensions. The law of change for quantum dynamical states is *Schrödinger's wave equation*

$$i\hbar \frac{\partial \psi}{\partial t} = \mathsf{H}\psi, \tag{1.2}$$

where H is the *Hamiltonian operator*† and $\hbar$ is $1/2\pi$ times Planck's constant. For a system of N point particles this operator is

$$\mathsf{H} = -\frac{\hbar^2}{2m} \sum_{i=1}^{3N} \frac{\partial^2}{\partial q_i^2} + U(q_1, \ldots, q_{3N}). \tag{1.3}$$

Since the partial differential eqn. (1.2) is of first order in the time, its solution is in principle fully determined by the form of the wave function ψ $(q_1, \ldots,$

† We use *sans serif* type for the linear operators of quantum mechanics.

$q_{3N}; t)$ as a function of the variables $q_1, \ldots, q_{3N}$ at any chosen time t_0. That is, if two systems have the same dynamical state at one particular time t_0, then (1.2) implies that they have the same dynamical state at all times. In this purely dynamical sense, quantum mechanics is just as deterministic as classical mechanics. Likewise, the observational states of a quantum system are indeterministic and are statistically but not individually predictable from previous observational states. Thus the basic principles of quantum statistical mechanics do not differ radically from those of classical statistical mechanics.

This similarity of quantum and classical statistical mechanics may seem surprising in view of the important conceptual differences between ordinary quantum and classical mechanics. However, these differences are all connected with the act of observation, tending to make the results of observation less predictable in quantum than in classical mechanics. In statistical mechanics the differences are minimized because here even classical observations are unpredictable (i.e. non-deterministic).

In order to have a sharp practical distinction between the dynamical and the observational descriptions of a physical system, it is necessary for the system to contain a very large number of molecules, although its spatial extent need not be large. For example, the solar system, although very large in spatial extent, is successfully treated in celestial mechanics as a system of only a few particles (10 if we consider only major planets and the sun). Observation can determine the positions and velocities of all these celestial "particles", and the recorded past observations are accurate enough to yield predictions good for many years ahead. In ordinary mechanics, therefore, because the systems considered always comprise only a few particles, we can discover the dynamical state of the system by observing its constituent particles individually. In statistical mechanics, on the other hand, the system always comprises a very large number of particles or molecules: a sewing needle, for example, contains some 10^{19} particles and therefore counts as "large" in statistical mechanics, even though it would seem infuriatingly small if lost in the proverbial haystack. When the number of particles in the system is as large as this, any attempt to discover its dynamical state by observing the particles individually would clearly be quite out of the question.

The mathematical device for obtaining simplifying approximations valid when some quantity is very large or very small is to use a limit process. In statistical mechanics this device takes the form of a limit process where N, the number of molecules in the system, tends to infinity.

Here we shall not take it as a postulate that N must tend to infinity, but it seems likely (see Chap. IV, § 7) that the postulates we do adopt can be satisfied exactly only in this limit. When the limit is used, the volume enclosed by the container holding the system may also be made indefinitely large so as to keep the number density N/V finite; in this case the limit $N \to \infty$ is called the *bulk limit*. The bulk limit is important in establishing the connection between statistical mechanics and thermodynamics (see Chap. V, § 4) since thermodynamic quantities such as temperature can be defined unambiguously for a dynamical system only in this limit. Moreover, since the walls of the container recede to infinity in the bulk limit, this operation eliminates some of the errors due to our oversimplified dynamical representation of the walls of the container (which would be interpreted as "surface effects" in thermodynamics).

Because of the arbitrariness in the choice of observational states in our model of observation it is sometimes convenient to consider a second limit process in which the experimental tolerance (error of observation) tends to zero. The sequence defining this limit may be thought of as a sequence of successive refinements of experimental technique, tending towards perfection. If this limit process is combined with the bulk limit $N \to \infty$ it is important to carry out the limiting processes in the right order. The well-known example from elementary analysis,

$$\lim_{x\to 0} \lim_{y\to 0} \frac{x-y}{x+y} = +1 \quad \text{but} \quad \lim_{y\to 0} \lim_{x\to 0} \frac{x-y}{x+y} = -1,$$

illustrates how the value of an expression involving two limit processes may be altered if their order is reversed. If we allow the experimental accuracy to approach perfection before we take $N \to \infty$, we get a theory of systems whose dynamical state can be observed perfectly, applicable as an approximation in the solar system example considered above, but quite distinct from statistical mechanics. Subsequently taking the bulk limit $N \to \infty$ will not basically alter this situation. The correct procedure is therefore to take the bulk limit first; the theory so obtained will be the statistical mechanics of very large, imperfectly observable systems. Subsequently taking a secondary limit, in which the width of the experimental tolerance intervals tends to zero, will not basically alter this fact.

It is particularly important to treat these limiting processes in the right order in quantum statistical mechanics. Let us represent the experimental tolerance in the measurement of energy by the symbol ΔE, and the separation of neighbouring energy levels by δE. Thus the number of energy levels compatible with any given observational state is roughly $\Delta E/\delta E$. To see whether this quantity is large or small we can use the two-limit processes just discussed. In the limit of perfect accuracy, the experimental tolerance ΔE vanishes, and in the bulk limit the energy levels become very dense and so δE vanishes.

For statistical mechanics, the bulk limit must be taken first, giving the estimate

$$\frac{\delta E}{\Delta E} \approx \lim_{\Delta E\to 0} \left[\lim_{\delta E\to 0} \frac{\delta E}{\Delta E} \right] = \lim_{\Delta E\to 0} [0] = 0, \tag{1.4}$$

so that $\Delta E/\delta E$, the number of energy levels per observational state, is extremely large in statistical mechanics. On the other hand, in ordinary mechanics the limit of perfect accuracy must be taken first, giving the estimate

$$\frac{\Delta E}{\delta E} \approx \lim_{\delta E \to 0} \left[\lim_{\Delta E \to 0} \frac{\Delta E}{\delta E} \right] = \lim_{\delta E \to 0} [0] = 0, \tag{1.5}$$

so that ΔE is much smaller than δE in ordinary mechanics. An observation to discover which energy level the system is in is therefore compatible with ordinary mechanics but incompatible with statistical mechanics. The comparison between the two theories, when the double limit is used, can also be written

$$\left.\begin{aligned} \delta E \ll \Delta E \ll 1 \quad & \text{(statistical mechanics)}, \\ \Delta E \ll \delta E \ll 1 \quad & \text{(ordinary mechanics)}. \end{aligned}\right\} \tag{1.6}$$

Although ΔE and δE are both vanishingly small their relative size is very different in the two types of mechanics.

In the remainder of this chapter we shall first formulate in more detail the dual description of a physical system by means of dynamical states changing with time according to a deterministic law and observational states changing according to a statistical law; afterwards, in § 5, we shall formulate our basic postulate about this statistical law. This postulate is a strong one with far-reaching consequences, and from it, together with the postulates specifying the dynamical and observational descriptions, and a postulate of a more technical nature described in Chap. IV, §§ 3 and 4, all the basic results of statistical mechanics can be deduced.

2. Dynamics

The purpose of this section is to show how a physical object can be represented approximately by a closed dynamical system of molecules. The model will be set up both in classical mechanics and in quantum mechanics, and its basic limitations will be discussed.

The Newtonian form (1.1) for the laws of classical mechanics has the disadvantage that it does not generalize easily to molecules with rotational degrees of freedom such as solid spheres; and it also obscures the correspondence with the corresponding quantum system. These disadvantages can be overcome by working instead with Hamilton's formulation of dynamics. The essence of Hamilton's method is to replace the velocity coordinates $\dot{q}_1, \dot{q}_2, \ldots$, used in describing a dynamical system, by a new set of momentum coordinates $p_1, p_2, \ldots$. If there are no magnetic forces, these coordinates are defined, for any system of F degrees of freedom, by

$$p_i \equiv \frac{\partial}{\partial \dot{q}_i} K(q_1, \ldots, q_F, \dot{q}_1, \ldots, \dot{q}_F) \quad (i = 1, 2, \ldots, F), \tag{2.1}$$

where K is the kinetic energy expressed as a function of the position and velocity coordinates. The *canonical coordinates* $p_1, \ldots, p_F, q_1, \ldots, q_F$ provide a new coordinate system in the $2F$-dimensional dynamical space; when this coordinate system is used, the dynamical space is called *phase space*. The simplest application of (2.1) is to a system of N point particles, each of mass m. Here the kinetic energy function $K(q_1, \ldots, \dot{q}_{3N})$ is $\sum_i \frac{1}{2}m\dot{q}_i^2$; consequently the canonical momenta are given by

$$p_i = \frac{\partial}{\partial \dot{q}_i} \sum_{j=1}^{3N} \frac{1}{2} m\dot{q}_j^2 = m\dot{q}_i \quad (i = 1, 2, \ldots, 3N) \tag{2.2}$$

so that in this case the momentum coordinate system in phase space differs from the velocity coordinate system only by a change of scale. For molecules with rotational degrees of freedom, however, the difference is more pronounced: see eqn. (2.10).

Hamilton† showed that the equations of motion for the canonical coordinates $p_1, \ldots, q_F$ are the system of $2F$ first-order differential equations

$$\left.\begin{aligned} \frac{dq_i}{dt} &= \frac{\partial H(p_1, \ldots, q_F)}{\partial p_i} \\ \text{and} \qquad \frac{dp_i}{dt} &= -\frac{\partial H(p_1, \ldots, q_F)}{\partial q_i} \end{aligned}\right\} \quad (i = 1, \ldots, F), \tag{2.3}$$

where $H(p_1, \ldots, q_F)$ is the *Hamiltonian function* which may be defined as the total energy of the system written as a function of the $2F$ canonical coordinates. The neat form of these equations makes canonical coordinates particularly useful in studying the geometry of phase space (Liouville's theorem: see Chap. III, § 3). For each value of i, the first equation of (2.3) recapitulates the kinematical relationship (2.1), and the second is a generalization of the corresponding Newtonian equation in (1.1). As an example let us consider once again a system of N interacting particles of mass m. The Hamiltonian function is obtained by writing the energy as a function of the canonical coordinates $p_1, \ldots, p_{3N}, q_1, \ldots, q_{3N}$ with the help of (2.2); this gives

$$H(p_1, \ldots, p_{3N}, q_1, \ldots, q_{3N}) = \sum_{i=1}^{3N} \frac{1}{2m} p_i^2 + U(q_1, \ldots, q_{3N}), \tag{2.4}$$

where U is the potential energy, which depends only on the *configuration* of the system (the set of all position coordinates). The Hamiltonian equations of motion (2.3) are now

$$\frac{dp_i}{dt} = -\frac{\partial U}{\partial q_i} \quad \text{and} \quad \frac{dq_i}{dt} = \frac{p_i}{m} \quad (i = 1, 2, \ldots, 3N). \tag{2.5}$$

By eliminating the p_is we can recover the Newtonian equations (1.1).

† See H. Goldstein, *Classical Mechanics* (Addison-Wesley, London, 1959), Chap. 7.

Since the dynamical systems considered in statistical mechanics are composed of molecules, we can divide their degrees of freedom into disjoint sets, one set for each molecule. We shall use the symbol $\boldsymbol{q}_i$ to represent the set of all the position coordinates belonging to the ith molecule; if this molecule has f_i degrees of freedom, then $\boldsymbol{q}_i$ is an f_i-dimensional vector. Likewise, we shall represent the momentum coordinates belonging to this molecule by an f_i-dimensional vector $\boldsymbol{p}_i$. Thus, for point particles, $\boldsymbol{q}_1$ stands for the set q_1, q_2, q_3; $\boldsymbol{p}_2$ for the set p_4, p_5, p_6; and so on. With the coordinates grouped in this way, the Hamiltonian of the system takes a particularly simple form, reflecting the fact that a system of N molecules is not the most general system with $F \equiv \sum f_i$ degrees of freedom. The simplest of all is the Hamiltonian for a system of non-interacting identical molecules, each having f degrees of freedom. This Hamiltonian is a sum of N terms, one for each molecule:

$$H(p_1, \ldots, q_F) = H(\boldsymbol{p}_1, \ldots, \boldsymbol{q}_N) = \sum_{i=1}^{N} H_{(1)}(\boldsymbol{p}_i, \boldsymbol{q}_i), \tag{2.6}$$

where $H_{(1)}(\boldsymbol{p}, \boldsymbol{q})$ is a function of $2f$ variables, called the *single-molecule Hamiltonian*. The equations of motion (2.3) now reduce to

$$\left.\begin{aligned} \frac{d\boldsymbol{q}_i}{dt} &= \frac{\partial H_{(1)}(\boldsymbol{p}_i, \boldsymbol{q}_i)}{\partial \boldsymbol{p}_i} \\ \frac{d\boldsymbol{p}_i}{dt} &= -\frac{\partial H_{(1)}(\boldsymbol{p}_i, \boldsymbol{q}_i)}{\partial \boldsymbol{q}_i} \end{aligned}\right\} \quad (i = 1, 2, \ldots, N), \tag{2.7}$$

where $d\boldsymbol{q}_1/dt$ stands for the f-component vector $(dq_1/dt, \ldots, dq_f/dt)$, $\partial H_{(1)}/\partial \boldsymbol{p}_1$ stands for $(\partial H_{(1)}/\partial p_1, \ldots, \partial H_{(1)}/\partial p_f)$, and so on. Since the right side of (2.7) depends only on the dynamical state $(\boldsymbol{p}_i, \boldsymbol{q}_i)$ of the ith molecule, the Hamiltonian (2.6) implies that the molecules move completely independently of each other. Hamiltonians of the form (2.6) are used to represent gases at densities low enough for the mutual interaction of the molecules to be negligible.

The single-molecule Hamiltonian is simplest when the molecules are particles. In this case f is 3, and by setting $U = 0$ in (2.4) and comparing with (2.6), we see that the appropriate single-molecule Hamiltonian is given by

$$H_{(1)}(\boldsymbol{p}_1, \boldsymbol{q}_1) = \frac{1}{2m}(p_1^2 + p_2^2 + p_3^2). \tag{2.8}$$

This single-molecule Hamiltonian is used for monatomic molecules such as helium, neon, and argon, and also for electrons and for monatomic ions. For diatomic molecules, the simplest treatment is based on the approximation that the two atoms are particles and that the distance between them is fixed. In this case f is 5; the 5 position coordinates of the first molecule may be taken to be the three Cartesian coordinates q_1, q_2, q_3 of the centre of gravity, the angle q_4 between the line of centres and a fixed direction, and the angle between the plane containing the angle q_4 and a fixed plane containing the fixed direction (in other words, q_4 and q_5 are the co-latitude and azimuth used in spherical polar coordinates). The kinetic energy of the first molecule is

$$\tfrac{1}{2}m(\dot{q}_1^2 + \dot{q}_2^2 + \dot{q}_3^2) + \tfrac{1}{2}I(\dot{q}_4^2 + \dot{q}_5^2 \sin^2 q_4), \tag{2.9}$$

where m is the total mass and I the moment of inertia. Consequently, by (2.1), the momentum coordinates associated with the line of centres are

$$p_4 = I\dot{q}_4 \quad \text{and} \quad p_5 = I\dot{q}_5 \sin^2 q_4, \tag{2.10}$$

and the single-molecule Hamiltonian is given by

$$H_{(1)}(\boldsymbol{p}_1, \boldsymbol{q}_1) = \frac{1}{2m}(p_1^2 + p_2^2 + p_3^2) + \frac{1}{2I}(p_4^2 + p_5^2/\sin^2 q_4). \tag{2.11}$$

Such a molecule is called a *rigid dumb-bell*. The method for more complicated molecules is the same except that it may be necessary to include a potential energy term in $H_{(1)}$ to allow for the interactions between different parts of the molecule.

An important generalization of (2.8) is the single-molecule Hamiltonian for an electrically charged particle in an electromagnetic field with sources outside the system. This Hamiltonian is

$$H_{(1)}(\boldsymbol{p}, \boldsymbol{q}) = \frac{1}{2m} \sum_{i=1}^{3} (p_i - A_i(\boldsymbol{q})\, e/c)^2 + e\varphi(\boldsymbol{q}), \tag{2.12}$$

where $\boldsymbol{A}$ and φ are the vector and scalar potentials, related to the electric and magnetic fields by the usual relations

$$\left.\begin{aligned} \boldsymbol{E} &= -\operatorname{grad} \varphi - c^{-1}\, \partial \boldsymbol{A}/\partial t \\ \boldsymbol{B} &= \operatorname{curl} \boldsymbol{A}, \end{aligned}\right\} \tag{2.13}$$

and e is the charge on the particle. The Hamiltonian equations of motion are now

$$\left.\begin{aligned} \dot{q}_i &= [p_i - A_i(\boldsymbol{q})\, e/c]/m \\ \dot{p}_i &= \sum_{j=1}^{3} (p_j - A_j(\boldsymbol{q})\, e/c)\, (\partial A_j/\partial q_i)\, e/mc - e\partial\varphi/\partial q_i. \end{aligned}\right\} \tag{2.14}$$

Taking the time derivative of the first equation and then using the second to eliminate $\dot{p}_i$, we obtain

$$m\ddot{q}_i = \sum_{j=1}^{3} \dot{q}_j(\partial A_j/\partial q_i - \partial A_i/\partial q_j)\, e/c - e(\partial\varphi/\partial q_i + c^{-1}\, \partial A_i/\partial t), \tag{2.15}$$

which is equivalent, by (2.16), to the Lorentz equation of motion

$$m\ddot{\boldsymbol{q}} = e\dot{\boldsymbol{q}} \wedge \boldsymbol{B}/c + e\boldsymbol{E}. \tag{2.16}$$

If there are interactions between the molecules, they can be allowed for by introducing additional terms into the Hamiltonian given by (2.6). It is simplest to assume that this energy of interaction is a sum of $\frac{1}{2}N(N-1)$ independent contributions, one from each of the $\frac{1}{2}N(N-1)$ pairs of molecules. Contributions involving 3 or more molecules could be included too, but they would complicate the mathematics without introducing anything essentially new. If we assume also that the interaction forces are not velocity-dependent, then the effect of the pair interactions is to change the Hamiltonian (2.6) to one of the form

$$H = \sum_i H_{(1)}(\boldsymbol{p}_i, \boldsymbol{q}_i) + \sum_{i<j} U_{(2)}(\boldsymbol{q}_i, \boldsymbol{q}_j), \tag{2.17}$$

where $U_{(2)}(\boldsymbol{q}, \boldsymbol{q}')$ is the potential energy of interaction between a pair of molecules whose configuration vectors are $\boldsymbol{q}$ and $\boldsymbol{q}'$.

If the molecules are point particles, then the interaction potential $U_{(2)}$ is a function of only the distance

$$r = [(q_1 - q_1')^2 + (q_2 - q_2')^2 + (q_3 - q_3')^2]^{1/2} \tag{2.18}$$

between the two atoms. The simplest interaction of this type is the Coulomb repulsion e^2/r between two particles of charge e. For neutral, chemically inert atoms such as argon or helium, the interaction potential is strongly repulsive at small separations and weakly attractive at large (Fig. 1); a

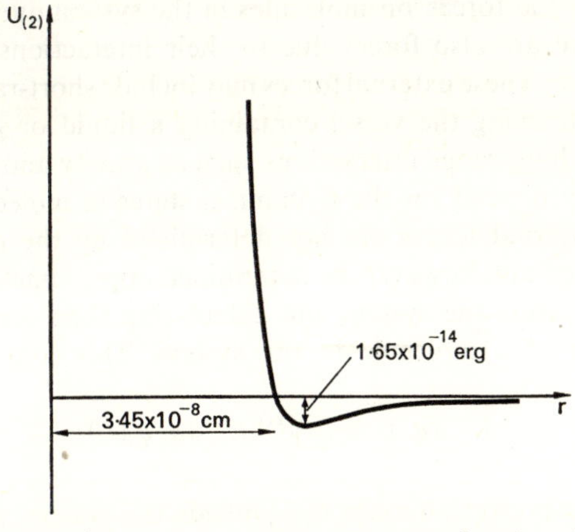

FIG. 1. Potential energy between two argon atoms as a function of their separation r

commonly used approximation for this type of potential is the *Lennard–Jones potential*

$$U_{(2)} = 4\varepsilon[(a/r)^{12} - (a/r)^6], \tag{2.19}$$

where ε and a are positive constants. For argon the value of a is about $3\frac{1}{2} \times 10^{-8}$ cm, which gives an idea of the size of an atom. For diatomic and larger molecules, $U_{(2)}$ is more complicated than a simple function of r. For example, if the molecules are electric or magnetic dipoles, their interaction includes a term of the form r^{-3} times a function of the angles the dipole axes make with each other and with the line joining them.

The equations of motion arising from (2.17) are

$$\left.\begin{aligned} \frac{d\boldsymbol{q}_i}{dt} &= \frac{\partial H_{(1)}(\boldsymbol{p}_i, \boldsymbol{q}_i)}{\partial \boldsymbol{p}_i}, \\ \frac{d\boldsymbol{p}_i}{dt} &= -\frac{\partial H_{(1)}(\boldsymbol{p}_i, \boldsymbol{q}_i)}{\partial \boldsymbol{q}_i} - \sum\nolimits_{j(\neq i)} \frac{\partial U_{(2)}(\boldsymbol{q}_i, \boldsymbol{q}_j)}{\partial \boldsymbol{q}_i}, \end{aligned}\right\} \tag{2.20}$$

where $\sum_{j(\neq i)}$ means a sum over the values $j = 1, 2, \ldots i-1, i+1, \ldots N$. The first line of (2.20), being identical with that of (2.7), shows that the kinematic relation between velocity and momentum is unaltered by this type of interaction. The second line shows that the total force on a molecule now includes a sum of forces, one due to every other molecule in the system. Since $U_{(2)}$ does not depend on the momenta, these interaction forces also do not depend on the momenta. The gain in generality that would be obtained by allowing $U_{(2)}$ to depend on the momenta as well as the positions would be more apparent than real, since the main type of velocity-dependent force, the magnetic interaction between two moving charged particles, cannot be disposed of by such a simple expedient.

In addition to the forces on molecules in the system due to their mutual interactions there are also forces due to their interactions with molecules outside the system. These external forces may include short-range interactions with molecules forming the vessel containing a liquid or gaseous physical system, and also long-range interactions such as gravity and external electric fields. Since they depend on the dynamical states of molecules outside the system, these external forces are not determined by the dynamics of the system alone; they can, however, be determined approximately by observing the molecules outside the system and calculating their contribution to the potential energy of a molecule in the system. This contribution may be written

$$U_{(1)}(\boldsymbol{q}, t) \equiv \sum\nolimits_j^{(\mathrm{ext})} U_{(2)}(\boldsymbol{q}, \boldsymbol{q}_j(t)), \tag{2.21}$$

where the sum goes over all molecules outside the system, and the fact that the molecules outside the system need not be at rest is allowed for in the notation. Although observation can never determine the individual positions $\boldsymbol{q}_j(t)$, it can sometimes determine the value of the sum on the right of (2.21) with some measure of accuracy. If so, then $U_{(1)}(\boldsymbol{q}, t)$ may be regarded as a known function of $\boldsymbol{q}$ and t, the *external potential*, and the effect of the external forces may be included in the dynamics of the system by using (2.17) with the modified single-molecule Hamiltonian

$$H_{(1)}(\boldsymbol{p}, \boldsymbol{q}, t) = H_{\text{free}}(\boldsymbol{p}, \boldsymbol{q}) + U_{(1)}(\boldsymbol{q}, t), \tag{2.22}$$

where $H_{\text{free}}(\boldsymbol{p}, \boldsymbol{q})$ is the Hamiltonian for a single molecule in the absence of external forces.

The most important service performed by the external potential $U_{(1)}$ is to represent the walls of a container enclosing the system. The force exerted by the walls will be zero for molecules well inside the container (that is, inside it by more than a few Angstrom units); therefore the contribution of the wall to $U_{(1)}$ may be taken as zero for any molecule that is well inside the container. Any molecule attempting to get out of the container will be met by an inward force; therefore the potential energy should rise rapidly as a molecule approaches the wall closely. A convenient mathematical model of such wall forces can be obtained by regarding the wall as a geometrical surface enclosing a region R of space, and to make

$$U_{(1)}(\boldsymbol{q}, t) \left\{ \begin{array}{l} \equiv 0 \text{ if a molecule with position } \boldsymbol{q} \text{ would be inside } R \\ \equiv +\infty \text{ if not.} \end{array} \right\} \tag{2.23}$$

Such a (model) wall we shall call a *perfectly hard* wall. If the molecules are point particles, they move quite freely within R, and every time they reach the boundary of R they bounce off it with perfect elasticity.

If the molecules carry electric charges or permanent magnetic moments, another contribution to $U_{(1)}$ comes from their interaction with charged or magnetized bodies outside the system. For example, if the molecules are dumb-bells with a permanent magnetic moment μ, such as O_2 molecules, and there is a uniform magnetic field $\boldsymbol{B}^{\text{ext}}$ arising from sources outside the system, then the magnetic contribution to $U_{(1)}(\boldsymbol{q})$ is $-\mu \boldsymbol{B}^{\text{ext}} \cos q_4$, in the notation of (2.9) with the spherical polar axis $q_4 = 0$ taken parallel to $\boldsymbol{B}^{\text{ext}}$. In general, since $U_{(1)}$ does not depend on $\boldsymbol{p}$, the force arising from it must be independent of velocity; thus $U_{(1)}$ cannot account for the Lorentz force exerted by an external magnetic field $\boldsymbol{B}^{\text{ext}}$ on charged particles within the system. Fortunately, however, this Lorentz force can be allowed for by the device of inserting a magnetic vector potential, as in (2.12), into the kinetic energy term of the single-molecule Hamiltonian.

If $H_{(1)}(\boldsymbol{p}, \boldsymbol{q}, t)$ is independent of t, then the entire Hamiltonian is independent of time and the system is said to be *isolated*. The energy of any isolated dynamical system is an invariant of the motion (that is, it does not change with time); for the rate of change of the energy of the system, calculated from the Hamiltonian equations of motion (2.3), is

$$\frac{d}{dt} H(p_1, \ldots, q_F; t) = \sum_{i=1}^{F} \left(\frac{dp_i}{dt} \frac{\partial H}{\partial p_i} + \frac{dq_i}{dt} \frac{\partial H}{\partial q_i} \right) + \frac{\partial H}{\partial t} = \frac{\partial H}{\partial t}, \tag{2.24}$$

and this vanishes if $H(p_1, \ldots, q_F)$ does not depend on t explicitly. It follows that $H_{(1)}$ may realistically be made independent of t only if energy does not enter or leave the system. This condition can be realized approximately by suitable experimental technique, for example, if the physical system is enclosed in a Dewar (thermos) flask.

If $H_{(1)}$ does depend on time, then (2.24) shows that the energy of the system can change, since $\partial H/\partial t \neq 0$. A convenient way of allowing for this time variation is to write the Hamiltonian in the form

$$H(p_1, \ldots, q_F; x_1, \ldots, x_n), \tag{2.25}$$

where $x_1, \ldots, x_n$ are a set of *external parameters*, which may depend on time in an arbitrary way and may be used to represent the currents in solenoids outside the system, the positions of pistons, etc. From (2.24) and (2.25) we obtain

$$\frac{dH}{dt} = \sum_{i=1}^{n} X_i \frac{dx_i}{dt}, \tag{2.26}$$

where $$X_1(p_1, \ldots, q_F; x_1, \ldots, x_n) \equiv \frac{\partial H}{\partial x_i} \quad (i = 1, \ldots, n) \tag{2.27}$$

is a dynamical variable called the *generalized force* associated with the parameter x_i. For example, if x_i is the displacement of a piston, then X_i is the force on that piston; if x_i is the volume enclosed by the walls containing the system, then X_i is the negative of the average pressure over the surface of the container. If the values of the external parameters are under the control of an experimenter, we shall use the term *mechanical operation* to denote a change in the values of these parameters. (The word "mechanical" is used in a generalized sense here, covering electrical and magnetic operations too.)

If the values of the external parameters are not all under the control of an experimenter, it should be possible to analyse the energy change described by (2.24) into two distinct parts in accordance with the first law of thermodynamics. One part would represent mechanical (or electrical or magnetic) work done on the system by external forces; the other would represent heat added to the system. In order to distinguish in a clear-cut way between heat and work we must, however, go beyond the purely dynamical considerations discussed in the present section. The method of making this distinction will be indicated in Chap. V, § 1; it depends on considerations of observability as well as dynamics.

The formalism can be extended to systems in which the particles are not all identical, as in mixtures of two or more chemical compounds, or an ionic solution. Suppose the N molecules fall into n different classes of identical molecules, so that $n = 2$ for a binary mixture, for example. We label the molecules $1, 2, \ldots, N$, as before; in addition we now assign to each molecule a *species coordinate* s_i ($i = 1, \ldots, N$) such that $s_i = s_j$ if and only if the ith and jth molecules are identical. The species coordinate is thus capable of just n distinct values, one for each kind of molecule in the system. These values may, but need not, be $s = 1, 2, \ldots, n$. To specify the dynamical state of the ith molecule we must now give not only $\boldsymbol{p}_i$ and $\boldsymbol{q}_i$ but also the species coordinate s_i.

Since the species coordinates are discrete, the phase space of a mixture consists of a number of disconnected parts, one corresponding to each possible set of values for the numbers $s_1 \ldots s_N$. The number of degrees of freedom of the ith molecule, and thence the number of components in the vectors $\boldsymbol{p}_i$ and $\boldsymbol{q}_i$, may depend on s_i, but this need not be shown explicitly in the general formulae. For example, if there are two types of molecule, one monatomic and one diatomic, and both are uncharged, then the free-molecule Hamiltonians given in (2.8) and (2.11) can be used to define $H_{(1)}(s, \boldsymbol{p}, \boldsymbol{q})$ for the mixture:

$$\left.\begin{aligned} H_{(1)}(1, \boldsymbol{p}_1, \boldsymbol{q}_1) &\equiv \frac{1}{2m_1}(p_1^2 + p_2^2 + p_3^2), \\ H_{(1)}(2, \boldsymbol{p}_1, \boldsymbol{q}_1) &\equiv \frac{1}{2m_2}(p_1^2 + p_2^2 + p_3^2) + \frac{1}{2I}(p_4^2 + p_5^2/\sin^2 q_4), \end{aligned}\right\} \tag{2.28}$$

and so the Hamiltonian for N free particles can still be written in the form analogous to (2.6), namely

$$H = \sum_{i=1}^{N} H_{(1)}(s_i, \boldsymbol{p}_i, \boldsymbol{q}_i). \tag{2.29}$$

In a similar way, $U_{(2)}$ will now depend on two species coordinates as well as two position coordinates; for example, the Coulomb interactions between the particles in an ionized monatomic gas can be written in the form $U_{(2)}(s, \boldsymbol{q}; s', \boldsymbol{q}') = e_s e_{s'}/r$, where r is defined in (2.18) and e_s is the electric charge on an ion of species s. Using this formalism, most parts of the statistical mechanics of a system of identical molecules can be easily generalized to mixtures; consequently in the following we shall only treat mixtures explicitly if there is some special point of interest or difficulty.

The formalism also generalizes easily to quantum mechanics. The quantum analogue of Hamilton's equations (2.3) is the Schrödinger wave equation

$$i\hbar \frac{\partial \psi}{\partial t} = H\psi. \qquad [(1.2)]$$

The wave function ψ is a function of t, of all the position coordinates, and, in the case of mixtures, of all the species coordinates as well. The Hamiltonian H has the same structure as the classical Hamiltonian given by (2.17) and (2.22):

$$\mathsf{H} = \textstyle\sum_i (\mathsf{H}_{i,\,\mathrm{free}} + \mathsf{U}_i) + \sum_{i<j} \mathsf{U}_{ij}, \tag{2.30}$$

where the operators U_i and U_{ij} are functions of position only:

$$U_i = U_{(1)}(\mathbf{q}_i; t),$$

$$U_{ij} = U_{(2)}(\mathbf{q}_i, \mathbf{q}_j), \tag{2.31}$$

and the operator $H_{i,\text{free}}$ is the Hamiltonian of the ith molecule travelling alone. In the cases where complications due to non-commutation and non-Cartesian coordinate systems do not arise, the operator $H_{i,\text{free}}$ can be obtained from its classical analogue by the standard recipe† of replacing $\mathbf{p}_i$ by $-i\hbar\,\partial/\partial\mathbf{q}_i$:

$$H_{i,\text{free}} = H_{\text{free}}(-i\hbar\,\partial/\partial\mathbf{q}_i, \mathbf{q}_i). \tag{2.32}$$

As an example, we may apply this recipe to the Hamiltonian (2.8) for a point particle: using (2.3) and (2.32) we obtain a quantum Hamiltonian of the form (1.3). The generalization of these formulae to mixtures is straightforward: it is only necessary to introduce species coordinates $s_1, s_2, \ldots$, as in the classical case, and to allow ψ to depend functionally on all the species coordinates and U_i, U_{ij}, and $H_{i,\text{free}}$ on the relevant ones:

$$\psi = \psi(s_1, \ldots, s_N; \mathbf{q}_1, \ldots, \mathbf{q}_N; t), \tag{2.33}$$

$$U_i = U_{(1)}(s_i, \mathbf{q}_i; t), \text{ etc.} \tag{2.34}$$

A quantum system may also contain discrete non-classical degrees of freedom such as electron spins. These can be formally allowed for by generalizing the species coordinate s to include all discrete degrees of freedom. Thus an electron gas can be treated kinematically just as if it were a mixture of two different types of electron, one with spin $s = +1$ and one with spin $s = -1$; and a metal would be a mixture of three kinds of particles: metallic ions, and two kinds of electrons. Thus the wave function will still have the form (2.33) and the Hamiltonian the form (2.30); the only change is that the various terms $H_{i,\text{free}}$, U_i and U_{ij} in the Hamiltonian will no longer be simply functions of s_i and s_j as in (2.34), but will involve matrices acting on the discrete variables s_i. As a result these variables need not be invariants of the motion as they are in classical mechanics.

The representation of a physical object by a dynamical system as discussed in this section has three main limitations. The first is the fact, already mentioned in connection with eqn. (2.21), that the interaction between the system and its surroundings can never be allowed for exactly. This is inevitable unless we treat the whole universe as a single dynamical system. In practice, however, the approximation of replacing the interactions with the surroundings by a fairly simple external potential $U_{(1)}$ works well as long as no heat enters or leaves the system; moreover, systems which heat does

† P. A. M. Dirac, *The Principles of Quantum Mechanics* (Oxford University Press, London, 1947), § 28.

enter or leave can usually be treated by regarding them as subsystems of a larger system with a simple external potential.

A second limitation of the method is that it is non-relativistic, so that all forces are assumed to propagate instantaneously. This limitation makes it impossible for us to treat exactly such electrodynamic effects as the interaction of light with matter and the electromagnetic interaction between moving electrons. The natural way to include these effects in the general theory would be to include the electromagnetic field in the dynamical description of the system.† This method works very well in deriving Planck's law of radiation, but as soon as we include particles as well as the electromagnetic field in our system, serious difficulties arise due to the fact that the number of degrees of freedom in a field is infinite. These difficulties include the "ultraviolet catastrophe" of classical radiation theory and the divergences of quantum electrodynamics.

A third limitation of dynamics is that it says nothing about the initial conditions for its differential equations of motion. It gives us an infinite class of possible motions for a dynamical system, but tells us nothing about which of these motions to expect on any particular occasion. In order to use the theoretical information provided by dynamics, therefore, one must always combine it with experimental information about the actual dynamical state of the system. The way in which observation is used to obtain this information will be the topic of the next section.

2.1. Exercises

1. Write down the equations of motion implied by the Hamiltonian (2.11) for the rigid dumb-bell molecule. Verify that p_5 and $p_4^2 + p_5^2 \cot^2 q_4$ are invariants of the motion.

2. Generalize your answer to question 1 to the case where the two particles in the dumb-bell have equal and opposite electric charges and there is a uniform external electric field E along the direction $q_4 = 0$. Find the generalized force belonging to E and give its physical interpretation.

3. Write down the Hamiltonian and the equations of motion for a free particle moving in a plane, using polar coordinates and classical mechanics. Verify that if the particle moves with constant velocity the equations of motion are satisfied.

3. Observation

The purpose of this section is to discuss in more detail the mathematical structure and physical interpretation of the idealized model proposed in § 1 for the observational properties of a physical system. The model is based on the assumption that an elementary observation is an instantaneous act, like

† The fundamental difficulties in the way of formulating an exact relativistic statistical mechanics without the use of fields are discussed by P. Havas in *Statistical Mechanics of Equilibrium and Non-equilibrium*, ed. J. Meixner (North-Holland, Amsterdam, 1965), p. 1.

taking a snapshot, with a limited set of possible outcomes which we call *observational states.*

We shall suppose these possible outcomes to form a countable set, and label them with positive integers, usually 1, 2, 3, ... Since the dynamical states form a continuum, there are in general many dynamical states to each observational state, and so it is not possible, in general, to deduce the dynamical state of a system from its observational state. The set of all dynamical states compatible with a given observational state A will be called the *dynamical image* of A and denoted by $\omega(A)$; in classical mechanics, the dynamical images may be thought of as a set of non-overlapping regions which together fill the whole of phase space. Although the observations are imperfect in the sense that they are inadequate to determine the dynamical state completely, they are assumed to be noiseless in the sense that if the dynamical state of a system is in the dynamical image $\omega(A)$ then an observation of that system is certain to produce the outcome A.

For a classical system, a convenient way to describe the dynamical images is to define for each observational state A a dynamical variable J_A, called the *indicator* of A, by

$$J_A(\alpha) \equiv \begin{Bmatrix} 1 \text{ if } \alpha \text{ is in } \omega(A) \\ 0 \text{ if not} \end{Bmatrix}, \tag{3.1}$$

where α represents the dynamical state $(p_1 \dots p_N, q_1 \dots q_N)$ of the system. That is, $J_A(\alpha)$ takes the value 1 at the instant t if and only if the observational state is A at that instant. Since the indicators take the values 0 and 1 only, they satisfy

$$J_A^2 = J_A \quad (A = 1, 2, \dots); \tag{3.2}$$

since the image regions do not overlap, they satisfy

$$J_A J_B = 0 \quad \text{if} \quad A \neq B; \tag{3.3}$$

and since the image regions together fill the whole of phase space, they also satisfy

$$\textstyle\sum_A J_A = 1, \tag{3.4}$$

just one term of the sum differing from zero at each phase space point.

The formal similarity between classical and quantum mechanics, so felicitously exploited by Dirac,† immediately provides the quantum analogue of our classical description of observation. Quantum dynamical variables, instead of being functions of the dynamical state as in classical mechanics, are Hermitian linear operators obeying the same algebraic relations as their classical counterparts. Thus the indicators $J_1, J_2, \dots$ of the observational states 1, 2, ... will in quantum mechanics be a set of Hermitian

† *Op. cit.*, §§ 21 and 28.

linear operators satisfying the algebraic relations (3.2), (3.3), and (3.4). The relation (3.2) implies that each J_A is a *projection operator* (a Hermitian operator whose only eigenvalues are 0 and 1), and the relation (3.3) implies that these projection operators commute.

The quantum dynamical image $\omega(A)$ of an observational state A consists,† as in classical mechanics, of all the dynamical states for which an observation of the dynamical variable J_A is certain to produce the result 1; that is, all the dynamical states α (usually written $|\alpha\rangle$) whose wave functions $\psi_\alpha(q_1, \ldots, q_F)$ satisfy the eigenvalue equation

$$J_A \psi_\alpha(q_1, \ldots, q_F) = \psi_\alpha(q_1, \ldots, q_F). \tag{3.5}$$

It is an immediate consequence of this equation that $\omega(A)$ is a *linear manifold:* that is if $|\alpha_1\rangle$ and $|\alpha_2\rangle$ are any two members of $\omega(A)$ and λ is any complex number, then $\lambda|\alpha_1\rangle$ and $|\alpha_1\rangle + |\alpha_2\rangle$ are also members of $\omega(A)$. The relation (3.3) implies that the linear manifolds $\omega(1)$, $\omega(2)$, ..., are mutually orthogonal: if $|\alpha\rangle$ is in $\omega(A)$ and $|\beta\rangle$ is in $\omega(B)$, with $A \neq B$, then by applying (3.5), the corresponding equation for $|\beta\rangle$, the Hermitian property of J_B, and finally (3.3), we deduce

$$\langle\beta|\alpha\rangle = \langle J_B\beta|J_A\alpha\rangle = \langle\beta|J_BJ_A\alpha\rangle = 0,$$

where $\langle\beta|\alpha\rangle$ denotes the *inner product* $\int \ldots \int \psi_\beta^*(q_1, \ldots, q_F)\,\psi_\alpha(q_1, \ldots, q_F) \times$ $\times\, dq_1 \ldots dq_F$ of the two quantum states $|\alpha\rangle$ and $|\beta\rangle$. Finally, the relation (3.4) implies that the linear manifolds $\omega(1)$, $\omega(2)$, ... together span the entire Hilbert space of dynamical states: an arbitrary dynamical state $|\gamma\rangle$, though not in general lying in any of the dynamical image manifolds $\omega(1)$, $\omega(2)$, ..., can be written as a superposition of dynamical states, each of which does lie in one of the dynamical image manifolds:

$$|\gamma\rangle = J_1|\gamma\rangle + J_2|\gamma\rangle + J_3|\gamma\rangle + \ldots$$

Any dynamical variable, classical or quantal, whose value can (in this model) be measured exactly, will be called an *observable*. By the definition of observational states, the value of an observable must be a function of the observational state of the system. The most elementary observables are the indicators themselves, and the rest can be expressed in terms of the indicators. Thus, in classical mechanics, if $G(\alpha)$ is any observable and G_A is (for $A = 1, 2, \ldots$) the value it takes when the observational state is A, then $G(\alpha)$ can be expressed in the form

$$G(\alpha) = \textstyle\sum_A G_A J_A(\alpha) \tag{3.6}$$

since only one term of the sum differs from zero whatever the dynamical

† This quantum-mechanical model of observation is due to J. von Neumann, *Mathematical Foundations of Quantum Mechanics* (Princeton University Press, Princeton, 1955), chap. V, § 4.

state of the system. Any function of an observable, or of more than one observable, is also an observable: for example, if $G'(\alpha)$ and $G''(\alpha)$ are two observables and $f(x, y)$ is any function of two variables, then (3.6) implies

$$f(G'(\alpha), G''(\alpha)) = \sum_A f(G'_A, G''_A)\, J_A(\alpha). \tag{3.7}$$

In quantum mechanics, the corresponding formulae are obtained by replacing the classical dynamical variables $G(\alpha)$, etc., by the corresponding operators throughout. Since all observables can, by definition, be measured simultaneously, their quantum-mechanical operators commute;† this fact may also be verified by operator algebra using (3.3) and the quantum version of (3.6).

Most of the dynamical variables we measure in physics are not observables in this sense; consequently they cannot be measured exactly. Some of them, however, can be measured approximately, and such an approximate measurement may be treated as an exact measurement of a new dynamical variable, approximating the original one, but differing from it in being an observable. For example if, for each observational state A, all the dynamical states in its image $\omega(A)$ have approximately the same energy E_A, then an approximate measurement of the energy (the value of the Hamiltonian) is equivalent to an exact measurement of the "observable Hamiltonian"

$$H_{\text{obs}} \equiv \sum_A E_A J_A. \tag{3.8}$$

Alternatively these approximate energy measurements may be described in terms of tolerance intervals, as outlined in § 1. According to this point of view, the range of possible energy values must be split into non-overlapping tolerance intervals, and it is assumed that observation cannot distinguish different energy values within the same tolerance interval.

For the two points of view to be consistent, it is sufficient that each tolerance interval should be the union of a set of observational states, all having the same value for the observable energy E_A. The dynamical image of such a tolerance interval is called an *energy shell*, and consists of all the dynamical states for which the exact Hamiltonian H satisfies the inequalities

$$E - \Delta E < H \leqq E,$$

where $E - \Delta E$ and E are the extremes of the tolerance interval.

Although the description of physical measurements in terms of dynamical images and observables may seem artificial it does correspond closely to the physical notion of an observational state. The ideal way of specifying, in physical terms, the observational state of an object would be to list all the measurable physical quantities relating to the object, with some indica-

† The physical justification of this assertion is discussed by von Neumann, *op. cit.*, and also by N. G. van Kampen, Quantum Statistics of Irreversible Processes, *Physica* **20**, 603 (1954). The meaning given here to the word "observable" should not be confused with the one used by Dirac, *op. cit.*, § 10. Dirac's "observables" do not all commute, for example.

tion of the accuracies of the measurements. Our scheme of description assumes, first, that all the measurable physical quantities are dynamical variables, and, second, that the inaccuracies in their measurement may be represented by dividing their ranges of variation into tolerance intervals. Each observational state then corresponds to a set of tolerance intervals, one for each of the measured dynamical variables, and its dynamical image is the intersection of the dynamical images of the tolerance intervals used to specify it. The first of these assumptions does not exclude the measurement of physical quantities that are not dynamical variables, such as temperature; for to make a temperature measurement we include a thermometer in the system and then measure some dynamical variable of the thermometer, such as the position of a mercury meniscus. The second assumption is admittedly an idealization but it has the virtue of representing the essential physical fact, that observations are never perfectly accurate, in a very simple way. Even the arbitrariness allowed by the theory in the choice of the dynamical images has its counterpart in the physical notion of an observational state; for the decision about which physical quantities are to be included in the list of "measurable physical quantities" referred to above contains a similar arbitrariness: it depends on the current state of experimental technique rather than on any basic physical law.

To complete the specification of this model of observation, we must specify the instants at which elementary observations may be made. Some restriction must be placed on these instants, for if observations are allowed at arbitrary instants the model fails in its objective of ruling out the possibility of perfectly accurate measurements. To see this, let us consider a classical system kept under continuous observation and suppose that some dynamical variable G is among those that are measured as part of the observation. Provided G is not a constant of the motion, its value will from time to time move from one tolerance interval to another, and at each moment when it does this its exact value (the value on the boundary between the two tolerance intervals) can be deduced from the observations. This possibility, if allowed, would nullify the original objective of the model, which was to rule out exact measurements. The obvious way to avoid this objectionable possibility is to forbid continuous observation, and the prohibition can be conveniently enforced by designating a discrete set of instants, to be called *instants of observation*, and to permit observations at these instants and no others.† For simplicity we shall assume the instants

† J. G. Kirkwood, The Statistical Mechanical Theory of Transport Processes, *J. Chem. Phys.* **14**, 180 (1946), makes the related assumption that instantaneous measurements are impossible and that one can only measure time averages of the form $\overline{G}(t) = (\Delta t)^{-1} \int_{t-\Delta t}^{t} G(t')dt'$, where $G(t')$ is the value of the dynamical variable G at time t'. Since a measurement of $\overline{G}(t)$ can be regarded as an instantaneous measurement of a new dynamical variable $\overline{G}$ at time t, however, Kirkwood's assumption is neither necessary nor sufficient for our purposes.

of observation to be equally spaced. The time interval Δt between successive instants of observation may be interpreted as a "dead time" which the measuring apparatus requires in order to assimilate one observation before it can proceed with the next.

Not every observation is an instantaneous "snapshot". It is also possible to combine a succession of instantaneous observations into a single *compound observation*,† in the same way that a succession of snapshots taken at different times combine to give a motion picture film. To incorporate compound observations into the theory, we must specify not only how the dynamical state affects the result of the observation, but also how the act of observation disturbs the dynamical state of the system. In classical mechanics with noiseless measurements the implicit assumption is that there is no disturbance: this was the assumption used in the foregoing discussion of time measurement. In quantum mechanics, however, even noiseless observations do in general disturb the system.

A well-known example is the experiment where a beam of electrons passes through a pair of slits in a screen and the diffraction pattern is recorded on a photographic plate. Any attempt to observe which slit the electrons go through destroys the phase relationship between the waves emanating from the two slits and so destroys the diffraction pattern too. We say that the earlier measurement (position in the plane of the screen) is *incompatible* with the later (position in the plane of the photographic plate). Such incompatibilities are not confined to the single-electron system; in principle they are possible in systems of any number of particles. With a macroscopic system, however, it is not possible in practice to maintain the phase relationships that are necessary for quantum interference effects ‡ and consequently the additional disturbance of phase relationships produced by making a measurement has no effect on the later measurements. The simplest way to incorporate this property of macroscopic systems into the theory is to make the assumption that any two observations made on the system at different times are compatible. This assumption will be called the *postulate of compatibility*. It is a generalization of our earlier assumption [see eqn. (3.3)] that any two observations made at the same time are compatible, and will be formulated mathematically in Chap. IV, § 1.

The fact that a system contains a large number of particles does not, of itself, ensure that the postulate of compatibility is satisfied: for example, a

† The use of compound observations in this model of observation appears to have been originated by U. Uhlhorn, On the Statistical Mechanics of Non-equilibrium Phenomena, *Ark. f. Fys.* **17**, 193 (1960).

‡ An interesting paper on this subject is that of A. Peres and N. Rosen, Macroscopic Bodies in Quantum Theory, *Phys. Rev.* **135**, B 1486 (1964). This points out some practical limitations precluding the observation of certain interference effects which are theoretically possible in quantum mechanics. The spirit of their work is similar to that of the theory of observation used here, which also emphasizes technical limitations not recognized by pure dynamics.

system consisting of an experimental apparatus designed to study quantum interference effects would contain many particles but would not satisfy the postulate. This system, however, falls outside the normal province of statistical mechanics in any case. Thus one of the functions performed by the postulate of compatibility is to provide an unambiguous rule for excluding such systems from consideration.

One further type of disturbance to the system, not this time specifically quantum-mechanical, must be allowed for in the theory. This is the disturbance produced by what may be called an *active observation*, which involves deliberate rather than unintentional interference with the system. An example of such an observation would be measuring the compressibility of a gas. This involves measuring the pressure for at least two different volumes. To change the volume we must move the walls enclosing the gas. This is represented in the dynamical model by a change in the external potential $U_{(1)}(\boldsymbol{q}, t)$ [see eqn. (2.21) and its immediate sequel], that is to say by a change in the Hamiltonian. If the time-dependence of the Hamiltonian is described by time-dependent external parameters, in accordance with (2.25), then active observations may be characterized as compound observations during which the external parameters are varied.

To preserve the spirit of our model of observation we shall assume that the external parameters can be changed only at the instants of observation. That is, just as the interactions between system and observer which constitute observations are confined, in this model, to a discrete set of equally spaced instants, so the interactions between observer and system which change the external parameters are also confined to these same special instants. At all other times, the system is assumed to be perfectly isolated, inaccessible both to observation and to interference changing its Hamiltonian. Thus any continuous change in H must be represented here as a succession of discontinuous jumps. If such a jump in H occurs, it does not cause a jump in the dynamical state of the system, since the equations of motion (2.3) or (1.2) ensure that the dynamical state always varies continuously with time. It is also natural to postulate that the indicator variables $J_1, J_2, \ldots$ specifying the observational states do not change their functional form when H changes. Consequently a jump in H also does not cause a jump in the observational state. Thus the effects of the jump in H are not felt immediately, either dynamically or observationally, but they do make themselves felt after a lapse of time, and can, therefore, be important for active observations.

Further insight into the theory of observation can be obtained by thinking of the dynamical state of the observed system as a "message" which the observer is trying to elicit from it, and the experimental apparatus used for his observations as a communication channel for transmitting this message. We can then borrow some of the ideas of the mathematical theory of communication.† In that theory a communication channel is said to be discrete if it only uses characters from a finite "alphabet" and transmits them at instants equally spaced in time; thus a teleprinter system uses an alphabet of thirty-two characters (including the twenty-six roman letters) and transmits about ten characters per second. A channel is said to be *noiseless* if its input and output are invariably identical; thus a teleprinter system is noiseless if it is functioning perfectly. Our model of observation represents the experimental apparatus by the simplest type of communication channel, a discrete channel without noise. Its alphabet is the finite set of observational states, its input is the succession of observational states occupied by the system at equally spaced instants, and its output faithfully reproduces this input.

The mathematical theory of communication provides a quantitative measure of the *information* in a message. The quantity of information is roughly proportional to the number of characters from the chosen alphabet required to encode the message, using the most efficient method of coding. Communication theory postulates an upper bound, called the *capacity* of a communication channel, to the rate at which that channel can transmit information. The starting point of statistical mechanics, the fact that observation cannot determine the dynamical state of a macroscopic physical system, can thus be rephrased in the language of communication theory: the quantity of information needed to specify the dynamical state of the system is far beyond the capacity of the channel available for communication between system and observer. For example, if the system is a cubic centimeter of gas it contains about 10^{19} particles and would therefore need at least 10^{20} decimal digits to specify its dynamical state even crudely; but a single real observation could hardly yield more than, perhaps, 10^3 decimal digits—that is, the capacity of the channel might be of the order of 10^3 decimal digits per observation. In this example, therefore, the system would need a message of at least 10^{17} characters (successive observational states) to communicate to the observer its dynamical state at any given instant. Thus no practicable sequence of observations can yield a significant fraction of the information required to specify the dynamical state of a macroscopic system.

More precisely, if there are M different observational states available to the system‡ and we make a compound observation comprising n instantaneous observations, then the number of different possible results of the compound observation is M^n. Consequently, if W is the number of effectively

† C. E. Shannon and W. Weaver, *The Mathematical Theory of Communication* (University of Illinois Press, Urbana, 1963). The connection between statistical mechanics and the theory of communication is also discussed in Chap. VI, § 4, below.

‡ The set of all observational states is normally infinite, but conservation of energy will be presumed to restrict any particular system, if isolated, to a finite subset.

distinct dynamical states available to the system, these states cannot all be distinguished by the observation if

$$W > M^n, \tag{3.9}$$

that is if

$$n < \frac{\ln W}{\ln M}. \tag{3.10}$$

According to the estimate made in the preceding paragraph, the right side of this inequality can be as large as $10^{20}/10^3$ for large systems, so that the condition is satisfied for any practical compound observation.

In the mathematical theory, however, there is the possibility (see Chap. IV, § 7) that contradictions may arise unless we can rule out even the theoretical possibility of a compound observation sufficiently lengthy to violate (3.10). It would be artificial to do this by putting an upper bound, say n_{max}, on the permitted number of observations on a system, because of the awkward discontinuity involved in allowing n_{max} observations but not $n_{max} + 1$. A more natural method, therefore, is to use the fact that the Hamiltonian of a large real system is never known with perfect accuracy. Because of this slight indefiniteness of the Hamiltonian, there is a slight indefiniteness in the relation between the dynamical states at different times, obtained by integrating the equations of motion. The longer the interval between these two times, the larger will be the accumulated effect of the indefiniteness in the Hamiltonian; and for sufficiently long time intervals, whose order of magnitude we denote by τ, the indefiniteness of the relation between the two dynamical states will be comparable with the size of the dynamical image of an observational state. It follows that nothing useful about the dynamical state of the system at time t can be deduced from observations made on it before $t - \tau$ or after $t + \tau$, so that the number of observations relevant to this dynamical state is about $2\tau/\Delta t$ where Δt is the time interval between successive instants of observation. Provided the condition

$$\tau < \left(\frac{\ln W}{\ln M}\right)\frac{\Delta t}{2} \tag{3.11}$$

is satisfied, therefore, the spirit of the condition (3.10) will also be satisfied.

The analogy with communication theory also draws attention to two ways in which our model of observation oversimplifies reality. First, noise is ignored; that is, the result of an observation (of a classical system) is assumed to depend only on the observational state of the observed system, which in turn is completely determined by the dynamical state of the system. In reality, other factors also affect the result of an observation, so that the result may differ from the true observational state. Such discrepancies would normally be described as "random errors of observation", though a thoroughgoing determinist might trace them to purely dynamical effects originating outside the observed system, for example in the experimental apparatus. One would expect this observational "noise" to be particularly important when the dynamical state of the system was near the boundary in phase space separating the images of two observational states. Its effect would be to blur these boundaries, which are unrealistically regarded as mathematical surfaces in the simplified model adopted here.

A second oversimplification in the theory is that both time and observational states are treated as discrete. One consequence of this is that the theory contains quantities, such as the time interval between successive observations and the tolerances in experimentally measurable variables such as the force on a piston, whose value is not precisely specified. In reality both time and most observable dynamical variables are continuous. In communication theory continuous messages (corresponding to the use of a telephone rather than a teleprinter) can be studied in a realistic way only if noise is taken into account, and this appears to be true of the theory of observation too.

It is clear from the last few paragraphs that our theory of observation is a highly idealized version of what really happens when an observation is made. Nevertheless, this lack of realism is not as important as it may look. The main goal of this theory of observation is not to provide a detailed and accurate description of the act of observation, but to provide a simple, well-defined mathematical model on the basis of which we can discuss the limitations of the observational description of a physical object and the relation between this description and the dynamical description.

3.1. Exercises

1. Verify that if an operator J satisfies the condition (3.2) then it has eigenvalues 0 and 1 only. Show by means of an example (e.g. a 2×2 matrix) that an operator can satisfy this condition without being a projection operator.

2. If J_1 and J_2 are commuting projection operators, show that $J_1 J_2$ is also a projection operator and that its linear manifold is the intersection of those of J_1 and J_2.

4. Probability

The observational description of physical systems gives by itself no indication how the observational states of the system at different times are related. We must now look for laws providing this relationship. These laws will supplement the purely descriptive treatment of observation given in § 3 where we thought of a system as passing through a more or less arbitrary discrete succession of observational states, in the same way that the laws of motion in mechanics supplement the purely descriptive kinematical treatment of motion as a more or less arbitrary continuous succession of dynamical states. Unlike the laws of motion, however, these new laws will not be deterministic but statistical.

The meaning of the term *statistical law* can be elucidated by considering what happens when a physical experiment is *replicated:* that is, when the experimental procedure is repeated at a different time, or in a different place at the same time. If the experiment is to measure a physical quantity or test a physical law, then ideally every replication gives the same outcome. Such an experiment, giving the same outcome at every replication, may be said to be *reproducible*. If the laws relating observational states at different times are deterministic, then any experiment which starts with the system in a known observational state is reproducible. For example, it is a reproducible experiment to project a heavy ball upwards in still air from a given point with given velocity and to measure its velocity when it hits the ground: the initial observational state (position and velocity) determines the later observational states with good accuracy, so that on any replication of the experiment the outcome will be reproduced.

When an experiment is not reproducible we may be able to turn it into a reproducible one by defining the experimental procedure more carefully, for example by controlling the temperature and neutralizing the earth's magnetic field. In other cases, however, the lack of reproducibility cannot be eliminated in this way because it springs from the fundamental lack of determinism of observational states. For example, with the observational techniques that are normally to hand, it is not a reproducible experiment to project a spinning coin (rather than a ball) upwards from a given point with given velocity, angular velocity, and angular orientation and to note which side is uppermost when the coin comes to rest. The difference from the experiment with the ball is that here the final observational state depends much more sensitively on the initial dynamical state, so that the accuracy with which this state can be determined from observation is insufficient for a firm prediction of the final observational state. Similarly, it is not a reproducible experiment to pass an electron of known energy through a narrow collimator and then a metallic crystal, recording photographically its direction of motion on emerging from the crystal: the Schrödinger wave of the electron is diffracted by the crystal and does not determine the position of the spot on the photographic plate, so that on different replications the spot indicates different outcomes. In this case the lack of reproducibility arises, not from limitations of experimental technique, but from the fundamental indeterminacy of quantum mechanics.

To avoid confusion with the physicist's customary usage, in which the word "experiment" means a reproducible experiment, we shall here use the word *trial* to mean any reproducible or non-reproducible experiment: that is, an experimental procedure which an observer may at any time attempt to apply to the system and which, if successful, leads to a well-defined outcome depending on the observational state of the system at one or more times during the experiment. We shall think of a trial as being defined by a set of instructions specifying both the system and the procedure that is to be applied to it. Any particular occasion on which the instructions are successfully carried out will be called a *replication* of the trial. A trial is *reproducible* if and only if every replication of it leads to the same outcome. The foregoing experiments with ball, coin, and electron are all examples of trials, but only the first is reproducible.

Although a trial does not in general, exhibit the perfect regularity which we call reproducibility, it may exhibit a less perfect *statistical regularity*. This statistical regularity manifests itself as a reproducible feature not of individual trials but of *statistical experiments*, performed by replicating the trial itself a large number of times. One of the ways of performing a statistical experiment is to use the same physical object (e.g. a coin) for all the replications (tosses of the coin), each new replication starting as soon as its predecessor is finished; an alternative way is to use a large number of replicas of the physical object (e.g. an electron) and to perform the same trial (diffracting the electron) on all the replicas at essentially the same time. There are many other ways, too, which need not be considered here but will be analysed in § 5. If the statistical experiment comprises $\mathcal{N}$ replications of the trial, and $\mathcal{N}(A)$ of them yield the outcome A, then the ratio $\mathcal{N}(A)/\mathcal{N}$ is called the *relative frequency* of A in that statistical experiment. For example, if a coin is tossed 10,000 times, yielding 5021 heads and 4979 tails, then the relative frequency of heads in this replication of the statistical experiment is 0·5021. Experience shows that for some trials the relative frequencies in statistical experiments with large $\mathcal{N}$ are approximately reproducible, the reproducibility being the better the larger the value of $\mathcal{N}$.

For example, if the trial is the toss of a coin, then experience shows that the relative frequency of heads in a statistical experiment consisting of a sufficiently long sequence of tosses is close to $\frac{1}{2}$; and if the statistical experiment itself is replicated a few times, then these replications also give relative frequencies close to $\frac{1}{2}$. Thus a statistical experiment comprising a large number of tosses of a coin yields reproducible relative frequencies, and we may say that the individual tosses are thereby showing statistical regularity. Likewise, if the trial is to pass a single electron of known energy through a collimator and then a thin crystal, recording photographically its direction of emergence, then the corresponding statistical experiment (which won G. P. Thompson the Nobel prize) is to pass a beam of such electrons through the crystal and photograph the diffraction pattern. The total amount of blackening in any region of the photograph is proportional to the number of electrons that have passed through this region and hence to the relative frequency of a particular outcome for the individual trial. Thus the reproducibility of photographs of diffraction patterns is another example of statistical regularity.

Not every trial exhibits statistical regularity. To show this we may consider the trial defined by the instructions "cast a cubical die and note which face is uppermost when it comes to rest". In a long sequence of throws of a cubical die the relative frequencies obtained will be influenced by the position of the centre of gravity within the die. Therefore if the statistical experiment is replicated using a differently loaded die, the relative frequencies will not reproduce themselves. That is to say, because the nature of the die is not adequately specified in the instructions defining the trial, the trial does not show statistical regularity. Just as we can sometimes turn a non-reproducible experiment into a reproducible one by specifying the experimental conditions (such as temperature) more carefully, so we can sometimes turn a trial without statistical regularity into one with it. In the present example we can give the trial statistical regularity by adding the additional instruction that the centre of gravity must be at the middle of the cube.

Experimentally, statistical regularity reveals itself in the approximate reproducibility of the relative frequencies in large statistical experiments. It is a natural idealization to assume that for suitable trials the relative frequencies could be made exactly reproducible by using a statistical experiment comprising an infinite number of trials. A trial having this property will be said to exhibit *statistical regularity*, or simply to be *statistically regular*. If A is one of the possible outcomes of a statistically regular trial $\mathfrak{T}$, then the relative frequency of A in an infinite statistical experiment, being exactly reproducible, depends only on A and the instructions defining $\mathfrak{T}$. This relative frequency is called the *physical probability* of the outcome A in the trial $\mathfrak{T}$, written $Pr(A|\mathfrak{T})$. A more precise formulation of this definition is

$$Pr(A|\mathfrak{T}) \equiv \lim_{\mathcal{N}\to\infty} \frac{n(A, \mathcal{N}, \mathfrak{T})}{\mathcal{N}}, \tag{4.1}$$

where $n(A, \mathcal{N}, \mathfrak{T})$ is the number of times the outcome A occurs during the first $\mathcal{N}$ replications of the trial $\mathfrak{T}$ in an infinite statistical experiment. This definition of probability hinges on the assumption that the trial $\mathfrak{T}$ is statistically regular, that is that the limit on the right-hand side of (4.1) exists and is the same at every replication of the infinite statistical experiment. For a trial that is not statistically regular, such as the rolling of an incompletely specified cubical die, (4.1) does not lead to well-defined probabilities.

It is natural to think of a physical probability as a permanent physical property of the observed system, analogous to mass.† Then the reproducibility of the relative frequencies observed in a statistical experiment can be

† This is the viewpoint taken by R. von Mises on p. 84 of his excellent book ***Probability, Statistics and Truth*** (George Allen & Unwin, London, 1957). The discussions given by H. Margenau in § 13.2 of *The Nature of Physical Reality* (McGraw-Hill, London, 1950) and by K. R. Popper, The Propensity Interpretation of Probability, *Brit. J. Philos. Sci.* **10**, 25 (1959), support essentially the same viewpoint.

understood by regarding each measurement of a relative frequency as an approximate measurement of this permanent physical property; in the same way the reproducibility of the ratio of the force on a body to its acceleration, in experiments carried out under widely differing circumstances, can be understood by regarding each measurement of this ratio as a measurement of the permanent physical property called mass. Just as in the experimental measurement of mass there is some error in the experimental measurement of probability, and much of it comes from the fact that the limit in (4.1) must in practice be replaced by the value of $n(A, \mathscr{N}, \mathfrak{T})/\mathscr{N}$ for some finite though large value of $\mathscr{N}$.

Another idealization used in defining physical probability is the assumption that the trial has ideal statistical regularity, so that the right-hand side of (4.1) is exactly reproducible. No trial performed with real physical objects could actually attain such perfection: however carefully the instructions defining the trial were drawn up, some factor (like the position of the centroid of a loaded die in the example considered earlier) would be inadequately controlled, and its variation from one replication of the statistical experiment to the next would spoil the exact reproducibility. This situation, too, has its analogue in the case of mass. Just as a trial has precisely defined probabilities only if it has ideal statistical regularity, so a body has a precisely defined mass only if it always consists of the same matter. In reality molecules evaporate or are rubbed off, while others are absorbed or settle on the surface as dust. This very slight variability of the mass of a body is usually neglected, being much less than the error of measurement; and likewise the deviation of a properly specified trial, such as the toss of a coin, from ideal statistical regularity is usually negligible, being much less than the error due to using a finite $\mathscr{N}$ in (4.1).

In this book the noun "probability" used without any qualifying adjective is always (except in the next four sentences) intended to mean physical probability, the measurable physical quantity defined by (4.1). Some statisticians and others find this definition of probability too narrow, since it applies only to statistically regular trials, and so leaves probabilities undefined in many situations where we should like to use them. For example, a person who is thinking of giving up smoking might be interested in the "probability" that smoking causes lung cancer, since he must weigh the pleasures of smoking against the risk of a painful death. This "probability" cannot be interpreted as a physical probability, since there is no trial which produces the outcome "smoking causes lung cancer" on some of its replications and "smoking does not cause lung cancer" on others; it can, however, be interpreted as a *reasonable degree of belief*. That is, if the person says this "probability" is $\frac{1}{2}$, he means that the evidence available to him lends equal support to the two alternative hypotheses, that smoking does and does not cause lung cancer.

The question how to assign actual numerical values to these reasonable degrees of belief is not easy to answer. One attractive proposal† is to define a "reasonable" person's degree of belief or *subjective probability*, that a particular event A will happen or that a particular assertion A is true, to be $o/(1 + o)$, where o is the odds that the person would barely be willing to offer in a monetary bet on A against not-A. This definition makes the subjective probability equal to the physical whenever the latter exists: a "reasonable" person would not offer odds greater than $o = p/(1 - p)$ on A against not-A, where A is a possible outcome of a statistically regular trial $\mathfrak{T}$ and p is its physical probability, for otherwise he would [by (4.1)] lose more money than he gained if his offer were accepted repeatedly during a very long sequence of replications of $\mathfrak{T}$. Unlike physical probabilities, subjective probabilities are also defined for events that are not outcomes of statistically regular trials; but for such events the subjective probabilities held by different people need not be the same. This lack of uniqueness is the main disadvantage of the subjective interpretation of probability.

In this book all the probability statements refer to statistically regular trials and can therefore be interpreted in terms of either physical or subjective probabilities. Since this is a book about physics rather than statistical inference we shall mainly use the physical interpretation, but the subjective interpretation (which under these conditions does not suffer from non-uniqueness) provides a useful alternative in some parts of the theory, particularly the part dealing with the relation between statistical mechanics and information theory (see Chap. VI, § 4). The value of the subjective interpretation for information theory lies in its explicit recognition of the fact that a person's subjective probabilities depend on the information available to him. With subjective probabilities defined in terms of betting odds, this dependence is evident from the way the odds on a horse-race change every time a new piece of information becomes available to the public concerning the condition of the turf, which horses have been doped, etc.

A third attitude towards probabilities, favoured by mathematicians, is to regard them as purely mathematical objects, like the points and lines of geometry, defined only through the axioms they satisfy. Such *formal probabilities* appear in their simplest form in the school textbook treatmentswhere the six faces of a die or the fifty-two cards in a pack are asserted (presumably for reasons of symmetry) to be "equally likely" without any statement of the practical consequences that arise from this equality. They appear in a more sophisticated form in the modern axiomatic treatments such as

† L. J. Savage, *The Foundations of Statistical Inference* (Methuen, London, 1962), p. 12. A "reasonable" person is one who (i) is always willing, whatever odds are proposed, to bet either for or against A, and (ii) is guided by the desire not to lose money in the long run.

For an alternative viewpoint, see H. Jeffreys, *Theory of Probability* (Oxford University Press, London, 1939), p. 15.

Kolmogorov's.† This formal definition of probability, if it is to be useful in practice, must be supplemented by some rule of interpretation, just as the axioms of geometry become useful in practical measurement only when supplemented by rules for associating the "points" and "lines" appearing in the axioms of geometry with the points and lines of everyday experience. Kolmogorov gives just such a rule of interpretation, part of which in our terminology would read: "One can be practically certain that if the trial $\mathfrak{T}$ is replicated a large number of times $\mathcal{N}$ then the ratio $n(A, \mathcal{N}, \mathfrak{T})/\mathcal{N}$ will differ very little from $Pr(A|\mathfrak{T})$." Apart from the intentional vagueness of the terms "practically certain", "large", "very little", this rule is very similar to our (4.1), so that the difference between Kolmogorov's interpretation of probability and the physical interpretation based on (4.1) would appear to be a matter more of emphasis than of substance.

5. The Markovian postulate

The definition (4.1) of physical probability hinges on the assumption that the trial $\mathfrak{T}$ is statistically regular, that is on the reproducibility of relative frequencies in statistical experiments comprising a very large number of replications of the trial. In applying this definition, therefore, it is important to have a criterion telling us which trials are statistically regular. So far we have only used a few examples of trials (the tossing of a coin, the casting of a die, the diffraction of an electron by a slit) which common experience has shown to be statistically regular. Now we shall formulate a physical assumption to provide a large class of statistically regular trials.

A convenient starting point is to analyse the notion of a trial a little further. In many cases, the experimental procedure that constitutes a trial can be split into three successive stages, which we shall call *preparation*, *development*, and *observation*. For example, in tossing a coin the preparation stage includes picking up the coin, examining it to see that it is not two-headed, and setting it into motion. The development stage is the coin's spinning flight through the air and coming to rest on the floor. The observation stage is seeing which face is uppermost at the end.

In general, the preparation stage consists of a sequence of observations and mechanical operations performed on the system. Both the duration of the preparation stage and the mechanical operations in it may depend on the results of previous observations. Thus in picking up the coin we must first observe its position, and the mechanical operation used to pick it up depends on the result of this observation; likewise, if the coin when ex-

† A. Kolmogorov, Grundbegriffe der Wahrscheinlichkeitsrechnung (*Ergebnisse der Mathematik* **2**, No. 3, Berlin, 1933); English translation *Foundations of the Theory of Probability* (Chelsea, New York, 1956).

amined is found to be two-headed, then the duration of the preparation stage must be prolonged until a genuine coin is found. Eventually, when observation shows that the system is ready, a decision is made to end the preparation stage and start the development stage. The instant when this happens will, in general, be different for different replications of the trial, but in discussing the instructions specifying the trial it is convenient, even so, to take the zero of time measurement at this instant (as if we measured time with a stop-watch which is started at the moment the preparation stage ends and the development stage begins).

The next stage of the trial, the development, is much simpler than the preparation stage. In many cases it consists merely in leaving the system isolated for a time interval whose duration, which we denote by t_1, is predetermined (and consequently the same at every replication). Thus if a tossed-up coin and the floor it lands on are regarded as a single physical system, then this system is essentially isolated between the time $t = 0$ when development is begun by setting the coin in motion and the time $t = t_1$ (say 5 sec later) when development is ended by observing the uppermost face. More generally, the development stage need not leave the system isolated; the time interval of predetermined duration t_1 may be occupied by a sequence of mechanical operations, themselves predetermined in the sense that the Hamiltonian must be the same function of t at every replication. For example, a variant of the coin-tossing trial would be to spin a coin suspended in a bearing, stop it at a predetermined instant t_1, and observe which side was uppermost at a later instant t_2. Here the coin would be stopped by a predetermined mechanical operation, say putting one's finger on it, at the time t_1. Thus the essential characteristics of the development stage are, first, that its duration must be predetermined, and second, that the external parameters fixing the form of the Hamiltonian must be predetermined functions of t.

The third, or observation, stage of the trial is again relatively simple. In many cases it consists merely of a single measurement performed at the time t_1. This measurement may not be sufficiently complete to reveal the observational state, but its result, the outcome of the trial, is determined by the observational state. More generally the observation stage could consist of a succession of measurements (a compound observation), beginning at time t_1.

Of these three stages of a trial, the preparation stage is potentially the most complicated because some of its important features, in particular its duration, are not predetermined but instead depend on the results of observations made during the preparation stage itself. In consequence the duration of the preparation stage, and possibly other features of it too, may be different at different replications of the same trial, depending on the observational history of the system during preparation. The analysis of the preparation stage can be much simplified by ignoring these "accidental" features of individual replications and concentrating on the predetermined features, which are evidently much closer to the essential nature of the trial.

For each of the simple trials we have been considering as examples, the preparation stage is distinguished by one salient predetermined feature, the observational state at the time $t = 0$. In tossing a coin, $t = 0$ is the instant when the coin begins its spinning flight, and provided the linear and angular velocity at this moment are not observably different at different replications, it is unimportant how the coin was set in motion. Likewise, if an electron is diffracted by a crystal, then provided it has the right momentum at the time $t = 0$ when it enters the crystal, the way it acquired this momentum is unimportant. Thus in each case the observational state at the time $t = 0$ is completely predetermined by the instructions specifying the trial, even though the way this observational state is reached is not.

The fact that for both the trials we have been using to exemplify statistical regularity the only really essential feature of the preparation stage is the observational state at $t = 0$ suggests that there may be a more general connection between statistical regularity and this observational state. In this book it will be assumed that such a connection does exist and that it can be expressed by the following postulate:

> If K is any observational state and $\mathfrak{T}$ is a trial whose preparation stage invariably ends with the system in the state K, then $\mathfrak{T}$ is statistically regular. (5.1)

This is a strong assumption, with many interesting implications. One of its implications is that the successive observational states of an isolated system constitute, in the language of probability theory, a Markov chain (see Chap. II, § 4). Accordingly (5.1) will be called the *Markovian postulate*.† This postulate may be thought of as an assumption that the observational description being used is complete in the sense that the current observational state embodies all the observational information about the past history of the system that is relevant to its observable future behaviour.

† The use of Markovian models for illustrative purposes in statistical mechanics dates back to the work of P. and T. Ehrenfest, Über zwei bekannte Einwände gegen das Boltzmannsche *H* Theorem, *Phys. Z.* **8**, 311 (1907). L. Onsager, Reciprocal Relations in Irreversible Processes, *Phys. Rev.* **38**, 2265 (1931), adopted a postulate equivalent to the restriction of (5.1) to systems that are isolated and unobserved for a long period of time preceding $t = 0$. J. G. Kirkwood, *J. Chem. Phys.* **14**, 180 (1946), applied the theory of Brownian motion (a branch of Markov chain theory) to very general problems in the theory of transport processes. The first explicit proposal to make a Markovian postulate a general principle of statistical mechanics appears to be that of M. S. Green, Markoff Processes and the Statistical Mechanics of Time-dependent Phenomena, *J. Chem. Phys.* **20**, 1281 (1952) and **22**, 398 (1954).

Whether or not the Markovian postulate is true depends upon two factors: the empirical properties of the physical system under consideration, and the details of the model used to represent observations made on it. These details are the choice of the dynamical images (or equivalently of indicators and observables) and the choice of the minimum time interval Δt between successive observations. In adopting the Markovian postulate we thus put a fairly stringent restriction on these choices which we have hitherto treated as arbitrary. For a given physical system, it is not usually a trivial task to find an observational description for which the Markovian postulate can reasonably be adopted; but there are indications that such a description can be found in many cases, if not all, particularly if we do not insist that all the observables appearing in the model should be instantaneously measurable. In the following paragraphs we consider some of the cases where a Markovian observational description is possible.

The simplest way for the Markovian postulate to be satisfied is for the observational states to be deterministic. Such determinism implies that the observational state at time t_1 is a unique function of the observational state at time 0; consequently, if the trial $\mathfrak{T}$ satisfies the premiss of (5.1) that the observational state at time 0 must be the same one, say K, at every replication, then the observational state at time t_1, which is the outcome of $\mathfrak{T}$, is also the same at every replication. That is, the trial $\mathfrak{T}$ is reproducible, and *a fortiori* any statistical experiment consisting of replications of $\mathfrak{T}$ is reproducible, so that $\mathfrak{T}$ itself is statistically regular, in conformity with the Markovian postulate (5.1). Moreover, since all the replications of $\mathfrak{T}$ composing the statistical experiment produce the same outcome, the probability of this outcome is 1 and that of all other hypothetical outcomes is 0.

One example where (5.1) is satisfied by virtue of determinism of the observational states is a classical system of only a few particles, such as the solar system. Here observational states and dynamical states are essentially identical, so that (5.1) is a consequence of the determinism of dynamical states. Another example is provided by the theory of heat conduction. If we regard the temperature distribution in a solid body as its observational state, then the heat conduction equation implies that the observational states are deterministic and hence Markovian. Likewise in the laminar hydrodynamics of an incompressible fluid the velocity field may be identified with the observational state, and the observational states so defined are deterministic and Markovian. Other branches of applied mathematics, such as elasticity and compressible hydrodynamics, supply similar examples, provided that in each case enough variables are used in defining the observational states to ensure that the equation of change for each one of them is of first order in the time variable.

To find an example where (5.1) is satisfied with probabilities other than 1 of 0, the nearest place to look is ordinary quantum mechanics. For a small quantum system, such as the electron in the diffraction experiment considered earlier, it is possible to make enough simultaneous measurements to determine the dynamical state (quantum state) of the system completely: in other words, the manifolds $\omega(A)$ defined in § 3 are one-dimensional. Consequently, if the preparation stage of a trial $\mathfrak{T}$ invariably ends with the system in the observational state K, then it also invariably ends with the system in the corresponding quantum state $|k\rangle$, with wave function $\psi_k(\boldsymbol{q}_1 \dots \boldsymbol{q}_N)$. Since dynamical states are deterministic the dynamical state at time t_1 is also uniquely determined (apart from an unimportant phase factor) by the conditions defining the trial; let us call it $|\alpha\rangle$. The probability that the observation completing the trial at time t_1 will show the system to be in, say, the observational state B is, according to the basic statistical postulate of quantum mechanics,

$$\langle\alpha|J_B|\alpha\rangle = \langle\alpha|\beta\rangle \langle\beta|\alpha\rangle = |\langle\beta|\alpha\rangle|^2; \tag{5.2}$$

here we have assumed that $\omega(B)$, like $\omega(A)$, is one-dimensional and have denoted a quantum state in it by $|\beta\rangle$, so that its projection operator is the outer product $|\beta\rangle \langle\beta|$. Both $|\beta\rangle$ and $|\alpha\rangle$ are assumed to be normalized, and $\langle\beta|\alpha\rangle$ denotes their inner product, already defined in § 3. Since quantum-mechanical probabilities are measurable physical quantities, the statistical experiments that measure them must be reproducible, so that the trial $\mathfrak{T}$ is statistically regular and the Markovian postulate is again satisfied.

A defect of the example just considered is that, although the Markovian postulate is satisfied, the postulate of compatibility is not. To avoid having to deal further with the postulate of compatibility, let us return to classical mechanics. A classical example where the observational states are Markovian but non-deterministic is given by the phenomenon of *Brownian motion*, discovered by the botanist Brown in 1827. Brown noticed that minute bodies, usually called particles although they are large enough to consist of many molecules, move erratically if suspended in a fluid. At first this motion was thought to be a manifestation of life, but further experiments made with particles of many different kinds, including ground-up stone from the Sphinx, soon made it clear that Brownian motion is caused by the impacts of molecules from the surrounding fluid against the Brownian particle. We owe the theory of Brownian motion to Einstein;† he based his work on the assumption that successive molecular impacts are essentially independent, so that if Δt is a time interval long compared with the duration of an impact (which is roughly 10^{-12} sec) the probability that the total impulse received by the particle during this time interval will have any specified value, say $\Delta \boldsymbol{p}$, depends on the history of the particle before the interval only through the velocity of the particle at the beginning of the interval.

† A. Einstein, Über die von der molekularkinetischen Theorie der Wärme geforderte Bewegung von in ruhenden Flüssigkeiten suspendierten Teilchen, *Ann. d. Phys.* **17** (4), 549 (1905).

Einstein's assumption is tantamount to assuming that if the observational states are defined in terms of small tolerance intervals for the velocity (or momentum) of the Brownian particle, and if Einstein's Δt and the time interval between successive observations are taken to be identical, then the Markovian postulate is satisfied. Incidentally, it should be noted that the system to which our definitions apply in this case comprises both the Brownian particle and the fluid it is suspended in: the Brownian particle alone cannot be taken as the system because its surroundings influence its motion strongly and unpredictably, so that its Hamiltonian cannot be regarded as even approximately determined by the instructions specifying the experimental procedure.

A disadvantage of this model of Brownian motion is that it treats the velocity of the Brownian particle, which cannot actually be observed, as an observable, whilst the position, which can be observed, is treated as unobservable. A more realistic model, which is also Markovian, can be obtained by treating both position and velocity as observables. That this model is Markovian is most easily seen if Δt is small enough to permit the neglect of its square. Then, if at time $t = 0$ the particle is in an observational state (x_0, v_0) corresponding to approximate values x_0 and v_0 for its observed position and velocity, its position at time Δt will be approximately $x_0 + v_0 \Delta t$, whilst the various possible values for the velocity at time Δt will have probabilities depending, as before, only on v_0. Thus, the probability for any observational state (x_1, v_1) at time Δt depends only on the observational state (x_0, v_0) at time 0, and so the Markovian postulate (5.1) is again satisfied.

Since the velocity of a Brownian particle cannot actually be observed, it is natural also to consider the model of observation in which the position of the particle is treated as the *only* observable. This model, however, is not Markovian unless Δt is relatively long (compared with the relaxation time of the velocity of the Brownian particle). If Δt is short, the particle moves approximately in a straight line, so that the observed positions at successive instants of observation are related by

$$x_1 = x_0 + (x_0 - x_{-1}) + \Delta x, \tag{5.3}$$

where x_{-1} is the observed position at time $-\Delta t$, and the term Δx, representing the effect of molecular impacts, has order of magnitude Δt times the order of magnitude of the velocity change due to molecular impacts during the time interval Δt. Let us consider a trial $\mathfrak{T}$ defined simply by the instruction that the observational state at time 0 is (x_0) and the outcome is the observational state at time Δt. If the Markovian postulate were valid here, the trial $\mathfrak{T}$ would be statistically regular. The simplest way to conduct this trial is for the experimenter to watch a cloud of Brownian particles until he observes one of them to be at the position x_0, and to note its position at the next instant of observation. Another way, which satisfies the definition of the trial equally well, is for him to choose an arbitrary value for x_{-1} and to wait until he observes one of the Brownian particles to have the positions x_{-1} and x_0 at two successive instants of observation before recording the position of this particle at the immediately following instant of observation. According to (5.3), an experimenter using this second method of conducting the trial can, if Δt is small enough to ensure that the term Δx in (5.3) does not outweigh the term $x_0 - x_{-1}$, exert a substantial influence on the outcome by his choice of x_1; hence, if he conducts a statistical experiment by replicating the trial many times, he can likewise exert an influence on the relative frequencies of the various possible outcomes of the trial through his choices of x_1 at the various replications. It follows that these relative frequencies need not be the same at every replication of the statistical experiment; that is, they are not reproducible, the trial is not statistically regular, and the Markovian postulate is not satisfied.

This last example suggests a method for converting a non-Markovian set of observational states into a set with a better chance of being Markovian. In the example, the lack of statistical regularity arises from the fact that the instructions defining the trial $\mathfrak{T}$ leave the value of x_{-1} at the experimenter's disposal. If, therefore, this freedom is taken away form the experimenter by including a specification of the observational state at time $-\Delta t$ as well as time 0 in the definition of the trial, then a much closer approach to statistical regularity will be achieved. Let us therefore define a new set of compound observational states to consist of pairs of successive observational states of the original type: that is, a system whose observational states at times $t - \Delta t$ and t are (x') and (x'') has in the new description the compound observational state (x', x'') at time t. We may define a trial $\mathfrak{T}'$ by the instructions that the compound observational state at time 0 is (x_{-1}, x_0) and that the outcome is the compound observational state (x_0, x_1) at time Δt. According to (5.3) this outcome is determined only by the initial compound observational states together with the random molecular impacts; accordingly it should be statistically regular. Thus the validity of the Markovian postulate has been restored by going from simple to compound observational states. In our example, each compound state is constructed from just two simple states, but it might be necessary to use three, or even more, in more complicated cases.

This method of converting a non-Markovian to a Markovian set of observational states has its counterpart in the theory of dynamical states. The necessary property of dynamical states corresponding to the Markovian property for observational states is the property of determinism—that is the differential equations of motion must be first order. One can set up a dynamical description of the system based on the positions of the particles only, without using their velocities. This description leads [eqn. (1.1)] to equations of motion of the second rather than the first order and is therefore non-deterministic; it can, however, be converted into a deterministic description by including the rates of change of position (the velocities) as well as the positions themselves in the set of quantities used to specify the dynamical state. This is essentially the same as using the positions at two slightly different times to specify the dynamical state at the later of the two times, and is thus an application of the method described in the preceding paragraph.

The examples we have considered indicate that there is a fairly large class of physical systems for which it is possible to choose observational states that satisfy the Markovian postulate. This class includes the deterministic models considered in the continuum treatments of matter and in thermodynamics; it also includes the models used in the standard discussions of fluctuation phenomena such as Brownian motion. These are just the systems to which we wish to apply statistical mechanics. Doubtless it is an oversimplification to assume, as we do here, that the postulate can be exactly satisfied by such physical systems (for further discussion of this point, see Chap. IV, § 7), but the consequences of the postulate are well worth investigating before trying anything more complicated.

5.1. Exercises

1. Show that the Markovian postulate is satisfied if the energy is the only observable [see eqn. (3.8) and its successor].

2. A system has two observational states, which we label 1 and 2, and its dynamical laws have the consequence that the observational states always follow the succession ...1122112211221122... Show that the Markovian postulate is not satisfied, but that a Markovian observational description can be set up using compound observational states.

6. Two alternative approaches

To support the position adopted in this book, that the physical hypothesis contained in the Markovian postulate is an essential ingredient in any complete account of the foundations of statistical mechanics, we consider in the present section two widely favoured alternative approaches to these foundations which do not make use of the Markovian postulate. These alternatives are the method based on the ergodic theorem and the method based on the hypothesis of equal *a priori* probabilities. We shall find that both methods encounter difficulties which can be traced back to the lack of some feature that the Markovian postulate can supply.

The first of these approaches is the attempt to base statistical mechanics on purely dynamical arguments, without any overt assumptions about observation or physical probabilities.† According to this approach, statistical

† A detailed review of this approach is given by I. Farquhar in *Ergodic Theory in Statistical Mechanics* (Wiley, New York, 1964).

mechanics is concerned only with the calculation of the *equilibrium values* of dynamical variables, which are defined as averages of these dynamical variables over an infinite time interval, calculated on the assumption that the system is isolated throughout. That is, if the dynamical state of a particular isolated system at time t is denoted by α_t, and $G(\alpha)$ is any dynamical variable, then the equilibrium value of G for that system is the time average

$$G_{eq} \equiv \lim_{T\to\infty} \frac{1}{T} \int_0^T G(\alpha_t)\, dt. \tag{6.1}$$

The point of the method is that, provided certain conditions are satisfied, the right-hand side of (6.1) can be evaluated without integrating the equations of motion. In classical mechanics this is accomplished by means of the *ergodic theorem* of general dynamics, according to which the time average on the right-hand side of (6.1) is, for all but a negligible class of initial states α_0, equal to an average over the *energy surface* defined as the surface in phase space whose equation is

$$H(\alpha) = E, \tag{6.2}$$

where $E \equiv H(\alpha_0)$ is the energy of the initial state. One way of writing this result is

$$G_{eq} = \lim_{\Delta E\to 0} \frac{\int_{\omega(E,\,\Delta E)} G(\alpha)\, d\alpha}{\int_{\omega(E,\,\Delta E)} d\alpha}, \tag{6.3}$$

where $\omega(E, \Delta E)$ means the region of phase space compatible with the condition

$$E - \Delta E < H(\alpha) \leq E. \tag{6.4}$$

Such a region is called an *energy shell* [see the discussion accompanying eqn. (3.8)].

The main condition of validity of the result (6.3) is that the energy surface must be *metrically transitive* (or *ergodic*) which means that there must be no dynamical variables, other than functions of the Hamiltonian, that are invariants of the motion. (If such invariants do exist the average in (6.3) must be confined to the part of the energy surface on which all the invariants take the same value as for the state α_0.) Until recently, the application of the ergodic theorem in statistical mechanics was justified by faith only, since there was no way of telling whether or not the energy surface of a given dynamical system was metrically transitive; but in 1963 Sinai† showed rigorously that the energy surface for a system of hard spheres is indeed

† Ya. G. Sinai, On the Foundations of the Ergodic Hypothesis for a Dynamical System of Statistical Mechanics, *Soviet Mathematics*, 4, 1818 (1963).

metrically transitive. Thus the rigorous calculation of equilibrium values for classical systems is at last a real possibility.

Despite these recent successes, the attempt to base statistical mechanics on purely dynamical arguments must face some fundamental difficulties. One of these is that there is no simple way of extending the method to quantum mechanics. In quantum mechanics the initial state does not in general have a definite energy, so that it is no longer possible to express time averages in terms of a single parameter representing the initial state (namely the energy of that state); the attractive simplicity and generality of the classical result (6.3) therefore do not survive transplantation to quantum mechanics.

Another difficulty, which arises in classical as well as in quantum mechanics, is that the method gives little information about non-equilibrium behaviour. Indeed, the validity of (6.3) is, by itself, not even sufficient to ensure that the system has any tendency to approach equilibrium. For example, the classical harmonic oscillator satisfies (6.3), but its dynamical variables never settle down to the "equilibrium" values given by (6.3); they are periodic functions of time and oscillate for ever. To ensure that the system is *dissipative*—that its dynamical variables do tend to approach their equilibrium values defined by (6.1)—we need a stronger condition than metrical transitivity. This stronger condition is known in ergodic theory as *mixing*.† For the hard-sphere system, Sinai's work proves that the energy surface has the mixing property as well as metrical transitivity.

A related difficulty is the so-called *paradox of irreversibility*. This "paradox" is the apparent contradiction between the symmetry of the laws of dynamics under the time reversal transformation $t \rightarrow -t$ and the asymmetry of the actual behaviour of material objects which is supposed to be derivable from these laws. Strictly speaking no paradox arises as long as the approach is used only for the calculation of equilibrium values in accordance with (6.1) or (6.3), for a state of equilibrium is essentially static (apart from fluctuations) and therefore symmetrical under time reversal.‡ On the other hand, the behaviour of systems that are far from equilibrium is not symmetrical under time reversal: for example, heat always flows from a hotter to a colder body, never from a colder to a hotter. If this behaviour could be derived from the symmetrical laws of dynamics alone, there would, indeed, be a paradox; we must therefore acknowledge the fact that some additional postulate, non-dynamical in character and asymmetrical under time reversal,

† For a rough indication of the meaning of "mixing", see p. 135. For a precise definition, see §8 of V. I. Arnold and A. Avez, *Problèmes Ergodiques de la Mécanique Classique* (Gauthier-Villars, Paris, 1967).

‡ The departure from time-reversal symmetry in using the limit operation $T \rightarrow +\infty$ in (6.1) rather than $T \rightarrow -\infty$ is only apparent, since both operations give the same limiting value for all but a negligible class of states α_0. This is a consequence of the ergodic theorem.

must be adjoined to the symmetrical laws of dynamics before the theory can become rich enough to include non-equilibrium behaviour.† In the present theory, this additional postulate is the Markovian postulate.

A second widely used approach to the fundamental problems of statistical mechanics is the method of *a priori probabilities*‡ (also called the method of *representative ensembles*). In this method the need for a non-dynamical postulate is explicitly recognized. This postulate, the *hypothesis of equal a priori probabilities*, asserts (in classical mechanics) that all sets of dynamical states compatible with the available observational information about the system, and occupying equal volumes in phase space, are equally likely. Let us suppose that the available information has been obtained by means of idealized observations of the type described in § 3, and denote by ω_t the set of all dynamical states compatible with the observational history of the system prior to time t. For example, if the observational history consists of a single measurement of the energy, indicating a value lying between the numbers $E - \Delta E$ and E, then ω_t is the energy shell defined in (6.4). Whatever the observational history of the system, the hypothesis of equal *a priori* probabilities implies that if a measurement of a dynamical variable G is made at the time t then the statistical expectation of its result is equal to the average value of G over the region ω_t, or, in symbols, that

$$\langle G\rangle_t = \int_{\omega_t} G(\alpha)\, d\alpha \Big/ \int_{\omega_t} d\alpha. \tag{6.5}$$

For the definition of statistical expectation, see Chap. II, § 2.

In this approach the difficulties encountered in the ergodic approach – that of distinguishing dissipative from non-dissipative systems and that of the paradox of irreversibility – no longer arise: the former because the theory is rich enough to deal with non-equilibrium as well as equilibrium situations, and the latter because the postulate (6.5) is not symmetrical under time reversal, ω_t being defined in terms of the behaviour observed before time t, not after. Unfortunately, however, a new difficulty takes their place. This difficulty concerns the physical interpretation to be placed on the quantity $\langle G\rangle_t$ in (6.5). For simplicity let us consider the case where $G(\alpha)$ is taken to be $J_A(\alpha)$, the indicator variable of an arbitrary observational state A. According to the mathematical rules of probability theory [see Chap. II,

† The point is clearly made by N. G. van Kampen on p. 173 of the volume *Fundamental Problems in Statistical Mechanics*, edited by E. G. D. Cohen (North-Holland, Amsterdam, 1962). He says that the paradox of irreversibility "makes it clear that there cannot be a rigorous mathematical derivation of the macroscopic equations from the microscopic ones. Some additional information or assumption is indispensable. One cannot escape from this fact by any amount of mathematical funambulism." Moreover, the additional assumption he chooses (p. 180) is that "the stochastic process used to describe the system macroscopically is a Markov process".

‡ R. C. Tolman, *The Principles of Statistical Mechanics* (Oxford University Press, London, 1938), §§ 23 and 24. See also H. Jeffreys, *Theory of Probability*, § 7.6.

eqn. (2.8)], the statistical expectation of an indicator variable is a probability, so that (6.5) becomes

$$Pr(A_t) = \int_{\omega_t} J_A(\alpha)\, d\alpha / \int_{\omega_t} d\alpha, \tag{6.6}$$

where $Pr(A_t)$ means the probability of observing A at time t. The difficulty we are discussing reduces here to a question of how to interpret the probability $Pr(A_t)$.

According to Tolman,† this probability is to be interpreted as the relative frequency of the result A_t on repeated trials of the same experiment—that is as a physical probability. We have seen in the previous section, however, that a physical probability has a definite value only if it refers to a statistically regular trial; before (6.6) becomes useful as a rule for calculating physical probabilities, therefore, it must be supplemented by a rule for determining whether or not the observational history defining the region ω_t in phase space is the preparation stage of a statistically regular trial. As we have seen in § 5, the simplest such rule is the Markovian postulate; but if the Markovian postulate is adopted then, as will be shown in Chapter IV, eqn. (6.5) becomes a theorem and therefore need no longer be taken as a postulate. To oversimplify somewhat, the proposed postulate (6.5) is either meaningless or unnecessary.

An alternative possibility, championed recently by Jaynes,‡ is to interpret the probability in (6.6) subjectively rather than physically. According to this point of view, (6.6) gives the subjective probability that the system is in state A at time t, calculated for a person who knows absolutely nothing about the system except the observational history which implies that its dynamical state is in the phase-space region ω_t. The assignment of equal subjective probabilities to equal volumes of phase space within the region ω_t is a form of the *principle of indifference*, according to which a person who sees no essential difference between two possible alternatives assigns them equal subjective probabilities.

An uncritical application of the subjective interpretation of (6.6) can lead to some paradoxical results, analogous to Bertrand's paradox in the classical theory of probability.* Consider, for example, a person who knows nothing whatever about some particular large physical system apart from its dynamical structure (its Hamiltonian function $H(\alpha)$) and the fact that its energy lies in the range between two values E_1 and E_2. He might reasonably estimate

† R. C. Tolman, *The Principles of Statistical Mechanics*, § 24.

‡ E. T. Jaynes, Information Theory and Statistical Mechanics, *Phys. Rev.* **106**, 171 (1957); A Katz, *Principles of Statistical Mechanics* (Freeman, San Francisco, 1967). Jaynes proposes a prescription more general than (6.6) for calculating subjective probabilities when given information about the system is available; but when this information is of the type considered here, comprising a specification of the observational states of the system at specified past times, then his method is equivalent to the method of *a priori* probabilities: see exercise 3 of Chap. II, § 7.1.

* R. von Mises, *Probability, Statistics and Truth*, p. 77.

that the probabilities for the energy to lie in each of the two halves of the range are roughly equal:

$$Pr\{E_1 < H(\alpha) < \tfrac{1}{2}(E_1 + E_2)\} \approx Pr\{\tfrac{1}{2}(E_1 + E_2) < H(\alpha) < E_2\}; \qquad (6.7)$$

but a calculation based on (6.6) (the calculation is simplest when the system is an ideal gas) leads to the much less plausible result that the probability for the half-range of higher energy is very nearly 1, so that

$$Pr\{E_1 < H(\alpha) < \tfrac{1}{2}(E_1 + E_2)\} \ll Pr\{\tfrac{1}{2}(E_1 + E_2) < H(\alpha) < E_2\}, \qquad (6.8)$$

unless $E_2 - E_1$ is comparable with the energy of a single molecule. Presumably one must conclude that a person who makes the reasonable probability estimates (6.7) does not really know "absolutely nothing" about the system, but has in some way made use of the information provided by his experience of similar systems on previous occasions. Thus the idealized mental state we have described as "knowing absolutely nothing" about the system is a poor approximation to the actual mental state of a person who has very little information about the system. The only way that comes to mind for guaranteeing that the subjective interpretation of (6.6) will not lead us into this type of difficulty is to restrict its application to the situations where the physical definition of probability also applies; but if this is done, then, as explained earlier, something like the Markovian postulate is once again needed in order to specify these situations.

CHAPTER II

Probability Theory

1. Events

In this chapter we shall explore those implications of the definition of probability and of the Markovian postulate that are independent of dynamics. Sections 1 and 2 will outline the formal consequences of the definition of probability; nothing in them will be new to readers familiar with the early chapters of such books as Feller's.† The remaining sections will be devoted to the consequences of the Markovian assumption. Even before introducing dynamics into the theory at all, we shall find that we can deduce important qualitative results, concerning in particular the nature of equilibrium and the way it is approached.

The elementary part of the theory of probability consists of the immediate consequences of the definition of probability itself. Let $\mathfrak{T}$ be any statistically regular trial. Since the outcome of $\mathfrak{T}$ is determined by observation and the set of observational states is countable, the set of possible outcomes of $\mathfrak{T}$ is also countable, and so the outcomes may be labelled by positive integers 1, 2, 3, ... In the mathematical theory of probability the set of all possible outcomes of a trial are regarded as an abstract space called the *sample space* of that trial. The number of points in the sample space may be finite or infinite, and if infinite may be countable (like the positive integers) or non-countable (like the real numbers). For our purposes, however, the part of the theory referring to finite and countably infinite sample spaces will suffice.

Any set of possible outcomes of $\mathfrak{T}$ is called an *event* of $\mathfrak{T}$: for example, if the trial is a throw of a cubical die, with outcomes 1, 2, 3, 4, 5, and 6, then one of the $2^6 = 64$ events of this trial is the event "the result of the throw is an even number", consisting of the outcomes 2, 4, and 6. At any particular replication of $\mathfrak{T}$ event E *occurs* if and only if some one out of the outcomes forming the set E occurs. Evidently each of the individual outcomes is also an event, called an *elementary event*; other examples are the *empty event*, denoted by O, which contains no outcomes at all and therefore never occurs, and the *complete event*, denoted by I, which includes all the possible outcomes and therefore occurs at every replication of the trial.

By a simple extension of the definition of probability [(4.1) of Chap. I], we can define a probability for each event E of the trial $\mathfrak{T}$:

$$Pr(E) = \lim_{\mathcal{N}\to\infty} \frac{n(E, \mathcal{N}, \mathfrak{T})}{\mathcal{N}}, \tag{1.1}$$

† W. Feller, *An Introduction to Probability Theory and its Applications* (Wiley, New York, 1957), Vol. 1. Feller's book is an excellent general reference for this whole chapter.

where $Pr(E)$ is an abbreviation for $Pr(E|\mathfrak{T})$, and $n(E, \mathcal{N}, \mathfrak{T})$ is the number of times E occurs in the first $\mathcal{N}$ replications of $\mathfrak{T}$ in an infinite statistical experiment. It is clear from this definition that all probabilities satisfy the inequalities

$$0 \leqq Pr(E) \leqq 1. \tag{1.2}$$

In special cases $Pr(E)$ can take one of the extreme values 0 or 1. For example, if E is the complete event I, which occurs at every replication, then its probability is 1. If an event has probability 1, however, we cannot conclude that the event is I, or even that it occurs at every replication; for example, by (1.1) an event has probability 1 if it occurs at all but a finite number of the replications in an infinite statistical experiment. An event with probability 1 is said to be *almost certain.* In the same way, the empty event has probability 0, but occurrences of an event whose probability is 0 are not necessarily impossible; an event with probability 0 may be said to be *almost impossible.* For example, if an arrow is shot at a circular target, the probability of the arrow (regarded as a geometrical point) landing at the precise geometrical centre of the target (or at any other specified point, for that matter) would generally be regarded as zero, even though this event could happen, without violating any principle, once or even a few times in a very long sequence of trials.

From the definitions of E and of $n(E, \mathcal{N}, \mathfrak{T})$ we have

$$n(E, \mathcal{N}, \mathfrak{T}) = \textstyle\sum_{A \in E} n(A, \mathcal{N}, \mathfrak{T}), \tag{1.3}$$

where the sum is over the outcomes A that together make up the event E. Substituting (1.3) into (1.1) we obtain

$$Pr(E) = \textstyle\sum_{A \in E} Pr(A), \tag{1.4}$$

where $Pr(A)$ means $Pr(A|\mathfrak{T})$; thus the probabilities of all possible events are determined by the probabilities of the elementary events. Equation (1.4) incidentally shows also that the limit in (1.1) does exist, provided the trial is statistically regular. Applying (1.4) to the complete event I, we obtain

$$\textstyle\sum_{A} Pr(A) = Pr(I) = 1, \tag{1.5}$$

where the sum is over all the possible outcomes of $\mathfrak{T}$. This equation is called the *normalization condition.* If the ratios of the numbers $Pr(1)$, $Pr(2)$, ..., are known but not their absolute values, then (1.5) can be used to find these absolute values.

From two or more different events we can construct further events by the operations of forming the *logical sum* and *logical product* of events. The logical sum, or *union,* of two events E_1 and E_2, written $E_1 \oplus E_2$ or $E_1 \cup E_2$, is the event consisting of all the elementary events that are in either E_1 or E_2. Thus, at any replication of the trial $\mathfrak{T}$, the event $E_1 \oplus E_2$ occurs if E_1

or E_2 or both occur, but does not occur if neither E_1 nor E_2 occurs. The logical product, or *intersection*, of two events E_1 and E_2, written E_1E_2 or $E_1 \cap E_2$, is the event consisting of all the elementary events that are in both E_1 and E_2. Thus, at any replication of the trial $\mathfrak{T}$, the event E_1E_2 occurs if E_1 and E_2 both occur, but does not occur if either E_1 or E_2 does not occur. For example, if the trial is the throw of a die, and E_1 is $\{2, 4, 6\}$ or in words "the outcome is even", and E_2 is $\{4, 5, 6\}$ or in words "the outcome exceeds 3", then $E_1 \oplus E_2$ is $\{2, 4, 5, 6\}$ or in words "the outcome is even or greater than 3", whilst E_1E_2 is $\{4, 6\}$ or in words "the outcome is even and greater than 3".

The point of the terminology and notation for logical sums and products is that the operations of logical addition and multiplication combine according to the commutative, distributive, and associative laws of ordinary algebra:

$$\left.\begin{aligned} &E_1 \oplus E_2 = E_2 \oplus E_1 \quad \text{and} \quad E_1E_2 = E_2E_1, \\ &E_1 \oplus (E_2 \oplus E_3) = (E_1 \oplus E_2) \oplus E_3 \quad \text{and} \quad E_1(E_2E_3) = (E_1E_2)E_3, \\ &E_1(E_2 \oplus E_3) = E_1E_2 \oplus E_1E_3. \end{aligned}\right\} \tag{1.6}$$

These laws, and also (1.7) and (1.8), are direct consequences of the definitions of logical sum and logical product. In this algebra of events† the events I and O play parts similar to those of 1 and 0 in the algebra of numbers:

$$E \oplus O = E, \quad EI = E, \quad EO = O. \tag{1.7}$$

The laws

$$E \oplus I = I, \quad E \oplus E = E, \quad EE = E \tag{1.8}$$

show, however, that the analogy with ordinary algebra is not complete. Another important difference between the algebra of events and the algebra of numbers is that the algebra of events contains no close analogues for subtraction and division. For example, if $E_1 = \{2, 4, 6\}$ and $E_2 = \{4, 5, 6\}$ as before, then with $E_3 = \{5\}$ we have $E_1 \oplus E_2 = E_1 \oplus E_3 = \{2, 4, 5, 6\}$ but $E_2 \neq E_3$; and with $E_4 = \{4, 6\}$ we have $E_1E_2 = E_1E_4 = \{4, 6\}$ but $E_2 \neq E_4$.

Although the algebra of events is different from the algebra of numbers, there are important similarities between the algebraic relationships connecting the events of a given trial and those connecting their probabilities. The simplest of these is the *addition law of probabilities*. Two events are said to be *mutually exclusive* if their logical product is the empty event O, so that

† It is called *Boolean algebra*, after G. Boole who first described it in *The Laws of Thought* (1854; Dover reprint, New York, 1951).

the occurrence of one event precludes the occurrence of the other. The addition law, which follows at once from (1.4), is

$$\left.\begin{aligned} &Pr(E_1 \oplus E_2) = Pr(E_1) + Pr(E_2) \\ &\text{if } E_1 \text{ and } E_2 \text{ are mutually exclusive.} \end{aligned}\right\} \tag{1.9}$$

For example, if E is any event and E' is its *complement*, defined to consist of all the elementary events not in E, so that E' occurs if and only if E does not occur, then $E \oplus E' = I$ and $EE' = O$. Consequently (1.9) gives

$$Pr(E') = 1 - Pr(E), \tag{1.10}$$

since $Pr(I) = 1$. The addition law is easily generalized to the logical sum of any number n of mutually exclusive events:

$$\left.\begin{aligned} &Pr(E_1 \oplus E_2 \oplus \cdots \oplus E_n) = Pr(E_1) + Pr(E_2) + \cdots + Pr(E_n) \\ &\quad \text{if } E_i E_j = O \text{ for all } i < j \leqq n. \end{aligned}\right\} \tag{1.11}$$

The normalization condition (1.5) is a special case of this law. If the events $E_1, \ldots, E_n$ are not mutually exclusive the addition law is replaced by the inequality

$$Pr(E_1 \oplus \cdots \oplus E_n) \leqq Pr(E_1) + \cdots + Pr(E_n) \tag{1.12}$$

which holds for arbitrary $E_1, \ldots, E_n$.

In order to supplement this addition law for probabilities by a multiplication law, we must first define *conditional probability*. If E_1 and E_2 are any two events of a trial $\mathfrak{T}$, then the conditional probability of E_1 given E_2, written $Pr(E_1|E_2\mathfrak{T})$ or in abbreviated notation $Pr(E_1|E_2)$, is defined by

$$Pr(E_1|E_2) = \frac{Pr(E_1E_2)}{Pr(E_2)} \quad \text{provided} \quad Pr(E_2) \neq 0. \tag{1.13}$$

Of course, if $E_2 = I$, then the conditional probability of E_1 is the same as its ordinary probability. By (1.1), the definition (1.13) is equivalent to

$$Pr(E_1|E_2\mathfrak{T}) = \lim_{\mathcal{N}\to\infty} \frac{n(E_1E_2, \mathcal{N}, \mathfrak{T})}{n(E_2, \mathcal{N}, \mathfrak{T})}. \tag{1.14}$$

This is the limiting relative frequency of the event E_1 in a modified statistical experiment consisting not of every member in a succession of replications of $\mathfrak{T}$ but only of those replications where E_2 occurs. Alternatively it can be regarded as the probability of E_1 in a modified trial, which may be called $E_2\mathfrak{T}$, performed by replicating $\mathfrak{T}$ repeatedly until the event E_2 occurs and counting only the outcome of this last replication. In the subjective interpretation of probability, $Pr(E_1|E_2)$ is interpreted as the subjective probability of E_1 for a person who knows that E_2 has occurred or will occur.

By virtue of their interpretation as probabilities for a modified trial, conditional probabilities satisfy laws analogous to the laws given earlier in this section for ordinary probabilities:†

$$0 \leqq Pr(E_1|E_2) \leqq 1, \tag{1.15}$$

$$Pr(E_1|E_2) = \textstyle\sum_{A\in E_1} Pr(A|E_2), \tag{1.16}$$

$$\textstyle\sum_A Pr(A|E_2) = 1, \tag{1.17}$$

$$Pr(E_1 \oplus E_2|E_3) = Pr(E_1|E_3) + Pr(E_2|E_3) \quad \text{if} \quad E_1E_2 = O, \tag{1.18}$$

$$Pr(E_1'|E_2) = 1 - Pr(E_1|E_2). \tag{1.19}$$

Alternatively, these results can be derived directly from the definition (1.10) of conditional probability. We can also obtain a generalization of (1.13), by considering the quotient $Pr(E_1E_2|E_3)/Pr(E_2|E_3)$. By the definition of conditional probability, this quotient is equal to $Pr(E_1E_2E_3)/Pr(E_2E_3)$ or $Pr(E_1|E_2E_3)$. Thus we derive

$$Pr(E_1E_2|E_3) = Pr(E_1|E_2E_3)\,Pr(E_2|E_3) \tag{1.20}$$

which may be called the *multiplication law of probabilities.*

Two events E_1 and E_2 are said to be *statistically* (or *stochastically*) *independent* if $Pr(E_1|E_2) = Pr(E_1)$. By (1.13) this definition may be written in the more symmetrical form

$$Pr(E_1E_2) = Pr(E_1)\,Pr(E_2). \tag{1.21}$$

In the same way, any number of events $E_1, E_2, \ldots, E_n$ are said to be statistically independent if

$$Pr(E_1E_2 \ldots E_n) = Pr(E_1)\,Pr(E_2) \ldots Pr(E_n). \tag{1.22}$$

For example, in the throw of a true cubical die (one for which the outcomes 1, 2, 3, 4, 5, 6 all have probability 1/6) the events $E_1 = \{2, 4, 6\}$ ("the outcome is even") and $E_2 = \{1, 2\}$ ("the outcome is less than 3") are statistically independent, since by (1.4) we have $Pr(E_1) = \frac{1}{6} + \frac{1}{6} + \frac{1}{6} = \frac{1}{2}$, $Pr(E_2) = \frac{1}{6} + \frac{1}{6} = \frac{1}{3}$, and $Pr(E_1E_2) = Pr(2) = \frac{1}{6}$. On the other hand, if the die were loaded the same events might well be statistically dependent. Thus statistical independence, unlike mutual exclusiveness, depends not only on

† Strictly, the condition (1.15) should be written

$$0 \leqq Pr(E_1|E_2) \leqq 1 \quad \text{if} \quad Pr(E_2) \neq 0$$

since $Pr(E_1|E_2)$ is not defined when $Pr(E_2) = 0$; likewise a corresponding proviso should be attached to every other formula involving conditional probabilities. To keep the formulae simple, this proviso is normally suppressed.

the definitions of the events concerned but also on the probabilities of the various outcomes. It is often possible to assert on physical grounds that two events are statistically independent, even though their probabilities are not known; some conditions that suffice to justify such assertions will be discussed in Chap. II, § 3, and Chap. VI, § 2.

1.1. Exercises

1. If E_1 and E_2 are statistically independent events, show that the pairs of events E_1' and E_2, E_1 and E_2', E_1' and E_2' are likewise statistically independent.

2. A room contains r people. Obtain an expression for the probability that at least two of them have the same birthday, ignoring leap years and assuming that all the elementary events have equal probabilities. With the help of the inequality $1 - n/365 < e^{-n/365}$, show that if $r \geqq 23$, then the probability is more than $\frac{1}{2}$.

2. Random variables

Any numerical quantity determined by the outcome of a trial is called a *random variable.*† If the value taken by a random variable X, when the trial outcome is A, is denoted by $X(A)$, then the set of numbers $X(1)$, $X(2)$, ... (which need not all be different), amount to a complete specification of the random variable X. For example, the square of the number of pips on top, after a throw of a cubical die, is a random variable, which can be specified by giving the 6 numbers

$$X(1) = 1,\ X(2) = 4,\ X(3) = 9,\ X(4) = 16,\ X(5) = 25,\ X(6) = 36.$$

The average value of a general random variable X in an infinite statistical experiment is called its *statistical expectation* or *mean.* It may be denoted by $\langle X \rangle$ or $\mathscr{E}(X)$ or, if we wish to exhibit the trial $\mathfrak{T}$ explicitly, by $\mathscr{E}(X|\mathfrak{T})$. More precisely, the definition is

$$\langle X \rangle \equiv \mathscr{E}(X|\mathfrak{T}) \equiv \lim_{\mathscr{N}\to\infty} \frac{1}{\mathscr{N}} \sum_{i=1}^{\mathscr{N}} X^{(i)}, \tag{2.1}$$

where $X^{(i)}$ is the value taken by X at the ith replication of $\mathfrak{T}$ in an infinite statistical experiment. The sum in (2.1) may be transformed by grouping the $\mathscr{N}$ replications according to their outcomes; in the notation of (1.1) it becomes

$$\textstyle\sum_i X^{(i)} = \sum_A X(A)\, n(A, \mathscr{N}, \mathfrak{T}), \tag{2.2}$$

where the sum ranges over all possible outcomes A. Substituting (2.2) into (2.1) we obtain

$$\langle X \rangle = \textstyle\sum_A X(A)\, Pr(A) \tag{2.3}$$

† If the outcome of the trial is determined by a single rather than a compound observation, the random variables are also observables.

or, if the trial $\mathfrak{T}$ is shown explicitly,

$$\mathscr{E}(X|\mathfrak{T}) = \sum_A X(A)\, Pr(A|\mathfrak{T}). \tag{2.4}$$

If two random variables are related by a linear equation, their statistical expectations are related in the same way:

$$\text{if } Y = aX + b \quad \text{then} \quad \langle Y\rangle = a\langle X\rangle + b. \tag{2.5}$$

This follows at once from (2.4) and the normalization condition (1.5). Useful particular cases of (2.5) are

$$\langle c\rangle = c \quad (c \equiv \text{const.}), \quad \langle\langle X\rangle\rangle = \langle X\rangle, \langle X - \langle X\rangle\rangle = 0. \tag{2.6}$$

The simplest non-trivial random variables are the *indicators* of the various events: the indicator of an event E is defined by

$$J_E \equiv \left\{\begin{array}{l} 1 \text{ if } E \text{ occurs} \\ 0 \text{ if } E \text{ does not occur.} \end{array}\right\} \tag{2.7}$$

In other words, $J_E(A) = 1$ if $A \in E$ and 0 if not. The expectation of any indicator is equal to the probability of its event; for with $X = J_E$ the sum in (2.1) is simply the number of occurrences of E in the first $\mathscr{N}$ replications, so that (1.1) gives

$$\langle J_E\rangle = Pr(E). \tag{2.8}$$

The same result can as easily be obtained from (2.3) and (1.4). The properties of events in the algebra of logical addition and multiplication are closely paralleled by properties of their indicators in the algebra of ordinary addition: thus

$$\begin{aligned} J_O &= 0, \\ J_I &= 1, \\ J_{EF} &= J_E J_F, \\ J_{E\oplus F} &= J_E \oplus J_F, \end{aligned} \tag{2.9}$$

$$\text{where } x \oplus y \equiv x + y - xy = \left\{\begin{array}{l} 0 \text{ if } x = 0 \quad \text{and} \quad y = 0 \\ 1 \text{ if } x = 1 \quad \text{or} \quad y = 1. \end{array}\right\} \tag{2.10}$$

It may be verified that for every law of the algebra of events given in (1.6)–(1.8) there is a corresponding relation of ordinary algebra between their indicators. For example, the middle equation of (1.8) corresponds to

$$\begin{aligned} J_E \oplus J_E &= J_E + J_E - J_E^2 \\ &= J_E, \quad \text{since} \quad J_E^2 = J_E. \end{aligned} \tag{2.11}$$

If X is a random variable, the function of a real variable x defined by

$$p_x \equiv Pr(X = x) \quad \text{for all real } x \tag{2.12}$$

will be called the *probability distribution* of X. Here we have used $(X = x)$ to stand for the event consisting of all outcomes for which the random variable X takes the value x; by (1.4) the value of p_x is the sum of the probabilities of all these outcomes. Since X can take at most a countable infinity of different values, the function p_x differs from zero only at these values. The probability distribution can be visualized as a distribution of mass along an infinite straight line, a mass equal to p_x being put at each point x on the line. The total mass is given as 1 by the normalization condition (1.5),

$$\textstyle\sum_x p_x = 1, \tag{2.13}$$

where the sum is over all values of x where $p_x \neq 0$. The position of the centroid of the distribution is

$$\textstyle\sum_x xp_x = \langle X \rangle, \tag{2.14}$$

derived by grouping together the terms of (2.3) according to their values of $X(A)$.

A measure of the spread or dispersion of the probability distribution of X about its mean value $\langle X \rangle$ is given by the analogue of the moment of inertia,

$$\textstyle\sum_x [x - \langle X \rangle]^2 p_x = \langle X^2 \rangle - \langle X \rangle^2, \tag{2.15}$$

where X^2 means the random variable whose value for each outcome is the square of the corresponding value of X. The quantity (2.15) is called the *variance* of X, and abbreviated to var (X). Since the sum in (2.15) is obviously non-negative, the variance of X is non-negative and has a real square root, called the *standard deviation* of X and often denoted by σ_X:

$$\sigma_X^2 = \text{var}\,(X) = \langle X^2 \rangle - \langle X \rangle^2. \tag{2.16}$$

The standard deviation has the same dimensions as X and is the analogue of a radius of gyration.

A probability distribution with small standard deviation tends to be highly concentrated near its mean; that is, the probability of X taking a value far away from $\langle X \rangle$ is small. This qualitative statement is made quantitative by *Chebyshev's inequality*

$$Pr(|X - \mu| \geqq \delta) \leqq \sigma_X^2/\delta^2 \quad \text{for any } \delta > 0, \tag{2.17}$$

where $\mu \equiv \langle X \rangle$. To prove this very useful result, let J_E be the indicator of the event $E \equiv (|X - \mu| \geqq \delta)$; that is, let

$$J_E \equiv \begin{cases} 0 \text{ if } \mu - \delta < X < \mu + \delta \\ 1 \text{ if not.} \end{cases} \tag{2.18}$$

Then (Fig. 2) J_E satisfies the inequality

$$J_E \leqq (X-\mu)^2/\delta^2 \quad \text{for all values of } X. \tag{2.19}$$

Taking the expectation of both sides and using (2.8) and (2.16), we obtain Chebyshev's inequality (2.17).

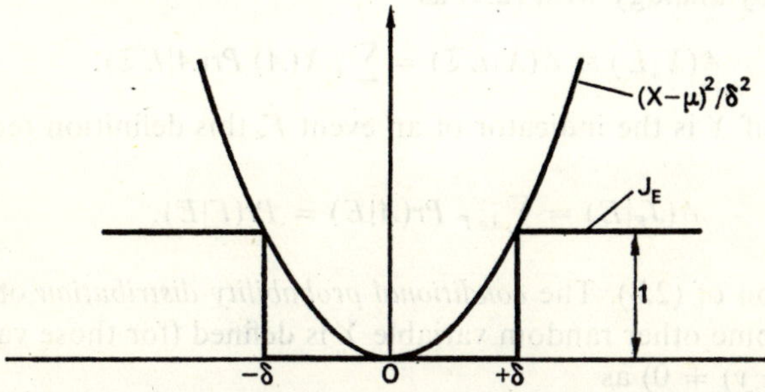

FIG. 2. The inequality (2.19).

The idea of a probability distribution is easily extended to two or more random variables. If X and Y are two random variables belonging to the same trial $\mathfrak{T}$, then their *joint probability distribution* is the function of two variables

$$p_{xy} \equiv Pr(X = x.Y = y), \tag{2.20}$$

where $(X = x.Y = y)$ stands for the logical product of the two events $(X = x)$ and $(Y = y)$. Such a joint probability distribution can be visualized as a distribution of mass in the xy-plane, a mass p_{xy} being placed at each point (x, y). The normalization condition is

$$\textstyle\sum_x \sum_y p_{xy} = 1. \tag{2.21}$$

By (1.4), the probability distribution of the single random variable X is

$$Pr(X = x) = \textstyle\sum_y p_{xy} \tag{2.22}$$

and similarly

$$Pr(Y = y) = \textstyle\sum_x p_{xy}. \tag{2.23}$$

The means, variances, and standard deviations of X and Y are defined, just as before, to be $\langle X\rangle$, $\langle X^2\rangle - \langle X\rangle^2$, etc. In addition we can now define a quantity corresponding to the product moment of the mass distribution about its centroid

$$\textstyle\sum_x \sum_y [x - \langle X\rangle][y - \langle Y\rangle]\, p_{xy} = \langle XY\rangle - \langle X\rangle\langle Y\rangle. \tag{2.24}$$

This is called the *covariance* of X and Y, abbreviated to covar (X, Y), and provides an indication of the amount of correlation between the two random variables: if covar (X, Y) is positive, then the deviations of X and Y away from their mean values tend to have the same sign, whereas if covar (X, Y) is negative, the deviations tend to have opposite signs.

The idea of conditional probability is also useful in connection with probability distributions. If X is a random variable and E an event, both belonging to the same trial $\mathfrak{T}$, then we may define the *conditional expectation* of X given E by analogy with (2.4) as

$$\mathscr{E}(X|E) \equiv \mathscr{E}(X|E\mathfrak{T}) = \sum_A X(A)\, Pr(A|E\mathfrak{T}). \tag{2.25}$$

For example, if X is the indicator of an event F, this definition reduces, by (1.16), to

$$\mathscr{E}(J_F|E) = \sum_{A \in F} Pr(A|E) = Pr(F|E), \tag{2.26}$$

a generalization of (2.8). The *conditional probability distribution* of X given the value of some other random variable Y is defined (for those values of y where $Pr(Y = y) \neq 0$) as

$$p_{x|y} \equiv Pr(X = x|Y = y) = p_{xy}\Big|\sum_x p_{xy} \tag{2.27}$$

by (1.13), (2.20), and (2.23). The conditional mean and variance of X, given $Y = y$, are related to $p_{x|y}$ by the natural generalizations of (2.14) and (2.15), for example

$$\mathscr{E}(X|Y = y) = \sum_x x p_{x|y}. \tag{2.28}$$

Using (2.4) and (1.4), the multiplication law (1.20) and finally (2.25) we can express the unconditional expectation of x in terms of its conditional expectations for the different values of Y:

$$\begin{aligned} \mathscr{E}(X) &= \sum_A X(A)\, Pr(A) = \sum_A X(A) \sum_y Pr(A.Y = y) \\ &= \sum_A \sum_y X(A)\, Pr(A|Y = y)\, Pr(Y = y) \\ &= \sum_y \mathscr{E}(X|Y = y)\, Pr(Y = y). \end{aligned} \tag{2.29}$$

The last expression can be regarded as the unconditional expectation of a new random variable whose value for the outcome A is $\mathscr{E}(X|Y = Y(A))$.

Two random variables X and Y are said to be *statistically independent* if $Pr(X = x|Y = y) = Pr(X = x)$ for all values of x and y such that $Pr(Y = y) \neq 0$; that is, if

$$Pr(X = x.Y = y) = Pr(X = x)\ Pr(Y = y) \quad \text{for all } x, y, \tag{2.30}$$

so that the joint probability distribution p_{xy} defined in (2.20) is the product of a function of x and a function of y. One consequence of this definition is that, by (2.24),

$$\text{covar}\,(X, Y) = 0 \tag{2.31}$$

if X and Y are statistically independent.

2.1. Exercise

1. A bottle holds 10^{22} molecules of air. Assuming that they are statistically independent of each other and that each of them has a probability $\frac{1}{2}$ of being in the lower half of the bottle, show that the variance of the number of molecules in the lower half is $2\cdot5 \times 10^{21}$. Using Chebyshev's inequality, find an upper bound on the probability that this number of molecules will deviate from its mean value, which is 5×10^{21}, by more than one part in a million.

3. Statistical independence

In this section we shall apply the calculus of probabilities outlined in §§ 1 and 2 to a more general type of trial than the simple type analysed in Chap. I, § 5 into preparation, development, and observation stages. The more general experimental procedure is to carry out two or more trials of this simple type, either successively or simultaneously; such a combination of simple trials, taken as a whole, may be regarded as a trial of a new type, which we shall call a *compound trial*.

The compound trial consisting of a performance of the simple trial $\mathfrak{T}_1$ followed by a performance of the trial $\mathfrak{T}_2$ may be denoted by $\mathfrak{T}_2\mathfrak{T}_1$. As usual in probability notation, the ordering of the symbols corresponds to their time order but with the earliest time at the right. Likewise, we may denote by B_2A_1 that outcome of the compound trial which consists of outcome A from $\mathfrak{T}_1$ followed by outcome B from $\mathfrak{T}_2$. This notation is consistent with the logical product notation of (1.6); for B_2A_1 is indeed the logical product of an event A_1, consisting of all outcomes of the compound trial in which the sub-trial $\mathfrak{T}_1$ gives the outcome A, with an event B_2, consisting of all outcomes for which $\mathfrak{T}_2$ gives the outcome B. For example, if the compound trial is to roll a die and then toss a coin, the outcome when the die shows a 6 and the coin a head may be denoted by H_26_1; this is the logical product of the event 6_1, consisting of all outcomes where the first sub-trial yields a 6, with the event H_2, consisting of all outcomes where the second sub-trial yields a head. The notation is easily generalized to compound trials consisting of three or more successive simple sub-trials.

A most important property of compound trials of this type, in which the first sub-trial ends before the second begins, is that the two sub-trials are *statistically independent:* this means that the two events A_1 and B_2 defined in the preceding paragraph are statistically independent in the sense defined in § 1, for all A and all B. To show this, consider the experimental procedure of repeatedly replicating $\mathfrak{T}_1$ until the first occurrence of the outcome A, and then immediately replicating $\mathfrak{T}_2$. We can look on this experimental procedure in two distinct ways: either as a single replication of $\mathfrak{T}_2$ preceded by a rather complicated but irrelevant preliminary manipulation of the system, or as number of abortive attempts to obtain a replication of $\mathfrak{T}_2\mathfrak{T}_1$ whose outcome belongs to the event A_1, followed at last by one successful attempt. (The word "abortive" is meant to imply that if a replication of $\mathfrak{T}_2\mathfrak{T}_1$ is started by replicating $\mathfrak{T}_1$, and the outcome of this replication of $\mathfrak{T}_1$ is found not to be A, then there is no need to complete $\mathfrak{T}_2\mathfrak{T}_1$ by replicating $\mathfrak{T}_2$: instead we proceed immediately to the next attempt.)

Consider now a statistical experiment conducted by replicating this experimental procedure a very large number of times, say $\mathcal{N}_2$, and let $\mathcal{N}(B)$ be the number of them at which the final outcome is B. According to the first of the two viewpoints just mentioned, each of these $\mathcal{N}_2$ replications of the experimental procedure is to be regarded as a replication of $\mathfrak{T}_2$; in our earlier notation, where $n(E, \mathcal{N}, \mathfrak{T})$ represents the number of occurrences of E in a statistical experiment comprising $\mathcal{N}$ replications of a trial $\mathfrak{T}$, we therefore have

$$n(B, \mathcal{N}_2, \mathfrak{T}_2) = \mathcal{N}(B)$$

from which it follows by (1.1) that

$$Pr(B|\mathfrak{T}_2) = \lim_{\mathcal{N}_2 \to \infty} \frac{\mathcal{N}(B)}{\mathcal{N}_2}. \tag{3.1}$$

According to the second viewpoint, on the other hand, each of the $\mathcal{N}_2$ replications of the experimental procedure is to be regarded as a succession of incomplete replications of $\mathfrak{T}_2\mathfrak{T}_1$, none of which if completed would yield an outcome belonging to the event A_1, followed by one complete replication whose outcome is of the form B_2A_1 and therefore does belong to the event A_1. If the total number of replications, both complete and incomplete (that is, the total number of replications of $\mathfrak{T}_1$), is denoted by $\mathcal{N}_1$, then, using the $n(E, \mathcal{N}, \mathfrak{T})$ notation again, we have

$$n(B_2A_1, \mathcal{N}_1, \mathfrak{T}_2\mathfrak{T}_1) = \mathcal{N}(B),$$

$$n(A_1, \mathcal{N}_1, \mathfrak{T}_2\mathfrak{T}_1) = \mathcal{N}_2,$$

from which it follows by (1.14) that

$$Pr(B_2|A_1\mathfrak{T}_2\mathfrak{T}_1) = \lim_{\mathcal{N}_1 \to \infty} \frac{\mathcal{N}(B)}{\mathcal{N}_2}. \tag{3.2}$$

To complete the proof of statistical independence, we invoke the definition (1.13) of conditional probability, according to which

$$Pr(B_2A_1|\mathfrak{T}_2\mathfrak{T}_1) = Pr(B_2|A_1\mathfrak{T}_2\mathfrak{T}_1)\, Pr(A_1|\mathfrak{T}_2\mathfrak{T}_1).$$

It follows, by (3.1) and (3.2), that

$$Pr(B_2A_1|\mathfrak{T}_2\mathfrak{T}_1) = Pr(B|\mathfrak{T}_2)\, Pr(A_1|\mathfrak{T}_2\mathfrak{T}_1). \tag{3.3}$$

Summing over all values of A_1 and using the addition law (1.11) and the normalization condition (1.5) we obtain

$$Pr(B_2|\mathfrak{T}_2\mathfrak{T}_1) = Pr(B|\mathfrak{T}_2), \tag{3.4}$$

so that (3.3) can be written

$$Pr(B_2A_1) = Pr(B_2)\, Pr(A_1) \tag{3.5}$$

all probabilities now referring to the compound trial $\mathfrak{T}_2\mathfrak{T}_1$. By the definition (1.21), this result shows that the events A_1 and B_2 are statistically independent, and so the successive sub-trials $\mathfrak{T}_1$ and $\mathfrak{T}_2$ may also be said to be statistically independent.

The argument which led to (3.5) can be extended to a compound trial $\mathfrak{T}_n \ldots \mathfrak{T}_1$ consisting of n successive (non-overlapping) trials; the result is

$$Pr(G_n \ldots B_2A_1) = Pr(G_n) \ldots Pr(B_2)\, Pr(A_1) \tag{3.6}$$

so that the successive sub-trials are again statistically independent.

One consequence of (3.6) is the *law of large numbers*, a theorem giving quantitative expression to the qualitative statement made in Chap. I, § 4 that large statistical experiments are approximately reproducible. If $\mathfrak{T}$ is a simple trial, the finite statistical experiment conducted by replicating this trial n times may be regarded as a compound trial $\mathfrak{T}_n\mathfrak{T}_{n-1} \ldots \mathfrak{T}_1$ whose n sub-trials are all identical. If A is a possible outcome for $\mathfrak{T}$, its relative frequency in this statistical experiment is a random variable equal to the average of the indicators of the events $A_1, A_2, \ldots, A_n$. This random variable is

$$\bar{J}_{A,n} \equiv \frac{1}{n} \sum_{i=1}^{n} J_{A,i}, \tag{3.7}$$

where

$$J_{A,i} \equiv \begin{Bmatrix} 1 \text{ if the outcome of } \mathfrak{T}_i \text{ is } A \\ 0 \text{ if not} \end{Bmatrix} \tag{3.8}$$

is the indicator of the event A_i.

By (2.8), the statistical expectation of $\bar{J}_{A,n}$ is

$$\begin{aligned} \langle \bar{J}_{A,n} \rangle &= \frac{1}{n} \sum_{i=1}^{n} \langle J_{A,i} \rangle \\ &= n^{-1} \textstyle\sum_i Pr(A_i | \mathfrak{T}_n \ldots \mathfrak{T}_1). \end{aligned} \tag{3.9}$$

Since the sub-trials $\mathfrak{T}_1, \ldots, \mathfrak{T}_n$ are all statistically independent and identical, we have (as in (3.4))

$$Pr(A_i | \mathfrak{T}_n \ldots \mathfrak{T}_1) = Pr(A | \mathfrak{T}_i) = Pr(A | \mathfrak{T}),$$

and so (3.9) reduces to

$$\langle \bar{J}_{A,n} \rangle = Pr(A | \mathfrak{T}). \tag{3.10}$$

The variance of $\bar{J}_{A,n}$, defined as in (2.16), is

$$\begin{aligned} \operatorname{var} \bar{J}_{A,n} &= \langle \bar{J}^2_{A,n} \rangle - \langle \bar{J}_{A,n} \rangle^2 \\ &= n^{-2} \textstyle\sum_i \sum_j [\langle J_{A,i} J_{A,j} \rangle - \langle J_{A,i} \rangle \langle J_{A,j} \rangle] \\ &= n^{-2} \textstyle\sum_i [\langle J^2_{A,i} \rangle - \langle J_{A,i} \rangle^2] \\ &\leqq 1/4n \end{aligned} \tag{3.11}$$

by (3.7), (2.31), and (2.8), since $J_{A,i}$ and $J_{A,j}$ are statistically independent if $i \neq j$. Using (3.10) and (3.11) in Chebyshev's inequality (2.17) we obtain, for any $\delta > 0$,

$$Pr\,\{|\bar{J}_{A,n} - Pr(A|\mathfrak{T})| \geqq \delta\} \leqq 1/4n\delta^2 . \tag{3.12}$$

Taking the limit $n \to \infty$ we obtain the *weak law of large numbers*

$$\lim_{n\to\infty} Pr\,\{|\bar{J}_{A,n} - Pr(A|\mathfrak{T})| \geqq \delta\} = 0 \quad \text{if } \delta > 0. \tag{3.13}$$

The meaning of this result is that for large n the value of the random variable $\bar{J}_{A,n}$ is unlikely to be far from its statistical expectation value $Pr(A|\mathfrak{T})$. If we measure the reproducibility of an experiment yielding a numerical outcome by the probability that this outcome will lie within a predetermined tolerance interval, then (3.13) asserts that the reproducibility of a statistical experiment to measure $Pr(A|\mathfrak{T})$ approaches perfection as $n \to \infty$ if the extremes of the tolerance interval are taken to be $Pr(A|\mathfrak{T}) - \delta$ and $Pr(A|\mathfrak{T}) + \delta$, with arbitrary positive δ.

The result (3.13) should not be confused with the *strong law of large numbers*,†

$$\lim_{n\to\infty} \lim_{k\to\infty} Pr\left\{\max_{r=0,\,1,\,\ldots,\,k} |\bar{J}_{A,n+r} - Pr(A|\mathfrak{T})| \geqq \delta\right\} = 0 \text{ for all } \delta > 0, \tag{3.14}$$

where the probability refers to the compound trial $\mathfrak{T}_{n+k} \ldots \mathfrak{T}_2\mathfrak{T}_1$. The weak law asserts that if n is large thne $\bar{J}_{A,n}$, the relative frequency of the outcome A in a statistical experiment of duration n, is likely to be near $P(A|\mathfrak{T})$; the strong law asserts that $\bar{J}_{A,n+r}$ is likely to remain near $P(A|\mathfrak{T})$ however many replications (r in number) after the nth are added to this statistical experiment. We shall not prove the strong law of large numbers since only the weak law is needed in the rest of this book.

3.1. Exercises

1. Prove (3.6) for $n = 3$.
2. A sequence of random variables, say $X_1, X_2, \ldots$, is said to *converge in probability* to a limiting numerical value, say x, if

$$\lim_{s\to\infty} Pr(|X_s - x| \geqq \delta) = 0 \quad \text{for all} \quad \delta > 0.$$

Thus the weak law of large numbers (3.13) asserts that the sequence $\bar{J}_{A,1}, \bar{J}_{A,2}, \ldots$ converges in probability to the limiting value $Pr(A|\mathfrak{T})$. Show that if X is any random variable associated with the simple trial $\mathfrak{T}$ then the sequence of random variables $\bar{X}_s \equiv s^{-1} \sum_{t=1}^{s} X(A_t)$ associated with the compound trials $\mathfrak{T}_s\mathfrak{T}_{s-1} \ldots \mathfrak{T}_1$ $(s = 1, 2, \ldots)$ converges in probability to $\langle X \rangle$.

† W. Feller, *An Introduction to Probability Theory and its Applications*, p. 190; R. von Mises, *Probability, Statistics, and Truth*, p. 127.

4. Markov chains

According to the Markovian postulate [Chap. I, eqn. (5.1)], the simplest type of statistically regular trial that can be performed with a given physical system is to prepare this system by any means in a definite observational state, say K, at time 0, and to make one further observation on it, say at time t. Between these two times the system may either be kept isolated or subjected to a specified sequence of mechanical operations (changes of Hamiltonian). The outcome of this trial is the observational state of the system at the instant of observation t. We shall denote the probability of the outcome A in this trial by $Pr(A_t|K)$, the form of the Hamiltonian not being shown explicitly. To simplify the notation, we shall use units of time where $\Delta t = 1$, so that the value of t must be a positive integer.

There is a useful conceptual device which simplifies the application of our basic definition of probability [Chap. I, eqn. (4.1)] to probabilities such as $Pr(A_t|K)$, which contain a time parameter t. According to the strict definition of physical probability, different values of t correspond to different trials and therefore to different statistical experiments. It is, however, possible to collapse all these statistical experiments into one, by taking a large collection of copies of the system, putting all of them simultaneously into the observational state K at a time 0, and measuring the fraction that are in each of the observational states $A = 1, 2, \ldots, M$ at each of the instants $t = 1, 2, \ldots$. The collection of copies of the system used for this portmanteau statistical experiment will be called a *statistical ensemble*. This idea of an ensemble has played an important part in the development of statistical mechanics.

Alternatively, one may interpret the probabilities $Pr(A_t|K)$ as subjective probabilities, but care is necessary to ensure that the right information is made available to the person whose degree of belief the probability $Pr(A_t|K)$ measures. The only relevant information available to this person must be a description of the system and the mechanical operations to be performed on it, together with the information that the observational state at time 0 is K; he must not be permitted to use the results of observations made at times later than 0. This is most simply achieved by regarding $Pr(A_t|K)$ as a subjective probability based only on the information available just after time 0. (If subjective probabilities are defined in terms of betting odds then $Pr(A_t|K)$ refers, whatever the value of t, to the odds for a bet placed just after the initial time 0.) Thus, in the subjective interpretation the dependence of $Pr(A_t|K)$ on t does not arise from the acquisition of new information as time proceeds, as in the case of horse-race betting odds mentioned in Chap. I, § 4, but rather from the fact that the different probabilities $Pr(A_t|K)$ refer to different events.

By making observations on a statistical ensemble at more than one time, we can measure probabilities relating to trials with compound observation stages. The simplest of these probabilities is $Pr(B_u A_t|K)$, the probability that a system prepared in state K at the time 0 passes through the state A at the instant t and then through the state B at the (later) instant u. (As in § 3, the notation is consistent with that for logical products, since the passage of the system through the state A at the instant t may be regarded as an event, denoted by A_t, in the sample space of the trial to which $Pr(B_u A_t|K)$ refers.) The most general such probability is $Pr(G_{t^n} \ldots B_{t^2} A_{t^1}|K)$, where n is an arbitrary positive integer, $t^n, \ldots, t^1$, are arbitrary instants of observation with $t^n > t^{n-1} \ldots > t^1 > 0$, and $G, \ldots, C, B, A, K$ are $n+1$ arbitrary observational states. As in the case $n = 1$, the Hamiltonian and its possible time-dependence are not shown.

An important consequence of the Markovian postulate [Chap. I, eqn. (5.1)] is that we can use it to express probabilities referring to more than two instants, such as $Pr(G_{t^n} \ldots A_{t^1}|K)$ with $n > 1$, in terms of two-instant probabilities such as $Pr(A_t|K)$. For simplicity we consider first the case $n = 2$. The multiplication law of probability, eqn. (1.20), gives

$$Pr(B_u A_t|K) = Pr(B_u|A_t K)\, Pr(A_t|K). \tag{4.1}$$

The conditional probability $Pr(B_u|A_t K)$ is measured by the fraction of occurrences of B_u in the sub-ensemble consisting of those members of the complete ensemble which pass through the state A at time t. Provided $u > t$, the procedure whereby this sub-ensemble is selected may be regarded as the preparation procedure of a new trial. This preparation procedure is to put a system into state K at time 0 and then at time t either to start the development stage if the system is in the state A or else to abandon the trial if the system is not in the state A. Since the development stage of this new trial invariably begins with the system in the state A, the Markovian postulate implies that it is statistically regular and that its probabilities are the same as for a trial whose preparation procedure is simply to put the system into state A at time t. Equating the probabilities in these two trials we obtain

$$Pr(B_u|A_t K) = Pr(B_u|A_t). \tag{4.2}$$

Finally, substituting into (4.1), we obtain

$$Pr(B_u A_t|K) = Pr(B_u|A_t)\, Pr(A_t|K). \tag{4.3}$$

A useful equation can be deduced by summing both sides of (4.3) over all values of A and using the addition law† (1.16):

$$Pr(B_u|K) = \textstyle\sum_A Pr(B_u|A_t)\, Pr(A_t|K). \tag{4.4}$$

This result is known as the *Chapman–Kolmogorov* equation.

† In quantum mechanics this application of the addition law in the form $\sum_A Pr(B_u A_t|K) = Pr(B_u|K)$ depends on the postulate of compatibility stated in Chap. I, § 3, which ensures that a measurement of A made on every system of the ensemble at the instant t does not affect the probability for B at the later time u. When quantum interference effects are important, it is not the probabilities but the probability amplitudes that satisfy this type of addition law. See R. P. Feynman, Space–Time Approach to Non-relativistic Quantum Mechanics, *Rev. Mod. Phys.* **20**, 267 (1948).

The argument that gave (4.2) can easily be generalized to the case where any number of observations are made during the trial. The generalized form of (4.2) is

$$Pr(G_{t^n}|F_{t^{n-1}} \dots A_{t^1}K) = Pr(G_{t^n}|F_{t^{n-1}}) \quad \text{for } n = 1, 2, \dots, \tag{4.5}$$

whence, by an inductive argument using the multiplication law of probability (1.20), we can obtain the generalized form of (4.3):

$$Pr(G_{t^n}F_{t^{n-1}} \dots A_{t^1}|K) = Pr(G_{t^n}|F_{t^{n-1}}) \dots Pr(B_{t^2}|A_{t^1})\, Pr(A_{t^1}|K) \quad \text{for } n = 1, 2, \dots. \tag{4.6}$$

A random sequence of states with probabilities satisfying (4.5) or (4.6) is called a *Markov chain:* this is our reason for calling eqn. (5.1) of Chap. I the Markovian postulate.†

The simplest Markov chains are those which are *stationary*, that is, those for which conditional probabilities such as $Pr(B_u|A_t)$ are invariant under time shifts:

$$Pr(B_{u+s}|A_{t+s}) = Pr(B_u|A_t) \quad \text{for all} \quad s \geqq -t. \tag{4.7}$$

For the Markov chains describing a physical system, this invariance comes about if the surroundings of the system do not alter, and in particular if the system is isolated. When (4.7) holds the quantity $Pr(B_u|A_t)$ is a function of $u - t$ only, so that we may write

$$Pr(B_u|A_t) = w_{BA}(u - t). \tag{4.8}$$

The case where $u - t = 1$ is particularly important. The quantity

$$w_{BA} \equiv w_{BA}(1) = Pr(B_1|A) \tag{4.9}$$

is the probability that a system reaching the state A at any instant of observation will make a transition to the state B in time for the next instant of observation. Accordingly the quantities w_{BA} are called the *transition probabilities* of the stationary Markov chain. It is a fundamental problem of statistical mechanics to express these transition probabilities in terms of the dynamical properties of the system. This problem will be tackled in Chapter IV.

Any of the probabilities relating to a stationary Markov chain can be written in terms of the probabilities $w_{BA}(t)$; according to (4.6) and (4.8) the formula is

$$Pr(G_{t^n}F_{t^{n-1}} \dots B_{t^2}A_{t^1}|K) = w_{GF}(t^n - t^{n-1}) \dots w_{BA}(t^2 - t^1)\, w_{AK}(t^1). \tag{4.10}$$

† A very complete account of the theory of finite Markov chains is given by M. Fréchet in *Traité du Calcul des Probabilités et de ses Applications*, by E. Borel and collaborators (Gauthier-Villars, Paris, 1938), Vol. I, Fascicule III, Book 2. See also chapter XV of Feller's *An Introduction to Probability Theory and its Applications* and § 2.2 of M. S. Bartlett's *Stochastic Processes* (Cambridge University Press, London, 1955).

In the special case where $t^1 = 1$, $t^2 = 2$, etc., this reduces to

$$Pr(G_n F_{n-1} \ldots B_2 A_1 | K) = w_{GF} \ldots w_{CB} w_{BA} w_{AK}. \tag{4.11}$$

Summing over all values for $F \ldots A$ and using the addition law (1.16) for conditional probabilities, together with (4.8), we obtain

$$w_{GK}(n) = Pr(G_n | K) = \sum_F \ldots \sum_A w_{GF} \ldots w_{CB} w_{BA} w_{AK}. \tag{4.12}$$

The right-hand side (containing n factors and $n-1$ summations) is just the nth power of the matrix $[w_{BA}]$; thus (4.12) may be rewritten in matrix notation

$$[w_{BA}(n)] = [w_{BA}]^n. \tag{4.13}$$

Using (4.10) together with (4.12) or (4.13) we can express any probability in the Markov chain in terms of the matrix w_{BA} of transition probabilities. From their definitions (4.8) and (4.9), the matrices $[w_{BA}]$ and $[w_{BA}(s)]$ have the properties

$$w_{BA} \geqq 0 \quad \text{and} \quad w_{BA}(n) \geqq 0 \quad \text{for all } B, A \text{ and all } n \geqq 1 \tag{4.14}$$

$$\textstyle\sum_B w_{BA} = 1 \quad \text{and} \quad \sum_B w_{BA}(n) = 1 \quad \text{for all } A \text{ and all } n \geqq 1 \tag{4.15}$$

by virtue of (1.15) and (1.17). Any matrix that shares with the matrices $[w_{BA}]$ and $[w_{BA}(n)]$ these properties of having non-negative elements and unit column sums is called a *stochastic matrix*.

A simple example of a stationary Markov chain is provided by the *Ehrenfest urn model*.† In the simplest form of this model, an urn contains N balls, some of which are black and the rest white. A ball is drawn at random from the urn and replaced by one of the opposite colour; this process is repeated many times. If the observational state is taken to be the number A of black balls, then the sequence of observational states constitutes a Markov chain, since the probability of drawing a black ball depends only on the number of balls of each colour in the urn at the time, not on the number at any preceding time. The transition probabilities for this model are

$$w_{BA} = \begin{Bmatrix} A/N \text{ if } B = A - 1 \\ \\ 1 - A/N \text{ if } B = A + 1 \\ \\ 0 \text{ otherwise} \end{Bmatrix} \tag{4.16}$$

since the probability of drawing a black ball is A/N and of drawing a white ball is $1 - A/N$.

† P. and T. Ehrenfest, Über zwei bekannte Einwände gegen das Boltzmannsche *H* Theorem, *Phys. Z.* **8**, 311 (1907). See also W. Feller, *An Introduction to Probability Theory and its Applications*, Vol. I, p. 313.

A variant of the Ehrenfest model is the "dog–flea model" in which well-trained fleas, numbered 1, 2, ..., N, live on two dogs. Numbers $\leqq N$ are called at random, and every time a flea's number is called he jumps to the other dog. If the observational state is taken to be the number of fleas on one of the dogs, this model has the same transition probabilities (4.16) as the Ehrenfest model. The model has played an important part in the development of statistical mechanics. It has the property that if N is large and A is $\ll N$, then A is practically certain to increase. This mimics irreversible physical processes such as heat conduction, in which the energy of the cooler of two bodies in thermal contact is practically certain to increase.

It is useful to write the equations for stationary Markov chains in a form that is also valid if the chain does not become stationary until some instant t^* later than the time 0 when the preparation stage of the trial ends. This corresponds to a physical situation where the system is subjected to a sequence of mechanical operations between the times 0 and t^* but is left isolated after the time t^*. We may call t^* the *initial of instant isolation.*

For a Markov chain of this type, eqns. (4.8) and (4.9) still hold for the times $t \geqq t^*$ when the chain is stationary, except that w_{BA} (1) need no longer be equal to $Pr(B_1|A)$. The Chapman–Kolmogorov equation (4.4) may therefore be written in the form

$$p_B(u) = \textstyle\sum_A w_{BA}(u-t)\, p_A(t) \quad \text{if} \quad t \geqq t^*, \tag{4.17}$$

where

$$p_A(t) \equiv Pr(A_t|K). \tag{4.18}$$

The time-dependent probability distribution $p_A(t)$ defined in (4.18) plays a very important part in the theory of Markov chains. For any t, this probability distribution satisfies the non-negativity and normalization conditions

$$\left.\begin{aligned} p_A(t) \geqq 0 \quad \text{for } A = 1, 2, \ldots, \\ \textstyle\sum_A p_A(t) = 1. \end{aligned}\right\} \tag{4.19}$$

There is no harm in suppressing the initial state K in the notation (4.18) since the operations performed between the instants 0 and t^*, which are just as important as the initial state, have already been suppressed. Apart from the conditions (4.19), we shall treat the probability distribution at the time t^* as completely arbitrary, since it is plausible that a set of operations prior to t^* can be found to produce any desired probability distribution at time t^*. Provided the distribution at t^* is known, the distribution at any later time is determined by (4.17). In particular, it can be determined from the equation obtained by setting $u = t + 1$ in (4.17):

$$p_B(t+1) = \textstyle\sum_A w_{BA} p_A(t) \quad (t \geqq t^*). \tag{4.20}$$

By subtracting $p_B(t)$ from both sides and using (4.15), we can rewrite this equation in the alternative form

$$\begin{aligned} p_B(t+1) - p_B(t) &= \sum_A [w_{BA} p_A(t) - w_{AB} p_B(t)], \\ &= \sum_A{}' [w_{BA} p_A(t) - w_{AB} p_B(t)], \end{aligned} \tag{4.21}$$

where $\sum_A'$ means a sum with the $A = B$ term omitted. The continuous-time analogue of (4.21), with a time derivative replacing the finite-difference expression on the left, is called the *master equation*. We shall use this name here for eqn. (4.21) as well. If the probabilities in it are interpreted in terms of an ensemble, the master equation expresses the increase in the number of systems in state B as the number of systems entering that state less the number leaving it.

If (4.20) has been solved for a general initial probability distribution $p_A(t^*)$, it can be used to find the higher transition probabilities $w_{BA}(t)$, since with $t^* = 0$ eqn. (4.18) gives

$$\left.\begin{aligned} w_{BA}(t) = p_B(t) \quad \text{if } p_A(0) &= \delta_{AK} \\ &\equiv \begin{cases} 1 \text{ for } A = K \\ 0 \text{ otherwise.} \end{cases} \end{aligned}\right\} \tag{4.22}$$

4.1. Exercises

1. A container holding gas is separated into two compartments by a diaphragm pierced by a hole much smaller than the mean free path in the gas. If N is the total number of gas molecules and A denotes the number of them that are in (say) the left-hand compartment, the probability that some molecule will pass through the hole to the right between the times t and $t+1$ is $cA(t)$, and the probability that one will pass to the left is $c[N - A(t)]$, where $A(t)$ denotes the value of A at time t and c is a small constant. Neglecting the probability that two or more molecules pass through the hole between the times t and $t+1$ and assuming that A is the only observable, obtain the matrix of transition probabilities.

2. In the Ehrenfest urn model, show that the conditional expectation of the observational state at time $t+1$, given that the observational state at time t was $A(t)$, is $N/2 + (1 - 2/N)[A(t) - N/2]$. Hence obtain a formula for the expectation of $A(t)$, given that $A(0) = 0$.

3. If two or more of the states in a Markov chain are lumped together into a single composite state, the resulting states are no longer Markovian, in general. Verify this statement for the case where the original Markov chain had three states 1, 2, and 3, with $w_{12} = w_{23} = w_{31} = 1$, and the states 2 and 3 are lumped together.

4. Show that if the compound observational state of the system at time t is defined to comprise the (simple) observational states at the times t and $t-1$, and the simple observational sates are Markovian, then the compound observational states are also Markovian.

5. Classification of observational states

A basic problem in statistical mechanics is to find a way of calculating the observable properties of systems in *equilibrium:* that is, of systems that have been isolated for a long time. Since the observational states of an isolated system form a stationary Markov chain, a prerequisite for dealing with this problem is to study the asymptotic behaviour of a stationary Markov chain at large values of the time t.

This asymptotic behaviour is simplest if the chain is *deterministic*. As explained in Chap. I, § 1, determinism means that the state at any time, say t_1, is uniquely determined by the state at any earlier time, say t_0, together with the form of the Hamiltonian specifying the structure of the system at all intermediate times. An example of a deterministic observational description is provided by the mathematical theory of heat conduction, according to which the temperature distribution in an isolated solid body at any time $t > t_0$ is uniquely determined by the distribution at time t_0; if the observational states of this solid body are taken to be the various possible temperature distributions, then they are deterministic. In general, we shall say that an observational state A is deterministic to mean that if the system is in the state A at any time t then its state at time $t+1$ is a unique function of A, which we write $U(A)$ or simply UA. If the state UA is also deterministic, then the state at time $t+2$ must be $U^2A \equiv U(UA)$, and so on. If the entire Markov chain is deterministic, its transition probability matrix w_{BA} and the related matrices $w_{BA}(s)$ are given by

$$\left.\begin{aligned} w_{BA} &= \begin{cases} 1 \text{ if } B = UA \\ 0 \text{ if not} \end{cases} \\ w_{BA}(s) &= \begin{cases} 1 \text{ if } B = U^sA \\ 0 \text{ if not.} \end{cases} \end{aligned}\right\} \tag{5.1}$$

Any stationary Markov chain can be represented by a diagram in which each state is represented by a point, or *vertex*, and the possible transitions by arrows from one vertex to another, as in Fig. 3. The diagrams representing deterministic chains are characterized by the fact that exactly one arrow emerges from each vertex. The long-time behaviour of any deterministic chain in can be read off immediately from such a diagram. Thus, for the chain illustrated by Fig. 3, a system started in state 1, 2 or 3 will continue round the cycle 1, 2, 3, 1, 2, 3, ..., indefinitely; a system started in state 4

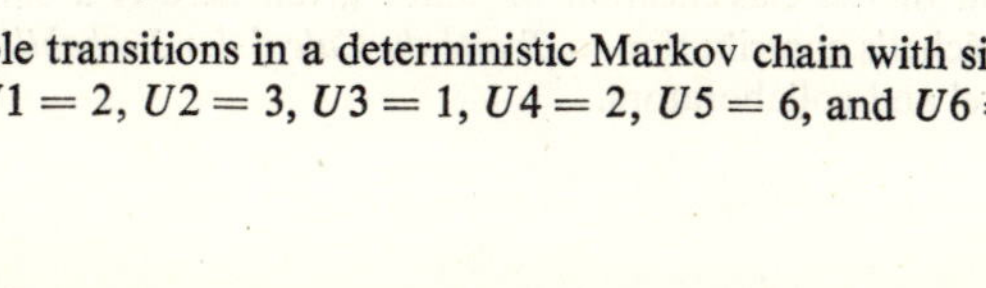

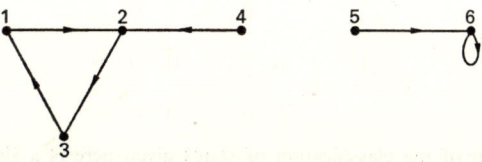

FIG. 3. The possible transitions in a deterministic Markov chain with six states for which $U1 = 2$, $U2 = 3$, $U3 = 1$, $U4 = 2$, $U5 = 6$, and $U6 = 6$.

will go to 2 and then cycle; a system started in state 5 will go to state 6 and stay there; and a system started in state 6 will stay there.

The states in this example can be classified into three types, in a way which generalizes easily to an arbitrary deterministic chain. Firstly, the states 4 and 5 which the system can be in at most once: we call these *transient* states. Secondly, the states 1, 2, 3 forming a cycle: we call these *periodic* states. And thirdly, the state 6 which cannot be left, once entered: we call this an *equilibrium* state (in the mathematical theory of Markov chains, such states are usually called *absorbing* states).

To make the corresponding classification for an arbitrary deterministic chain, we define the *period* of an arbitrary state A as the smallest integer (if any) with the property

$$U^s A = A. \tag{5.2}$$

If $s = 1$, then a system reaching the state A will stay there indefinitely, and we call A an equilibrium state. If $s \geqq 2$, then a system reaching the state A will pass through $s - 1$ other states $UA, U^2A \ldots U^{s-1}A$ before recommencing the cycle by returning to A at the sth step, and we call A a periodic state. If no value of s satisfies (5.2) then a system reaching the state A will never return to it, and we call A a transient state. A state is said to be *persistent* if it is not transient; that is, if it is either periodic or an equilibrium state. A transition from a persistent to a transient state is impossible, since if A (being persistent) satisfies (5.2) then so do UA, U^2A, etc. In the physical example of heat conduction used above, the non-uniform temperature distributions correspond to transient observational states, the uniform distributions to equilibrium states, and there are no periodic states.

This classification can be extended to non-deterministic Markov chains.† A physical example of such a chain is provided by the observational states (velocities, or positions and velocities: see the discussion in I § 5) of a Brownian particle. A mathematical example is provided by the Ehrenfest urn model, with transition probabilities given by (4.16). To illustrate the classification of states, an even simpler model will suffice. Consider the Markov chain whose transition probability matrix is

$$[w_{BA}] = \begin{bmatrix} \frac{1}{3} & 0 & 0 & 0 & 0 & 0 \\ \frac{1}{3} & 0 & 0 & 0 & 1 & 0 \\ 0 & 0 & \frac{1}{3} & 0 & 0 & 1 \\ \frac{1}{3} & 0 & 0 & 1 & 0 & 0 \\ 0 & 1 & 0 & 0 & 0 & 0 \\ 0 & 0 & \frac{2}{3} & 0 & 0 & 0 \end{bmatrix}.$$

† The treatment of the classification of states given here is a simplified version of pp. 187–201 of Fréchet's contribution to *Traité du Calcul des Probabilités et de ses Applications*, by E. Borel and collaborators.

The diagram representing this Markov chain is shown in Fig. 4. As in Fig. 3, each arrow represents a possible transition, that is, a non-vanishing element of the matrix w_{BA}; but now the vertices 1 and 3 have more than one arrow emerging, since more than one transition is available to a system in

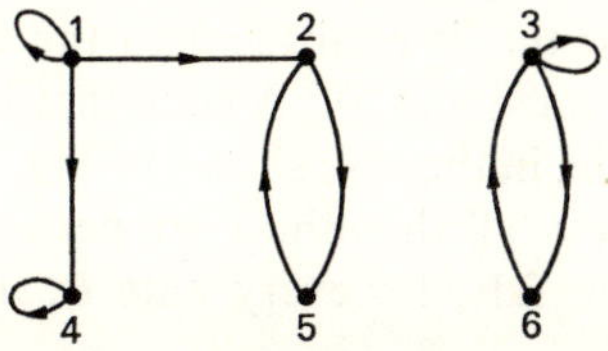

FIG. 4. The possible transitions in the non-deterministic Markov chain whose matrix of transition probabilities is given in the text.

state 1 or 3. Even if numerical values of the non-vanishing transition probabilities are not shown in such a diagram, the qualitative behaviour of the Markov chain can be read from it. Thus, for the chain illustrated in Fig. 4 a system started in state 4 will stay there, a system started in state 2 or 5 will oscillate regularly between states 2 and 5, and a system started in state 1 will either finish permanently in state 4 or oscillate regularly for ever between states 2 and 5. A system started in state 3 or 6 will oscillate irregularly for ever between states 3 and 6. The fact that this chain is non-deterministic is evident from the feature that some vertices have more than one arrow emerging, indicating that the future behaviour of a particle in state 1 or 3 is not determined. The same fact can equally easily be seen by looking at the matrix of transition probabilities $[w_{BA}]$, in which elements other than 0 and 1 appear.

As in the case of deterministic chains, the states of a general stationary Markov chain can be classified into transient states, persistent states, and so on. If the chain is not deterministic, however, we can no longer define the single-valued function $U(A)$ on which the classification for deterministic chains was based. Instead we introduce the new symbol $B \leftarrow A$ to mean that $w_{BA}(s) > 0$ for some s; or, in words, that the state B can be reached from the state A. In the contrary case, where $w_{BA}(s) = 0$ for all s, we shall write $B \nleftarrow A$. For deterministic chains, $B \leftarrow A$ means $B = U^s A$ for some s. The relation '←' has the property of *transitivity*:

$$\text{if } C \leftarrow B \quad \text{and} \quad B \leftarrow A \quad \text{then} \quad C \leftarrow A; \tag{5.3}$$

this is obvious from the diagram representation, and it may be derived more pedantically from the relation

$$w_{CA}(s+t) = \sum_B w_{CB}(s)\, w_{BA}(t) \tag{5.4}$$

which is itself an immediate consequence of (4.4) and (4.8).

A state A of a general Markov chain is said to be *transient* if there exists a state B that can be reached *irreversibly* from A: that is, if the system can get from A to B but not back again:

$$B \leftarrow A \quad \text{and} \quad A \not\leftarrow B. \tag{5.5}$$

(For deterministic chains this is equivalent to the definition already given, since the state $B = UA$ satisfies (5.5).) A state that is not transient is said to be *persistent*. For example, in the case shown in Fig. 4 the state 1 is transient, since $2 \leftarrow 1$ and $1 \not\leftarrow 2$, but all the others are persistent. If a state A is persistent, then (5.5) must be false for every state B: that is,

$$\text{if } A \text{ is persistent and } B \leftarrow A, \text{ then } A \leftarrow B. \tag{5.6}$$

The converse of this statement is also true:

$$\text{if } B \leftarrow A \text{ implies } A \leftarrow B, \text{ then } A \text{ is persistent,} \tag{5.7}$$

since the first part of (5.7) implies that (5.5) is false for every B.

If A is a persistent state the set of all states B reachable from A will be denoted by $Z(A)$. By the normalization condition (4.15), this set must contain at least one state; it follows from (5.6), together with the transitivity relation (5.3), that $A \leftarrow A$ so that A itself is a member of $Z(A)$. If it is the only member we call A an *equilibrium state* (or absorbing state.) Let B be any member of $Z(A)$, so that $B \leftarrow A$. If follows from the transitivity relation (5.3) that

$$B \leftarrow A \quad \text{implies} \quad Z(B) \subset Z(A), \tag{5.8}$$

where $\subset$ means "is a subset of". Moreover, since A is persistent, we also have $A \leftarrow B$ [by (5.6)] and hence, by (5.8) with A and B exchanged, $Z(A) \subset Z(B)$, so that $Z(A)$ and $Z(B)$ are identical. It follows that the persistent states may be separated into one or more distinct sets (*ergodic sets*) of states such that one persistent state can be reached from another if and only if they are in the same ergodic set. Thus, once the system reaches a persistent state it will remain for ever in the ergodic set containing that state, visiting each of the states in the set many times but never going outside. In Fig. 4 the state 4 by itself constitutes an ergodic set, the states 2 and 5 a second ergodic set and the states 3 and 6 a third (the remaining state, 1, is transient). In the case of the urn model (4.16), there is just one ergodic set, comprising all the states of the system.

Any observable that is a constant of the motion, such as the observable energy, must take the same value for all the states in an ergodic set. Thus the different ergodic sets may be labelled by different sets of values of the observable constants of the motion. In the simplest case, the energy is the only observable constant of the motion, and each distinct value for the observable energy corresponds to a distinct ergodic set. This is the situation that is assumed to exist in the thermodynamics of fluids. For solids, on the other hand, there are other observables which, although not exact constants of the motion like the energy, can remain effectively constant over very long periods of time. One example is the number of screw dislocations in a crystal; another is the magnetic flux through a superconducting ring. (These additional "constants of the motion" differ from the energy in that they are not defined for all observational states: if the system is given enough energy to melt the crystal, or to turn the superconductor into a normal metal, the long-range order on which the definition of the constant of the motion depends will disappear.)

Ergodic sets fall into two classes: periodic and aperiodic. In the chain illustrated in Fig. 4 the ergodic set consisting of states 2 and 5 is said to be periodic because a system starting in state 2 and returning to it must do so in an even number of steps; therefore $p_2(t)$ oscillates indefinitely with period 2 and does not approach a limit as $t \to \infty$. Periods larger than 2 are also possible; for example, a chain with three states having $w_{12} = w_{23} = w_{31} = 1$ and all other transition probabilities zero would consistute a single ergodic set with period 3. Periodic ergodic sets would be expected to occur in the observational description of physical systems with observable undamped periodicities, such as a planet going round the sun, a purely reactive electrical network, or an ideal harmonic oscillator. It is gratifying that the framework of the theory is wide enough to include such cases, but they are of little importance in the applications of statistical mechanics.

An ergodic set which is not periodic is said to be *aperiodic*. An example is the set consisting of states 3 and 6 in Fig. 4. If the system starts in state 3 it can be in state 3 again after one step, or after two steps (via state 6), or after three (via 3 and 6) or, indeed, in any number of steps, so that we would not expect $p_3(t)$ to oscillate indefinitely with time. For our purposes, the most convenient general definition is that an ergodic set is aperiodic when there exists an integer s such that

$$w_{BA}(s) > 0 \quad \text{for all } A, B \text{ in the ergodic set.} \tag{5.9}$$

For example, in the ergodic set consisting of states 3 and 6 in Fig. 4, any value of $s \geqq 2$ satisfies (5.9). An ergodic set for which no value of s satisfies (5.9) will be called *periodic*.

As in the special case of deterministic chains, there is one-way traffic between transient and persistent states: once in a persistent state, the system will always be in a persistent state. That is to say, if A is persistent and $B \leftarrow A$ then B is persistent. To prove this, let C be any state such that $C \leftarrow B$. Then by the transitivity relation (5.3) we have $C \leftarrow A$, and consequently, by (5.6), $A \leftarrow C$. Combining this with $B \leftarrow A$ we obtain $B \leftarrow C$. That is, B can be reached from any state C that can be reached from B and so, by (5.7), B is persistent.

Since a system can get from a transient to a persistent state but not back again it is plausible that the system will almost certainly be eventually "absorbed" by the persistent states. We can show, in fact, that

$$\text{if } A \text{ is transient, then } \lim_{t\to\infty} p_A(t) = 0. \tag{5.10}$$

To prove (5.10), let Q be the set of all states that can be reached irreversibly from A; that is, the set of all states B satisfying (5.5). Since A is transient, the set Q is not empty; on the other hand, Q does not contain A itself. The set Q has the property that all the states reachable from any state in Q are also in Q:

$$\left.\begin{array}{l} \text{if } C \leftarrow A \text{ and } A \nleftarrow C \text{ (i.e. } C \in Q) \\ \qquad \text{and } B \leftarrow C \\ \text{then } B \leftarrow A \text{ and } A \nleftarrow B \text{ (i.e. } B \in Q). \end{array}\right\} \tag{5.11}$$

The first part of the conclusion follows at once from the transitivity property (5.3), and the second part follows by *reductio ad absurdum*, since if it were false (i.e. if $A \leftarrow B$) then by transitivity we should have $A \leftarrow C$, contradicting the first line of (5.11).

Let $q(t)$ be the probability that the state of the system is in Q at time t; that is,

$$q(t) \equiv \sum_{B\in Q} p_B(t). \tag{5.12}$$

According to (4.17) the change of $q(t)$ with time is given by

$$\begin{aligned} q(t+s) &= \sum_{B\in Q}\sum_{C} w_{BC}(s)\, p_C(t) \\ &\geqq \sum_{B\in Q} w_{BA}(s)\, p_A(t) + \sum_{B\in Q}\sum_{C\in Q} w_{BC}(s)\, p_C(t) \end{aligned} \tag{5.13}$$

since A is not in Q and all terms are positive. Since every state reachable from a state in Q is itself in Q, it follows that $w_{BC}(s) = 0$ if C is in Q and B is not; consequently we may, without altering its value, extend the second sum over B in (5.13) to cover all states instead of just those within Q. It then follows, by the normalization condition (4.15) on $w_{BA}(s)$ and the definition (5.12) of q, that

$$q(t+s) \geqq \sum_{B\in Q} w_{BA}(s)\, p_A(t) + q(t) \tag{5.14}$$

for all $s \geqq 1$ and $t \geqq 0$. Hence $q(t)$ is a non-decreasing function of time. Being a probability it is bounded above by 1 and must therefore tend to a limit as $t \to \infty$. Taking the limit $t \to \infty$ on both sides of (5.14) we obtain, since all terms in the sum are non-negative,

$$0 = \lim_{t\to\infty} \sum_{B\in Q} w_{BA}(s)\, p_A(t) \tag{5.15}$$

for all integers $s \geqq 1$. By the definition of Q, there must exist values of s for which $\sum_{B \in Q} w_{BA}(s) > 0$, and the proof of (5.10) is immediately completed.

If the Markov chain is finite (comprises a finite number of states) then it follows from (5.10) and the addition law of probabilities, eqn. (1.11), that the probability of the system's being in a transient state tends to 0 as t tends to ∞; the probability of its being in a persistent state therefore tends to 1. This result implies, incidentally, that a finite Markov chain must have at least one persistent state. If the Markov chain is infinite, on the other hand, these deductions cannot be made, since an infinite sum each of whose terms tends to 0 need not itself tend to 0. In order to apply these deductions to the observational states of a physical system, which may form an infinite (though countable) set, we shall make the assumption [also used in Chap. I, eqn. (3.9)] that conservation of energy restricts the number of states available to the system in any particular trial to a finite subset of the observational states. The Markov chain can then be taken to comprise only the states in this subset, and the deductions just mentioned are valid.

5.1. Exercises

1. Draw the diagram corresponding to Fig. 4 for the Ehrenfest urn model with $N = 5$.

2. Show that the Ehrenfest urn model has period 2, whatever the value of N. How many ergodic sets are there?

6. Statistical equilibrium

The concept of equilibrium described in the previous section, based on equilibrium observational states which cannot be left once entered, is restricted to systems that do possess equilibrium states. If, however, the observational description is sufficiently refined to reveal some of the fluctuations due to thermal molecular motion, then this observational description is non-deterministic and in general does not possess equilibrium states. An example is the case of a fluid containing a Brownian particle: the Brownian particle never settles into an equilibrium observational state, but goes on dancing about indefinitely. To describe the behaviour of systems that have been isolated for a long time in a way that allows for such thermal fluctuations, a new and more general concept of equilibrium is necessary.

This new concept of equilibrium applies whenever the probabilities $p_A(t)$ approach limiting values as t becomes large. Under this condition, the ensemble as a whole will tend towards a *statistical equilibrium* in which the fraction of its members in any specified observational state A becomes effectively constant, even though the composition of this fraction is for ever changing as individual members of the ensemble enter and leave the state A. If the observational states are deterministic, this new concept of equilibrium is equivalent to the old one, since all the systems in the ensemble behave the same way and $Pr(A_t|K)$ becomes independent of time as soon as they all reach an equilibrium observational state; but the new concept applies to non-deterministic states too.

The main difference between statistical equilibrium and the simpler notion based on equilibrium states is that statistical equilibrium is a property not of an individual system but of an ensemble. To say whether or not a particular system is in equilibrium it is no longer sufficient to consider that system alone, observing it and seeing whether or not it is in an equilibrium observational state; instead we must consider the system as part of an imaginary ensemble.† This ensemble, if it existed, would consist of a large collection of copies of the system, all prepared at time 0 in the same observational state K and subjected thereafter to the same mechanical operations. Except in the case of deterministic (or nearly deterministic) observational states, for which a single system is as good as an ensemble since all (or nearly all) systems in the ensemble behave identically, it is perhaps unwise to try to answer the question "Is this individual system in equilibrium now?"; if one does wish to give an answer, however, that answer depends only on the way the system was prepared at some specified initial time, and on the mechanical operations it has suffered since, and is completely independent of the present observational state of the system. This increased subtlety and complexity of the concept of equilibrium is the price we must pay for the possibility of treating fluctuations as an integral part of the theory.

To apply the theory of stationary Markov chains to the study of statistical equilibrium, we assume for simplicity that the initial state K is persistent; this entails no great loss of generality since, as shown at the end of § 5, the system almost certainly reaches a persistent state eventually. If the ergodic set containing K is denoted by $\mathbf{Z}$, the system can never escape from $\mathbf{Z}$, so that

$$w_{BA} = 0 \quad \text{if } A \in Z \quad \text{and} \quad B \notin Z \tag{6.1}$$

and

$$p_A(t) = 0 \quad \text{if} \quad A \notin Z. \tag{6.2}$$

Any probability distribution over $\mathbf{Z}$ which is independent of time will be called an *equilibrium probability distribution* over $\mathbf{Z}$ and denoted by π_1, π_2, ... (instead of $p_1(t)$, $p_2(t)$, ...). From (4.20) we obtain the system of equations

$$\pi_B = \sum_A w_{BA}\pi_A \tag{6.3}$$

which any time-independent probability distribution must satisfy. By virtue of (6.2) and the normalization condition (4.19), an equilibrium probability

† R. C. Tolman, in *The Principles of Statistical Mechanics*, calls it the *representative* ensemble of that system.

distribution over Z will also satisfy the conditions

$$\sum_A \pi_A = 1 \tag{6.4}$$

$$\pi_A = 0 \text{ unless } A \in Z. \tag{6.5}$$

The three conditions (6.3–5) suffice to determine a unique equilibrium probability distribution $\pi_1, \pi_2, \ldots$ for Z. To show this† we first prove that (6.3) has a non-vanishing solution satisfying (6.5). Consider the two systems of linear equations

$$x_B = \sum_{A \in Z} w_{BA} x_A \quad \text{for all } B \text{ in } Z \tag{6.6}$$

$$y_A = \sum_{B \in Z} y_B w_{BA} \quad \text{for all } A \text{ in } Z. \tag{6.7}$$

These can be written in matrix form

$$0 = (W - I)x \quad \text{and} \quad 0 = y(W - I), \tag{6.8}$$

where W is the square matrix formed from the rows and columns of $[w_{AB}]$ that refer to states in Z, and I is a unit $(M \times M)$ matrix, where M is the number of states in Z. By virtue of the normalization condition (4.15), the second system of equations has the solution

$$y_A = 1 \quad \text{for all } A \text{ in } Z.$$

Consequently the matrix $W - I$ must be singular, and so the first system of equations must also have a non-vanishing solution.

Let $\{x_A\}$ be any non-vanishing solution of (6.6). Then we can show that $\{|x_A|\}$ is also a solution. It is an immediate consequence of (6.6) that

$$|x_B| \leqq \sum_{A \in Z} w_{BA} |x_A| \tag{6.9}$$

since all transition probabilities are non-negative. Summing both sides of (6.9) over B we obtain

$$\sum_B |x_B| \leqq \sum_B \sum_A w_{BA} |x_A|, \tag{6.10}$$

all sums being confined to the ergodic set Z.

By the normalization condition (4.15), the right-hand side of (6.10) is equal to $\sum_A |x_A|$ and is therefore equal to the left-hand side. Consequently every one of the inequalities (6.9) that were summed to give (6.10) must be an equality:

$$|x_B| = \sum_{A \in Z} w_{BA} |x_A|. \tag{6.11}$$

† I am indebted to Professor G. E. H. Reuter for help in simplifying this proof.

This equation has the consequence that if one of the quantities x_A is positive, then they all are. To prove this we note first that a matrix equation $\xi = W\xi$ implies $\xi = W^s\xi$ for any positive integer s, so that, by (4.13), (6.11) implies

$$|x_B| = \sum_{A \in Z} w_{BA}(s)\,|x_A| \quad \text{for} \quad s = 1, 2, \dots. \tag{6.12}$$

Since not all the numbers in the set $\{x_A\}$ are zero, we can find an A in Z such that $|x_A| > 0$. Now let B be any member of Z. It follows from the definition of Z that $B \leftarrow A$, which means that $w_{BA}(s) > 0$ for some s. Using this value of s in (6.12), we deduce that $|x_B| > 0$, since no term of the sum can be negative, and the one belonging to the chosen value of A is positive. That is,

$$\text{if } x_A \neq 0 \text{ holds for some } A \text{ in } Z, \text{ then it holds for all } A \text{ in } Z. \tag{6.13}$$

This result can be strengthened by considering the set of numbers $x'_A \equiv x_A + |x_A|$. By adding (6.6) and (6.12) we see that the set $\{x'_A\}$ is a solution of (6.6) and is therefore subject to the condition (6.13). Since $x'_A \neq 0$ is equivalent to $x_A > 0$, we conclude that

$$\text{if } x_A > 0 \text{ holds for some } A \text{ in } Z, \text{ then it holds for all } A \text{ in } Z. \tag{6.14}$$

We now define

$$\pi_A \equiv \begin{cases} |x_A| / \sum_{B \in Z} |x_B| & \text{if } A \in Z, \\ 0 & \text{if not.} \end{cases} \tag{6.15}$$

The numbers $\pi_1, \pi_2, \dots,$ form an equilibrium probability distribution over Z: they satisfy (6.4) and (6.5) trivially, and because of (6.1) and (6.10) they also satisfy (6.3). Moreover, because of (6.14), they must be positive:

$$\pi_A > 0 \quad \text{for all } A \text{ in } Z. \tag{6.16}$$

To show that the set of equilibrium probabilities defined by (6.15) is unique, let us suppose that $\pi'_1, \pi'_2, \dots$ is also a solution of the conditions (6.3, 4 and 5), and define $x''_A \equiv \pi'_A - \pi_A$ for all A in Z. Since the sets $\{\pi_A\}$ and $\{\pi'_A\}$ both satisfy (6.3), it follows by (6.5) that the set x''_A satisfies (6.6). Consequently (6.14) tells us that if $\pi'_A > \pi_A$ holds for some A in Z, then it holds for all A in Z. Since this conclusion is incompatible with the normalization conditions (6.4) on the sets $\{\pi_A\}$ and $\{\pi'_A\}$, the premiss that $\pi'_A > \pi_A$ holds for some A must be false. In a similar way, we can show that $\pi_A > \pi'_A$ cannot hold for any A: thus the only possibility is $\pi_A = \pi'_A$ for all A, which proves the uniqueness of the equilibrium probability distribution over the ergodic set Z.

The conditions under which the probability distribution tends to the equilibrium distribution for large t will be established in the next two sections, but before embarking on these calculations we prove the weaker result that the average of the probability distribution over a time interval tends to the equilibrium distribution as this time interval becomes infinite. This time average is defined by

$$\bar{p}_A(s) \equiv \frac{1}{s} \sum_{t=1}^{s} p_A(t). \tag{6.17}$$

Applying (4.20), we have

$$\textstyle\sum_A w_{BA}[\bar{p}_A(s) - s^{-1}p_A(s)] = \bar{p}_B(s) - s^{-1}p_B(1). \tag{6.18}$$

Rearranging and taking the limit $s \to \infty$ gives

$$\lim_{s\to\infty} [\textstyle\sum_A w_{BA}\bar{p}_A(s) - \bar{p}(s)] = 0. \tag{6.19}$$

The definition (6.17) also implies

$$\left.\begin{aligned} \bar{p}_A(s) &= 0 \quad \text{if} \quad A \notin Z \\ \text{and} \qquad \textstyle\sum_A \bar{p}_A(s) &= 1 \quad \text{for all } s. \end{aligned}\right\} \tag{6.20}$$

It follows from (6.19) and (6.20) that every limit point of the sequence of probability distributions $\bar{p}_A(s)$ satisfies the equations (6.3), (6.4), and (6.5). Since these equations define a unique steady distribution π_A, there is only one limit point, and it is given by

$$\lim_{s\to\infty} \bar{p}_A(s) = \pi_A \quad \text{for all } A \text{ in } Z. \tag{6.21}$$

6.1. Exercises

1. A *doubly stochastic matrix* (compare with eqns. (4.14) and (4.15)) is one whose row sums and column sums are all equal to 1, and whose elements are all non-negative. If the matrix of transition probabilities is doubly stochastic, what can be deduced about the equilibrium probabilities?

2. Show that the equilibrium probabilities for the Ehrenfest urn model form a binomial distribution.

3. In a *random walk with reflecting barriers*, a man takes steps to right or left at random along a corridor, his probabilities for stepping in either direction being equal. If his attempted step brings him up against the wall at either end of the corridor, however, he does not move. Find the matrix of transition probabilities and the equilibrium probability distribution.

7. The approach to equilibrium

Having proved the existence of an equilibrium probability distribution π_A over each ergodic set Z, we now wish to show that if Z is aperiodic then any probability distribution $p_A(t)$ over Z will approach this steady distribution in the limit $t \to \infty$. We shall do this by introducing a numerical measure of the deviation of the distribution $p_A(t)$ from the steady distribution π_A, and showing that this measure of deviation decreases to zero as $t \to \infty$.

The argument will make use of the mathematical theory of convex functions. Accordingly we first set out the relevant part of this theory.† A function $\varphi(x)$ is said to be *convex* if its graph never lies above any chord, and *strictly convex* if the graph lies below every chord (see Fig. 5). For example, the function x^2 is both convex and strictly convex; the function x is convex

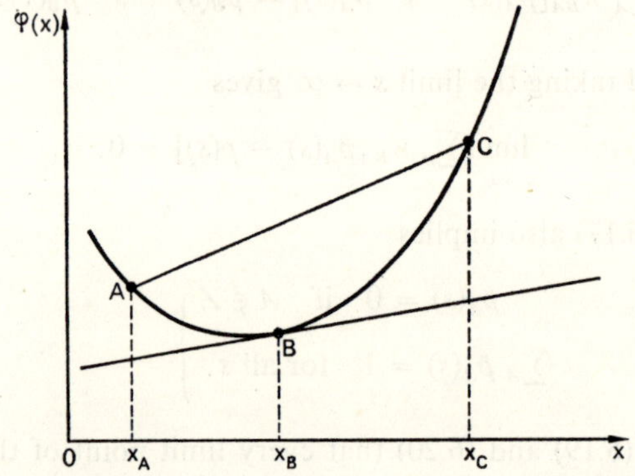

FIG. 5. Graph of a strictly convex function $\varphi(x)$. Note that the graph lies below the chord AC and above the tangent at B.

but not strictly convex; and the function x^3 is neither unless the variable x is restricted to non-negative values. This defining property of a convex function may equally well be expressed as an inequality

$$\varphi(x_B) \leqq \frac{x_C - x_B}{x_C - x_A}\varphi(x_A) + \frac{x_B - x_A}{x_C - x_A}\varphi(x_C) \tag{7.1}$$

holding whenever $x_A < x_B < x_C$. If φ is strictly convex, the sign $\leqq$ in (7.1) may be replaced by $<$.

For simplicity we shall assume that $\varphi(x)$ is twice differentiable. The defining property (7.1) of convex functions is then equivalent to

$$\varphi''(x) \equiv \frac{d^2\varphi}{dx^2} \geqq 0. \tag{7.2}$$

This is intuitively obvious from Fig. 5; a rigorous proof is given by Hardy, Littlewood, and Polya.‡ The corresponding condition for strictly convex functions is

$$\varphi''(x) > 0. \tag{7.3}$$

† For more details of the theory of convex functions, see G. H. Hardy, J. E. Littlewood, and G. Polya, *Inequalities* (Cambridge University Press, London, 1959), chapter III.
‡ *Inequalities*, § 3.10.

An important property of convex functions is that the graph of such a function never lies below any tangent; for the equation of the tangent at $x = x_0$ is

$$y = \varphi(x_0) + (x - x_0)\,\varphi'(x_0) \tag{7.4}$$

and Taylor's theorem combined with (7.2) implies

$$\begin{aligned}\varphi(x) &= \varphi(x_0) + (x - x_0)\,\varphi'(x_0) + \tfrac{1}{2}(x - x_0)^2\,\varphi''(x_\theta)\\ &\geqq \varphi(x_0) + (x - x_0)\,\varphi'(x_0),\end{aligned} \tag{7.5}$$

where x_θ is some number between x_0 and x. For strictly convex functions, the corresponding result is that the curve lies above the graph everywhere except at the point of contact.

A corollary of (7.5) is that the centre of gravity of any set of masses placed on the graph of $\varphi(x)$ lies above or on the curve. In symbols, this property is

$$\sum m_i \varphi(\bar{x}) \leqq \sum m_i \varphi(x_i), \tag{7.6}$$

where

$$\bar{x} \equiv \sum m_i x_i / \sum m_i, \tag{7.7}$$

$$m_i \geqq 0 \quad \text{for all } i,$$

and the sums go over any set of indices, say from 1 to n. To prove (7.6) we apply (7.5) and then (7.7) to its right-hand member:

$$\begin{aligned}\sum m_i \varphi(x_i) &\geqq \sum m_i [\varphi(\bar{x}) + (x_i - \bar{x})\,\varphi'(\bar{x})]\\ &= \sum m_i \varphi(\bar{x}).\end{aligned} \tag{7.8}$$

If φ is strictly convex, the appropriately strengthened version of (7.5) shows that (7.6) becomes an equality only if $x_i = \bar{x}$ whenever $m_i > 0$.

As in the previous section, we consider a stationary Markov chain whose initial state is persistent, so that its probability distribution $p_A(t)$ is permanently restricted to a single ergodic set $\mathbf{Z}$. To obtain a measure of the deviation of this probability distribution from the equilibrium distribution π_A, we define† the quantity (not to be confused with the Hamiltonian)

$$H(t) \equiv \sum_{A \in Z} \pi_A \varphi(p_A(t)/\pi_A), \tag{7.9}$$

† This device originated with Boltzmann, who used it for a more complicated problem, discussed in Chap. V, § 8 of this book. It was applied to Markov chains, in the particular case where $w_{AB} = w_{BA}$, and using the particular convex function $x \ln x$, by W. Pauli in his article "Über das H Theorem vom Anwachsen der Entropie vom Standpunkt der neuen Quantenmechanik" in the book *Probleme der Moderne Physik*, edited by P. Debye (Hirzel, Leipzig, 1928). The observation that any convex function may be used is due to M. J. Klein, Entropy and the Ehrenfest Urn Model, *Physica* **22**, 569 (1956). See also A. Rényi, On Measures of Entropy and Information, in *Proceedings of the Fourth Berkeley Symposium on Mathematical Statistics and Probability* (Editor J. Neyman, University of California, Berkeley and Los Angeles, 1961), Vol. I, p. 547; P. A. P. Moran, Entropy, Markov Processes and Boltzmann's H theorem, *Proc. Camb. Phil. Soc.* **57**, 833 (1961); T. Morimoto, Markov Processes and the H Theorem, *J. Phys. Soc. Japan* **18**, 328 (1963).

where φ is any strictly convex function. At the present stage there is no reason to prefer any one strictly convex function φ to any other. Later, when the connection between probability and dynamics has been established, we shall find that if φ is chosen properly then $-H$ can be identified with the entropy of the system.

Applying (7.6) with $m_A \equiv \pi_A$ and $x_A \equiv p_A(t)/\pi_A$, we obtain from (7.9) the lower bound

$$H(t) \geqq \varphi(1), \tag{7.10}$$

the two sides being equal only if all of p_A/π_A are equal, that is, if $p_A(t)$ is the equilibrium probability distribution. Thus the excess of H over its minimum value $\varphi(1)$ provides a measure of the deviation of the distribution p_A from the steady distribution. For example, if we take $\varphi(x) = x^2$, then (7.9) becomes $H = \sum_A p_A^2/\pi_A$. In view of the normalization conditions on p_A and π_A this is equivalent to $H = 1 + \sum_A (p_A - \pi_A)^2/\pi_A$, from which the truth of (7.10) for this example is evident.

In order to apply the basic property of convex functions, the inequality (7.6), to a discussion of the time variation of $H(t)$, it is convenient to make the definitions

$$x_A \equiv p_A(t)/\pi_A \quad \text{for all } A \text{ in } Z \tag{7.11}$$

$$\bar{x}_B \equiv p_B(s+t)/\pi_B \quad \text{for all } B \text{ in } Z, \tag{7.12}$$

where s and t are arbitrary positive integers. With these definitions eqn. (4.17), which governs the change with time of the probability distribution and therefore of H, takes the form

$$\bar{x}_B = \textstyle\sum_A m_{BA} x_A, \tag{7.13}$$

where

$$m_{BA} \equiv \pi_B^{-1} w_{BA}(s)\, \pi_A \tag{7.14}$$

is a square matrix with a row and a column for each member of the ergodic set Z. In (7.13) and throughout the rest of this section, the summation is confined to states within Z since the probabilities of the remaining states vanish for all t. By virtue of the normalization condition (4.15), the matrix m_{BA} satisfies the condition

$$\pi_A = \textstyle\sum_B \pi_B m_{BA}. \tag{7.15}$$

Also, combining (6.3) with (4.12), we obtain $\pi_B = \sum_A w_{BA}(s)\, \pi_A$, which is equivalent to

$$\textstyle\sum_A m_{BA} = 1. \tag{7.16}$$

By (6.16) and (4.14) all elements of the matrix m_{BA} are finite and non-negative.

The change in H with time is given, according to (7.9), (7.11), and (7.12), by

$$H(t) - H(s+t) = \sum_A \pi_A \varphi(x_A) - \sum_B \pi_B \varphi(\bar{x}_B). \qquad (7.17)$$

To make use of the strict convexity of $\varphi(x)$ we transform the first sum on the right into a double sum using (7.15) and the second using (7.16). This gives

$$\begin{aligned} H(t) - H(s+t) &= \sum_B \sum_A \pi_B m_{BA}[\varphi(x_A) - \varphi(\bar{x}_B)] \\ &= \sum_B \sum_A \pi_B m_{BA}[\varphi(x_A) - \varphi(\bar{x}_B) - (x_A - \bar{x}_B)\,\varphi'(\bar{x}_B)]. \end{aligned} \qquad (7.18)$$

The part added in the last line, whose form is suggested by (7.8), vanishes by virtue of (7.13) and (7.16). By the property (7.5) of convex functions, every term of the double sum in (7.18) is non-negative, so that

$$H(t) - H(s+t) \geqq 0; \qquad (7.19)$$

that is, H is a non-increasing function of time. According to (7.10) this non-increasing function is bounded below, and therefore it tends to a limit as $t \to \infty$.

To obtain the limiting form of the probability distribution, we take the limit $t \to \infty$ on both sides of (7.18). This gives

$$\lim_{t\to\infty} \sum_A \sum_B \pi_B m_{BA}[\varphi(x_A) - \varphi(\bar{x}_B) - (x_A - \bar{x}_B)\,\varphi'(\bar{x}_B)] = 0.$$

Since every term in the sum is non-negative, and $\pi_B > 0$ by (6.16), it follows that

$$\lim_{t\to\infty} m_{BA}[\varphi(x_A) - \varphi(\bar{x}_B) - (x_A - \bar{x}_B)\,\varphi'(\bar{x}_B)] = 0. \qquad (7.20)$$

Applying the first line of (7.5), we may write (7.20) in the form

$$\lim_{t\to\infty} m_{BA}[\tfrac{1}{2}(x_A - \bar{x}_B)^2\,\varphi''(x_{AB})] = 0, \qquad (7.21)$$

where x_{AB} is some number between x_A and $\bar{x}_B$. The strict convexity of $\varphi(x)$ implies, by (7.3), that $\varphi''(x)$ has a positive lower bound on any bounded interval of the x-axis. Denoting this lower bound for the interval $0 < x < \max_B(1/\pi_B)$ by a, we deduce from (7.21) that

$$\frac{a}{2} \lim_{t\to\infty} m_{BA}(x_A - \bar{x}_B)^2 = 0.$$

That is, either $m_{BA} = 0$ or else $\lim (x_A - \bar{x}_B) = 0$. Using (7.11), (7.12), and (7.14) to convert this statement back to the original notation, we obtain the statement

$$\left.\begin{aligned} &\text{either} \qquad && w_{BA}(s) = 0 \\ &\text{or} \qquad && \lim_{t\to\infty}\left[\frac{p_A(t)}{\pi_A} - \frac{p_B(s+t)}{\pi_B}\right] = 0. \end{aligned}\right\} \tag{7.22}$$

If the ergodic set Z is aperiodic, then we may rule out the first alternative by giving s a value satisfying (5.9). Multiplying the remaining alternative by π_B and summing over all values of B, we obtain

$$\lim_{t\to\infty}\left[\frac{p_A(t)}{\pi_A}\sum_B \pi_B - \sum_B p_B(s+t)\right] = 0.$$

By the normalization conditions on π_B and $p_B(s+t)$, this reduces to

$$\lim_{t\to\infty} p_A(t) = \pi_A \quad \text{if } A \in Z \text{ and } Z \text{ is aperiodic}. \tag{7.23}$$

Thus, if the initial state of a stationary Markov chain is aperiodic, then the statistical ensemble that would be used to measure $p_A(t)$ approaches a statistical ensemble in which the probabilities are independent of time, even though the observational state of any individual member of the ensemble never ceases to change.

7.1. Exercises

1. Discuss the behaviour of the probability distribution $p_A(t)$ for large t in a stationary Markov chain where the initial state is transient. Assume that all the ergodic sets are aperiodic.

2. If $\varphi(x)$ is convex for $x > 0$, show that the function $x\varphi(1/x)$ is also convex for $x > 0$.

3. The "information-theory" approach to statistical mechanics† is based on the principle that the subjective probability density in phase space, denoted here by $D(\alpha)$, is to be chosen so as to give the *fine-grained entropy*, defined by

$$S_{\text{fine}} \equiv k \int D(\alpha) \ln [c/D(\alpha)]\, d\alpha$$

with k and c constants ($k > 0$), the largest value consistent with the available information about the system. If the available information at time t is equivalent to the statement that the system must be somewhere in a given region ω_t of phase space (so that $D(\alpha)$ vanishes outside this region, is non-negative inside, and satisfies the normalization condition $\int D(\alpha)\, d\alpha = 1$), verify that this prescription leads to expression (6.5) of Chap. I for the expectation

$$\langle G\rangle_t \equiv \int G(\alpha) D(\alpha) d\alpha$$

of any dynamical variable $G(\alpha)$ at time t.

† E. T. Jaynes, *Phys. Rev.* **106**, 171 (1957).

8. Periodic ergodic sets

An ergodic set that is not aperiodic according to the definition (5.9) is said to be *periodic*. In the present section, the parts of the theory that were restricted to aperiodic ergodic sets in previous sections will be generalized to the periodic case. This material is included for completeness only, not as an essential part of the exposition; a reader who skips this section will not suffer for it in his reading of the later sections.

Let Z be any ergodic set and (A, B) any ordered pair of states of Z. If u is any positive integer, we shall say "u is *a period of* (A, B)" to mean that there exists a positive integer v such that the inequality

$$w_{BA}(v + ku) > 0 \tag{8.1}$$

holds for all sufficiently large positive integers k. For example, in the Markov chain whose possible transitions are shown in Fig. 6, it is possible to get from state 2 to state 1 in 5 steps, or in 8 steps (via state 7), or in 11 steps, etc.; consequently we have $w_{12}(t) > 0$ if $t = 5$ or 8 or 11 or .... Two of the many ways of satisfying the definition of a period of (2,1) are to take $v = 5$ and $u = 3$, and to take $v = 2$ and $u = 6$; thus both the numbers 3 and 6 are periods of the ordered pair (2, 1).

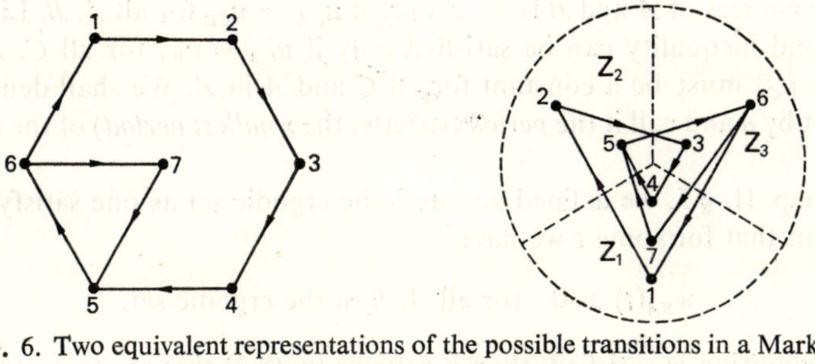

FIG. 6. Two equivalent representations of the possible transitions in a Markov chain whose non-vanishing transition probabilities are $w_{21} = w_{32} = w_{43} = w_{54} = w_{65} = w_{57} = 1$, $w_{16} = w_{76} = \frac{1}{2}$. The period of this chain is 3. The subsets Z_r are also shown.

To show that any ordered pair (A, B) does have at least one period, we use the inequality

$$w_{CA}(s + t) \geqq w_{CB}(s)\, w_{BA}(t) \tag{8.2}$$

which follows immediately from eqn. (5.4), and holds for any positive integers s and t and any states A, B, C. Consider first ordered pairs of the form (A, A). Since A belongs to an ergodic set Z, the transition $A \leftarrow A$ must be possible; that is, there exists an s such that $w_{AA}(s) > 0$. By putting $C = B = A$ in (8.2) and then successively putting $t = s$, $t = 2s$, $t = 3s$, etc., we see that $w_{AA}(ks) > 0$ for any positive integer k. It follows that eqn. (8.1) with $B = A$ can be satisfied for all k by taking $v = u = s$, and thus s is a period of (A, A). Consider now a general ordered pair, say (A, B), of states belonging to Z. Since A and B are in the same ergodic set, there exists a t such that $w_{BA}(t) > 0$, and hence, replacing C, B, A, s, and t in (8.2) by B, A, A, t, and ks, we see that $w_{BA}(t + ks) > 0$ for any positive integer k. Consequently, (8.1) can be satisfied by taking $v = t$ and $u = s$, and so s is a period of (A, B).

Since every ordered pair (A, B) of states in Z has at least one period, it has a smallest period. We denote this smallest period by u_{BA}. To relate the smallest periods of different ordered pairs, let A, B, C be any three states in Z. By suitably choosing v and v' we can, by (8.1), ensure that $w_{CB}(v + ku_{CB}) > 0$ and $w_{BA}(v' + k'u_{BA}) > 0$ for all sufficiently large k and k'. It therefore follows by (8.2) that

$$w_{CA}(v + v' + ku_{CB} + k'u_{BA}) > 0. \tag{8.3}$$

Consequently, by the definition (8.1), both u_{CB} and u_{BA} are periods of (A, C), and so the smallest period of (A, C) satisfies

$$u_{CA} \leqq u_{CB} \quad \text{and} \quad u_{CA} \leqq u_{BA}. \tag{8.4}$$

For a given choice of C, the first of these inequalities can be satisfied for all possible choices of A and B from Z only if $u_{CA} = u_{CB}$ for all A, B. Likewise, the second inequality can be satisfied only if $u_{CA} = u_{BA}$ for all C, B in Z. That is, u_{CA} must be a constant for all C and A in Z. We shall denote this constant by q and call it the *period* (strictly, the *smallest period*) of the ergodic set Z.

In Chap. II, § 5, we defined an aperiodic ergodic set as one satisfying the condition that for some t we have

$$w_{BA}(t) > 0 \quad \text{for all } A, B \text{ in the ergodic set.} \tag*{[(5.9)]}$$

If this condition is satisfied, then it follows by (8.2) that $w_{CA}(s + t) > 0$ for any positive integer s, since there must be at least one state B in the ergodic set for which $w_{CB}(s) > 0$. That is to say, if eqn. (5.9) holds for some value of t, then it also holds for all larger values. Consequently, by (8.1), any pair of states in an aperiodic ergodic set have the period 1, and therefore the period q of the ergodic set itself is 1. On the other hand, if the ergodic set is periodic, then there is no value of t for which (5.9) holds; consequently there must be at least one ordered pair, say (A, B), such that $w_{BA}(t)$ vanishes for an infinite number of values of t. Such an ordered pair cannot, by (8.1), have 1 as a period, and the value of q for a periodic ergodic set is therefore at least 2.

Although the same value of u may be used in (8.1) for every ordered pair (A, B), the same is not in general true for the value of v; in fact the values of v for the various ordered pairs may be used for a sub-classification of the states within the ergodic set Z. Let us therefore define v_{BA} to be the smallest positive integer such that

$$w_{BA}(v_{BA} + kq) > 0 \tag{8.5}$$

holds for all sufficiently large integers k. The value of v_{BA} must lie in the range

$$1 \leqq v_{BA} \leqq q, \tag{8.6}$$

for if v is a number which satisfies the condition that $w_{BA}(v + kq) > 0$ for all sufficiently large k and which also exceeds q, then $v - q$ is positive and also satisfies this condition, so that v, not being the smallest positive integer satisfying the condition, cannot be v_{BA}.

For ordered pairs of the special form (A, A) we can evaluate v_{AA} explicitly: we find that v_{AA} is a period of (A, A) and must therefore be equal to q, the only period compatible with (8.6). To show this, we start from (8.5) with $B = A$, and make successive applications of (8.2) to show that each of the inequalities

$$\left.\begin{aligned} w_{AA}(v_{AA} + kq) &> 0, \\ w_{AA}(2v_{AA} + kq) &> 0, \\ w_{AA}(3v_{AA} + kq) &> 0, \\ &\cdots\cdots \\ w_{AA}(qv_{AA} + kq) &> 0, \end{aligned}\right\} \tag{8.7}$$

holds for sufficiently large k. There must therefore exist a number, say K, such that the q inequalities (8.7) are all true for all $k > K$. Now let n be any integer greater than $q(K/v_{AA} - 1)$, and let l and $m - 1$ be the quotient and remainder when $n - 1$ is divided by q, so that $n = m + lq$ holds with $1 \leqq m \leqq q$ and $l > K/v_{AA}$. Then we have

$$w_{AA}(nv_{AA}) = w_{AA}(mv_{AA} + kq), \tag{8.8}$$

where $k \equiv lv_{AA} > K$. It follows by (8.7) that $w_{AA}(nv_{AA}) > 0$ and hence by (8.1) that v_{AA} is a period of (A, A). Since q is the smallest period of (A, A) we conclude, by (8.6), that

$$v_{AA} = q \tag{8.9}$$

for any state A in Z.

For a general ordered pair, say (A, B), of states from Z, the basic property of the quantity v_{BA} is given by the following selection rule:

$$\text{if } w_{BA}(t) > 0 \quad \text{then} \quad t = v_{BA} + M(q), \tag{8.10}$$

where the symbol $M(q)$ stands for an unspecified integral multiple of q or for zero. To prove (8.10), we start with (8.2) and by appropriate substitution obtain, for any integer k,

$$w_{AA}(t + v_{AB} + kq) \geqq w_{AB}(v_{AB} + kq)\, w_{BA}(t). \tag{8.11}$$

By (8.5) and the premiss of (8.10), both factors on the right of (8.11) are positive for all sufficiently large k; consequently, if r is any number differing from $t + v_{AB}$ by a multiple of q, we have

$$w_{AA}(r + k'q) > 0 \tag{8.12}$$

for all sufficiently large k'. Let us choose r in the range $1 \leqq r \leqq q$; that is, let us make $r - 1$ the remainder when $t + v_{AB} - 1$ is divided by q. Then, by the definition (8.5) we have $r \geqq v_{AA}$, and it follows by (8.9) that $r = q$. Recalling the definition of r, we deduce that

$$t + v_{AB} = M(q). \tag{8.13}$$

To complete the proof of (8.10) we must relate v_{AB} to v_{BA}. According to (8.5) we have $w_{BA}(v_{BA} + kq) > 0$ for some k; consequently $v_{BA} + kq$ is an acceptable value for t in (8.10), and it follows by (8.13) that $v_{BA} + v_{AB} + kq = M(q)$, or more simply

$$v_{AB} + v_{BA} = M(q) \tag{8.14}$$

with the convention that $M(q)$ can stand for different multiples of q in different formulae. Combining (8.14) with the original (8.13) completes the proof of (8.10).

An extension of the addition law (8.14) can be obtained by setting

$$t = v_{BA} + kq \quad \text{and} \quad s = v_{CB} + k'q$$

in (8.2) and then using (8.5) to show that the right-hand side of (8.2) is positive for large enough k and k'; the left-hand side is then positive, too, and so (8.10) gives $s + t = v_{CA} + M(q)$. Eliminating s and t we obtain

$$v_{CB} + v_{BA} = v_{CA} + M(q). \tag{8.15}$$

In view of the condition (8.6) the numerical value of $M(q)$ in (8.15) must be either 0 or q.

These results make it possible to classify the states of a periodic ergodic set Z into q subsets, which we may call $Z_1, Z_2, \ldots, Z_q$. Let K be any state of Z, say the state of the system at time 0. Then we define the subsets Z_r $(r = 1, 2, \ldots, q)$ by

$$A \in Z_r \text{ if and only if } A \in Z \text{ and } v_{AK} = r. \tag{8.16}$$

It then follows from (8.15) that $v_{CB} = 1$ if $C \in Z_{r+1}$ and $B \in Z_r$, or if $C \in Z_1$ and $B \in Z_q$, but not otherwise. Consequently, by (8.10) with $t = 1$, each transition carries the system from its current subset, say Z_r, to the next one, Z_{r+1}, or back to Z_1 if $r = q$. An illustration of this cyclic behaviour is shown in Fig. 6.

Because of this cyclic alternation of the subsets $Z_1, \ldots, Z_r$, which continues indefinitely, the probabilities $p_A(t)$ in a periodic ergodic set oscillate in the limit $t \to \infty$ instead of approaching equilibrium values as they do for an aperiodic ergodic set. To describe this oscillatory behaviour in more detail we return to the result proved in the previous section, which applies to any ergodic set:

$$\left.\begin{array}{ll}\text{either} & w_{BA}(s) = 0 \\ \text{or} & \lim\limits_{t\to\infty}\left[\dfrac{p_A(t)}{\pi_A} - \dfrac{p_B(s+t)}{\pi_B}\right] = 0.\end{array}\right\} \qquad [(7.22)]$$

By choosing $s = v_{BA} + kq$ with k sufficiently large we can ensure, through (8.5), that the first alternative in (7.22) is ruled out. Multiplying the second alternative in (7.22) by π_B and summing over all the states B in one of the subsets, say Z_r, we obtain

$$\lim_{t\to\infty} [p_A(t) - \pi_A f(t)] = 0, \qquad (8.17)$$

where
$$f(t) \equiv \textstyle\sum_B^{(r)} p_B(s+t) / \sum_B^{(r)} \pi_B \qquad (8.18)$$

and $\sum^{(r)}$ denotes a summation over all states in Z_r. The numerator on the right of (8.18) is the probability that the system is in a state belonging to Z_r at time $s + t$. The integers r and s in it are connected by the relation

$$r - s = v_{BK} - [v_{BA} + M(q)] = v_{AK} + M(q)$$

(in whose derivation (8.16) and (8.15) have been used) but are otherwise arbitrary.

If the Markov chain is stationary, that is, if $t^* = 0$, then the formula for $f(t)$ can be greatly simplified; for now (8.10) and (8.16) show that the system is, at time $s + t$, in some state B belonging to Z_r if and only if

$$t = -s + v_{BK} + M(q) = -s + r + M(q) = v_{AK} + M(q),$$

and consequently (8.18) reduces to

$$f(t) = \left\{\begin{array}{ll} 1/\sum_B^{(r)} \pi_B & \text{if} \quad t = v_{AK} + M(q) \\ 0 & \text{if} \quad \text{not}. \end{array}\right\} \qquad (8.19)$$

To evaluate the denominator in this expression for $f(t)$, we note that the value of r does not appear explicitly in the formula (8.17), and since $p_A(t)$ and π_A are both independent of r, $f(t)$ must also be independent of r. By (8.19), therefore, the value of $\sum^{(r)}\pi_B$ must be independent of r. Writing the normalization condition for equilibrium probabilities [eqn. (6.4)] in the form

$$\sum_{r=1}^{q} \left(\textstyle\sum^{(r)} \pi_B\right) = 1 \qquad (8.20)$$

we deduce that $\sum^{(r)}\pi_B = 1/q$ holds for each of the q possible values of r. Thus (8.19) finally reduces to

$$f(t) = \begin{cases} q & \text{if } t = v_{AK} + M(q), \\ 0 & \text{if not.} \end{cases} \tag{8.21}$$

When substituted into (8.17), this gives the asymptotic behaviour of the probability distribution for a periodic ergodic set in a stationary Markov chain and so generalizes the earlier result (7.23) which was restricted to aperiodic ergodic sets. Because of the selection rule (8.10), the probability distribution for a periodic ergodic set cannot approach a limiting steady form; instead it approaches a cyclic alternation of q different distributions, each one confined to one of the subsets $Z_1, \ldots, Z_q$, and proportional, within its own subset, to the equilibrium distribution.

8.1. Exercises

1. Show that if all the ergodic sets of a system have the same period q and the interval between successive observations is changed from 1 to q (as if the system were observed with a stroboscope), then each of the original periodic ergodic sets splits into q distinct aperiodic ergodic sets. What are the equilibrium probabilities in these new ergodic sets?

2. If the Markov chain is stationary for times after t^* but not between 0 and t^*, and the probability distribution at time t^* is concentrated on a single periodic ergodic set but not on any particular state within this set, what conditions must this probability distribution satisfy in order that the probabilities tend to their equilibrium values as $t \to \infty$?

9. The weak law of large numbers

The weak law of large numbers, proved in Chap. II, § 3, for a succession of independent simple trials, asserts that the fraction of occurrences of a given outcome A has a probability distribution which becomes infinitely sharp as the sequence of trials is made infinitely long. The purpose of the present section is to generalize this result to stationary Markov chains. For simplicity we shall assume, as in the previous three sections, that the initial state K is persistent; moreover, we shall prove the law of large numbers only for the case where the ergodic set Z containing K is aperiodic, although the law is true for periodic ergodic sets as well.

The fraction of occurrences $\bar{J}_{A,s}$ of a given state A in a Markov chain of arbitrary duration s is given by a formula analogous to the formula (3.7) for independent trials:

$$\bar{J}_{A,s} \equiv \frac{1}{s} \sum_{t=1}^{s} J_{A,t}, \tag{9.1}$$

where $J_{A,t}$ is now the indicator of the event A_t that the observational state of the system is A at time t. The basic property (2.8) of indicators, applied to (9.1), shows that the expected fraction of occurrences of A is

$$\langle \bar{J}_{A,s} \rangle = \frac{1}{s} \sum_{t=1}^{s} p_A(t) \equiv \bar{p}_A(s), \tag{9.2}$$

where the definition of $\bar{p}_A(s)$ recapitulates (6.17). It follows by (6.21) that

$$\lim_{s\to\infty} \langle \bar{J}_{A,s} \rangle = \pi_A . \tag{9.3}$$

The magnitude of a typical deviation of the random variable $\bar{J}_{A,s}$ from its mean value π_A is given by its standard deviation, the square root of the variance. As in the corresponding proof for independent trials, we can show that the probability distribution of $\bar{J}_{A,s}$ becomes infinitely sharp in the limit $s \to \infty$, by showing that the variance tends to 0.

According to (9.1) and the definition (2.16) of variance, the variance of $\bar{J}_{A,s}$ is

$$\begin{aligned} \operatorname{var}(\bar{J}_{A,s}) &= \langle \bar{J}_{A,s}{}^2 \rangle - \langle \bar{J}_{A,s} \rangle^2 \\ &= \frac{1}{s^2} \sum_{t=1}^{s} \sum_{u=1}^{s} \{\langle J_{A,t} J_{A,u} \rangle - \langle J_{A,t} \rangle \langle J_{A,u} \rangle\} \\ &\equiv v_1 + v_2 + v_3 , \end{aligned} \tag{9.4}$$

where v_1, v_2, v_3 are the contributions to the double sum from the terms with $u = t$, $u > t$, and $u < t$ respectively. The first of these contributions is

$$\begin{aligned} v_1 &= \frac{1}{s^2} \sum_{t=1}^{s} [\langle J_{A,t}{}^2 \rangle - \langle J_{A,t} \rangle^2] \\ &= s^{-2} \sum p_A(t)\,[1 - p_A(t)] \leqq \tfrac{1}{4}s \end{aligned} \tag{9.5}$$

which tends to 0 as $s \to \infty$, just as in the corresponding calculation (3.11) for independent trials. The remaining contributions v_2 and v_3 are equal by symmetry, but they do not vanish as for independent trials. Using first (2.8) and (2.9), then (1.13) and (4.18), and finally (4.8), we can put v_2 in the form

$$\begin{aligned} v_2 &= \frac{1}{s^2} \sum_{t=1}^{s} \sum_{u=t+1}^{s} \{Pr(A_u A_t|K) - Pr(A_t|K)\, Pr(A_u|K)\} \\ &= \frac{1}{s^2} \sum_{t=1}^{s} \sum_{u=t+1}^{s} p_A(t)\, \{Pr(A_u|A_t K) - p_A(u)\} \\ &= \frac{1}{s^2} \sum_{t=1}^{s} p_A(t) \sum_{u=t+1}^{s} \{w_{AA}(u - t) - p_A(u)\}. \end{aligned} \tag{9.6}$$

Since $p_A(t) \leqq 1$ it follows that

$$|v_2| \leqq \frac{1}{s^2} \sum_{t=1}^{s} \sum_{u=t+1}^{s} |w_{AA}(u-t) - p_A(u)| . \tag{9.7}$$

We have seen in eqn. (7.23) that $p_A(u) \to \pi_A$ as $t \to \infty$ since the initial state is persistent and aperiodic; and it follows by eqns. (4.8) and (4.18) that $w_{AA}(u-t) \to \pi_A$ as $u \to \infty$. Therefore when s is large most of the $s(s-1)/2$ terms in the double sum (9.7) are small, and we may expect v_2 to tend to 0 as $s \to \infty$.

To prove that $v_2 \to 0$, we derive from (9.7) the upper bound

$$\begin{aligned} |v_2| &\leqq \frac{1}{s^2} \sum_{t=1}^{s} \sum_{u=t+1}^{s} \{|w_{AA}(u-t) - \pi_A| + |\pi_A - p_A(u)|\} \\ &= \frac{1}{s^2} \sum_{t=1}^{s} \sum_{r=1}^{s-t} |w_{AA}(r) - \pi_A| + \frac{1}{s^2} \sum_{t=1}^{s} \sum_{r=t+1}^{s} |\pi_A - p_A(r)| \\ &\leqq \frac{1}{s} \sum_{r=1}^{s} \{|w_{AA}(r) - \pi_A| + |\pi_A - p_A(r)|\} . \end{aligned} \tag{9.8}$$

We wish to show that for any positive ε the inequality $|v_2| \leqq \varepsilon$ holds for all sufficiently large s. Since both $w_{AA}(r)$ and $p_A(r)$ approach the limit π_A as $r \to \infty$, there exists an n such that the summand in (9.8) is less than $\frac{1}{2}\varepsilon$ for all $s \geqq n$. Also, since all of $w_{AA}(r)$, $p_A(r)$, and π_A lie between 0 and 1, the summand cannot exceed 2. It follows, for $s \geqq n$, that

$$|v_2| \leqq \frac{1}{s} \sum_{r=1}^{n} 2 + \frac{1}{s} \sum_{r=n+1}^{s} \frac{1}{2} \varepsilon \leqq \frac{2n}{s} + \frac{\varepsilon}{2} .$$

Thus the inequality $|v_2| \leqq \varepsilon$ holds whenever both $s \geqq n$ and $s \geqq 4n/\varepsilon$ hold, and by the definition of a limit we conclude that

$$\lim_{s \to \infty} v_2 = 0 . \tag{9.9}$$

Since $v_3 = v_2$ by symmetry, it follows from (9.4), (9.5), and (9.9) that

$$\lim_{s \to \infty} \operatorname{var}(\bar{J}_{A,s}) = 0 \tag{9.10}$$

and hence, by (9.3) and Chebyshev's inequality (2.17), that

$$\lim_{s \to \infty} Pr(|\bar{J}_{A,s} - \pi_A| \geqq \delta) = 0 \quad \text{for all} \quad \delta > 0 . \tag{9.11}$$

That is, the random variable $\bar{J}_{A,s}$, representing the fraction of occurrences of A in the first s observational states after preparation, has a probability distribution which in the limit $s \to \infty$ becomes infinitely sharply concentrated about its mean value π_A. This result is called the *weak law of large numbers*; although the proof given here applies only if Z is aperiodic, the law is true also when Z is periodic. We shall use this law in Chapter IV to express the steady probability distribution π_A and the transition probability matrix w_{AB} in terms of dynamical properties of the system.

It is also possible to prove a *strong law of large numbers* analogous to eqn. (3.14); but the strong law will not be needed in this book.

The result (9.11) can be extended to give information about the rates of occurrence of pairs, triplets, etc., of states. For example, if A and B are any two states in Z we may define

$$\bar{J}_{BA,s+1} = \frac{1}{s} \sum_{t=1}^{s} J_{BA,t+1}, \tag{9.12}$$

where

$$J_{BA,t+1} \equiv J_{B,t+1} J_{A,t} \tag{9.13}$$

is the indicator of the event that A occurs at time t, immediately followed at time $t + 1$ by B. To estimate the probability distribution of $\bar{J}_{BA,s+1}$ we need its mean and variance. The mean is

$$\begin{aligned} \langle \bar{J}_{BA,s+1} \rangle &= \frac{1}{s} \sum_{t=1}^{s} Pr(B_{t+1} A_t | K) \\ &= \frac{1}{s} \sum_{t=1}^{s} w_{BA} p_A(t) \\ &= w_{BA} \langle \bar{J}_{A,s} \rangle \end{aligned} \tag{9.14}$$

so that

$$\lim_{s \to \infty} \langle \bar{J}_{BA,s+1} \rangle = w_{BA} \pi_A \tag{9.15}$$

by (9.3). For the variance of $\bar{J}_{BA,s+1}$ we have

$$\text{var}\,(\bar{J}_{BA,s+1}) = \frac{1}{s^2} \sum_{t=1}^{s} \sum_{u=1}^{s} \{\langle J_{BA,t+1} J_{BA,u+1} \rangle - \langle J_{BA,t+1} \rangle \langle J_{BA,u+1} \rangle\}. \tag{9.16}$$

As in (9.4), this can be written in the form $v_1' + v_2' + v_3'$; here v_1' is the contribution of terms with $u = t$ or $t \pm 1$, v_2' that of terms with $u > t + 1$, and v_3' that of terms with $u < t - 1$. The sum in v_1' contains $s + 2(s - 1)$ terms, none of which exceeds 1 in absolute value, and therefore we have

$$|v_1'| \leqq \frac{1}{s^2} [s + 2(s - 1)] < \frac{3}{s},$$

which tends to 0 as $s \to \infty$.

The contribution v'_2 reduces, by an extension of the method used for (9.6), to

$$v'_2 = \frac{1}{s^2} \sum_{t=1}^{s} \sum_{u=t+2}^{s} [w_{BA} w_{AB}(u - t - 1) - w_{BA} p_A(u)] \, w_{BA} p_A(t) \quad (9.17)$$

so that

$$|v'_2| \leqq \frac{1}{s^2} \sum_{t=1}^{s} \sum_{u=t+2}^{s} |w_{AB}(u - t - 1) - p_A(u)| \quad (9.18)$$

which tends to 0 for $s \to \infty$, as in the treatment of (9.7). Since $v'_3 = v'_2$, we conclude that

$$\lim_{s \to \infty} \operatorname{var} (\bar{J}_{BA,\, s+1}) = 0 \quad (9.19)$$

and so, by Chebyshev's inequality and (9.15), that

$$\lim_{s \to \infty} Pr(|\bar{J}_{BA,\, s+1} - w_{BA} \pi_A| \geqq \delta) = 0 \quad \text{for all} \quad \delta > 0. \quad (9.20)$$

Similar results follow for the rates of occurrence of successive triplets, quadruplets, etc., of states; for example, the result for triplets is

$$\lim_{s \to \infty} Pr(|\bar{J}_{CBA,\, s+2} - w_{CB} w_{BA} \pi_A| \geqq \delta) = 0 \quad \text{for all} \quad \delta > 0 \quad (9.21)$$

where

$$\bar{J}_{CBA,\, s+2} \equiv \frac{1}{s} \sum_{t=1}^{s} J_{C,\, t+2} J_{B,\, t+1} J_{A,\, t}. \quad (9.22)$$

Like the law of large numbers itself, the generalizations (9.20) and (9.21) hold for periodic as well as aperiodic ergodic sets.

A corollary of the weak law of large numbers is the *weak ergodic theorem*. Let X be any observable and let $X(A)$ be the value taken by X when the observational state is A. Then it follows from the definition (2.7) of J_A that

$$X = \textstyle\sum_A X(A) \, J_A. \quad (9.23)$$

In analogy with (9.1) we define the time average

$$\bar{X}_s \equiv \frac{1}{s} \sum_{t=1}^{s} X_t, \quad (9.24)$$

where X_t is the value taken by X at the time t, that is, the random variable

$$X_t = \textstyle\sum_A X(A) \, J_{A,\, t}. \quad (9.25)$$

Substituting (9.25) in (9.24) and using the definition (9.1), we have

$$\bar{X}_s = \textstyle\sum_A X(A) \, \bar{J}_{A,\, s}. \quad (9.26)$$

The weak law of large numbers leads us to expect that in the limit $s \to \infty$ the probability distribution of the right side will concentrate near its mean value $\sum_A X(A)\,\pi_A$. To prove this, let E denote the event $|\overline{X}_s - \sum_A X(A)\,\pi_A| > \delta$ and E_A the event $|\overline{J}_{A,s} - \pi_A| > \delta/\sum_A X_A(A)\,(A = 1, 2, \ldots)$. From the inequality

$$|\overline{X}_s - \textstyle\sum_A X(A)\,\pi_A| \leqq \textstyle\sum_A X(A)|\overline{J}_{A,s} - \pi_A|, \tag{9.27}$$

obtained by subtracting the quantity $\sum_A X(A)\,\pi_A$ from both sides of eqn. (9.26) and then taking absolute values, it follows that if $J_E = 1$ then at least one of the quantities J_{E_A} is 1, and hence that

$$J_E \leqq \textstyle\sum_A J_{E_A}. \tag{9.28}$$

Taking expectations on both sides we obtain

$$Pr(E) \leqq \textstyle\sum_A Pr(E_A). \tag{9.29}$$

In the limit $s \to \infty$, all the probabilities on the right tend to 0, by (9.11), and hence the one on the left does so too. This completes the proof of the weak ergodic theorem,

$$\lim_{s\to\infty} Pr(|\overline{X}_s - \textstyle\sum_A X(A)\,\pi_A| > \delta) = 0 \quad \text{if} \quad \delta > 0. \tag{9.30}$$

The meaning of this theorem is that for large s the time average of the values taken by the observable X at the first s instants of observation is likely to be very close to the expectation of this observable calculated with the equilibrium probabilities. It has points both of similarity and difference with the ergodic theorem of general dynamics quoted in eqn. (6.3) of Chap. I. The two theorems are similar in that both relate time averages to averages taken over a suitable stationary probability distribution; but they differ not only in that (9.30) refers to Markov chains and observables whereas eqn. (6.3) of Chap. I refers to dynamical systems and dynamical variables but also in that eqn. (6.3) of Chap. I is an ergodic theorem of the strong rather than the weak type (analogous to the strong law of large numbers mentioned in § 3).

The result (9.11) can be used to calculate the *mean recurrence time*† of a persistent state A, defined as the mean interval between successive occurrences of A in a system that is isolated over a long time. According to (9.11), if the system is isolated for a long time s then the number of occurrences of A is close to $\pi_A s$, and so the mean recurrence time is $s/\pi_A s = 1/\pi_A$.

† See W. Feller, *An Introduction to Probability Theory and its Applications*, Vol. 1, p. 353.

Smoluchowski made valuable use of the notion of mean recurrence time in his classic discussion of irreversible physical processes.† He argued that a physical process (by which we shall understand here a sequence of observational states in an isolated system) should be regarded as reversible or irreversible according as the recurrence time of its initial state is short or long in comparison with the duration of an experiment. The idea is that if the recurrence time is short, the state A is likely to recur many times (roughly $\pi_A \tau$) during an experiment of duration τ, and on each of these occasions the process that initially took the system away from the state A may be said to have been reversed; on the other hand, if the recurrence time is long, compared with τ, then we expect few or no reversals of the process. In the limiting case where $\tau = \infty$, Smoluchowski's criterion amounts to defining an *irreversible process* as one whose initial state is transient. This is in keeping with the result proved in § 5, that a system started in a transient state will (with probability 1) eventually make the irreversible transition to a persistent state.

For finite values of τ, Smoluchowski's criterion classes a process as irreversible if its initial state is *quasi-transient*, by which we mean that it is either transient or else is persistent and has $\pi_A < 1/\tau$. This classification suffers from the mathematical defect that the value of τ is only vaguely specified, but it is physically reasonable, since a persistent state A with very small π_A behaves approximately like a transient state. To prove this, we may start from the fact that the equilibrium probability distribution, being a steady solution of the evolution eqn. (4.20) for Markov chains, must also satisfy the more general evolution eqn. (4.17) (with A and B interchanged, and t in place of $u - t$)

$$\pi_A = \sum_B w_{AB}(t)\, \pi_B. \tag{9.31}$$

Since the sum contains no negative terms, this equation implies

$$\pi_A \geqq w_{AB}(t)\, \pi_B,$$

that is,

$$w_{AB}(t) \leqq \pi_A/\pi_B \quad \text{for all} \quad t. \tag{9.32}$$

In view of the normalization condition (6.4), there must be at least one state B in Z, the ergodic set containing A, for which $\pi_B \geqq 1/M$, where M is the number of states in Z; and so (9.32) implies that this state B also has the property

$$w_{AB}(t) \leqq M\pi_A \quad \text{for all} \quad t. \tag{9.33}$$

The result (9.33) shows that, if π_A is very small, then there is a state B such that the system can get from A to B (since B and A are in the same ergodic set) but it is unlikely to return in any specified number of steps. In other

† M. von Smoluchowski, Molekulartheoretische Studien über Umkehr thermodynamisch irreversibler Vorgänge und Wiederkehr abnormaler Zustände, *SB Akad. Wiss. Wien* (IIa) **124**, 339 (1915).

words, the probabilities nearly satisfy the condition (5.5) for A to be transient. This is our reason for calling a state quasi-transient if its equilibrium probability is very small. We may also define a *quasi-irreversible transition* as one for which the probability of the reverse transition is very small. Equation (9.32) shows that a sufficient condition for the transition from A to B to be quasi-irreversible is $\pi_A/\pi_B \ll 1$. Since it is practically impossible to decide experimentally whether a probability is zero or merely very small, there is no practical distinction between quasi-transient and strictly transient states, or between quasi-irreversible and strictly irreversible transitions. For mathematical simplicity, however, we shall use only the strict classification (corresponding to $\tau = \infty$) in this book.

9.1. *Exercises*

1. A *quasi-equilibrium state* may be defined as a state, say B, for which $\pi_B \cong 1$: one which at equilibrium is overwhelmingly the most probable in its ergodic set. Given that a system is in a quasi-equilibrium state at time 0, show that the probability of finding it in some other state at any given later time is small (of order $1 - \pi_B$).

2. A *recurrence* of the state A may be defined as an occurrence of A that is not immediately preceded by an occurrence of A. Show† that the mean interval of time between successive recurrences of A (not quite the same thing as the mean recurrence time as defined above) is $1/\pi_A(1 - w_{AA})$.

3. Show that the weak law of large numbers also holds for periodic ergodic sets. (Hint: use the result of exercise 1 of Chap. II, § 8.)

4. Use the result of exercise 4 of Chap. II, § 4 to provide an alternative derivation of (9.20).

† See M. S. Bartlett, *Stochastic Processes*, p. 183.

CHAPTER III

The Gibbs Ensemble

1. Introduction

In the previous chapter we examined the consequences of the Markovian model of observation, without making any use of the dynamics of the observed system. The next step on our path, taking us beyond the realm of probability theory into that of statistical mechanics, will be to combine these purely statistical results with the further information provided by dynamics. This will make it possible to express the fundamental quantities of the statistical theory, the transition probabilities w_{BA}, in terms of purely dynamical properties of the system.

As explained in Chap. II, § 4, the time-dependent probability distribution $p_A(t)$ can be interpreted physically as the fraction of systems in a large statistical ensemble that are in the observational state A at time t. In order to relate these probability distributions to dynamics, we therefore begin by extending the dynamical description of matter, which in its elementary form applies only to individual systems, to ensembles of systems.

This idea of applying the laws of dynamics to " a great number of independent systems, identical in nature but differing in phase, that is in their condition with respect to configuration and velocity", was introduced into physics by Willard Gibbs.† Such a collection of systems is now called a *Gibbs ensemble.* The systems comprising a Gibbs ensemble do not interact;‡ they have identical Hamiltonians but their dynamical states need not be identical. If all the dynamical states do happen to be identical we speak of a *pure ensemble*; otherwise, of a *mixed ensemble.*

The idea of a Gibbs ensemble is more general than that of a statistical ensemble, for every statistical ensemble can in principle be realized in a statistical experiment, but there are some Gibbs ensembles which cannot be so realized owing to the limitations on experimental technique that are formalized in the model of observation described in Chap. I, § 3. As examples of realizable Gibbs ensembles we may take the ensemble of pennies produced by tossing a large number simultaneously, or the ensemble of iron bars produced by bringing about a specified temperature distribution in each one of a large number of replicas of a given iron bar. An example of

† J. W. Gibbs, *Elementary Principles in Statistical Mechanics* (Yale University Press, New Haven, 1902; Dover reprint, New York, 1960), p. 5. The wording of Gibbs's definition applies only in classical mechanics, but we shall treat the quantum case too.

‡ If there is a weak interaction between the systems in the collection it is called not an ensemble but an *assembly.* See exercise 4 of Chap. VI, § 1.

a Gibbs ensemble that cannot be realized is the one that would be obtained if we could simultaneously reverse the velocities of all the particles in the ensemble of iron bars just mentioned. In this hypothetical ensemble, heat would flow from cold to hot, a violation of the law of heat conduction which shows that, in general, the thought-operation of reversing all the velocities is not among the operations that it is actually possible to perform on a physical system.† The question which Gibbs ensembles can be realized in statistical experiments is connected with the limitations on experimental procedures and therefore with the theory of observation. This question will exercise us again in Chapter IV, but in the rest of the present chapter we shall be concerned, as Gibbs was, only with the formal and dynamical properties of arbitrary Gibbs ensembles, and not with how, if at all, these ensembles can be realized. Since the limitations on observation and experimentation are not considered here, the distinction made in Chap. I, § 1, between "large" and "small" systems disappears; the theory in this chapter applies to any Gibbs ensemble, realizable or non-realizable, consisting of replicas of any dynamical system, large or small.

The fundamental operation in the theory of Gibbs ensembles is that of taking the *ensemble average* of a dynamical variable. Let G be any dynamical variable referring to a single system of the ensemble; in classical mechanics G is represented by a function $G(\alpha)$ over the dynamical states α of the system, and in quantum mechanics by an operator in its Hilbert space. Since the restrictions imposed in our theory of observation are irrelevant here, we may define the ensemble average in terms of *ideal* measurements of G (measurements of perfect accuracy) carried out separately on each system of the ensemble. The ensemble average of G, written $\langle G \rangle$, is defined as the average of the results of these measurements:

$$\langle G \rangle = \lim_{n\to\infty} \frac{1}{n} \sum_{i=1}^{n} G^{(i)}, \tag{1.1}$$

where $G^{(i)}$ is the result of the measurement of G in the ith system. The limit process $n \to \infty$ is a mathematical idealization of Gibbs's requirement that the number of systems in the ensemble be "large". The use of the bracket notation $\langle\rangle$ for ensemble averages is a natural extension of its earlier use in Chap. II, eqn. (2.1), for the expectations of random variables; the earlier definition of $\langle G \rangle$ may be thought of as a special case of (1.1), applying whenever the Gibbs ensemble is also a statistical ensemble and the dynamical variable G is an observable.

† J. M. Blatt, An Alternative Approach to the Ergodic Problem, *Prog. Theor. Phys.* **22**, 745 (1959), points out that there is an experiment—the spin-echo experiment—where such a velocity reversal *is* possible. Some resulting theoretical difficulties are interpreted by Blatt as evidence of fundamental shortcomings in the model of observation advocated in Chap. I, § 3, but these difficulties can equally well be attributed to the fact that the observational description he works with is non-Markovian.

The main purposes of this chapter are, first, to show how to specify Gibbs ensembles in such a way that ensemble averages can be calculated easily and, secondly, to study the dynamical behaviour of Gibbs ensembles. Because both dynamics and the theory of ideal measurements in classical mechanics are very different from their quantum-mechanical counterparts, the two types of mechanics will be treated separately, although the final results for the two types of mechanics can be put into close correspondence. We consider classical mechanics first.

2. The phase-space density

As in Chap. I, §§ 1 and 2, we may visualize the dynamical state α of a single system as a point in a dynamical space, which in classical mechanics is a $2F$-dimensional phase space with the coordinate system $p_1, ..., p_F, q_1, ..., q_F$. If a Gibbs ensemble comprises n systems with dynamical states $\alpha^{(1)}, \alpha^{(2)}, ..., \alpha^{(n)}$, the dynamical state of the whole ensemble may be visualized as a "dust" consisting of the n points $\alpha^{(1)}, ..., \alpha^{(n)}$ in the phase space of a single system. When the limit $n \to \infty$ is taken, as in (1.1), the "dust" in phase space may be imagined to coalesce into a continuous medium whose density is a function of position in phase space (that is, of the dynamical state).

To describe the density of this continuous medium quantitatively, we need a definition of *volume in phase space*. The natural definition is that the volume of a phase-space region R is the $2F$-fold integral

$$\int_R dp_1 \dots dp_F \, dq_1 \dots dq_F, \tag{2.1}$$

where $p_1, ..., p_F, q_1, ..., q_F$ is any canonical coordinate system (a system of F position coordinates $q_1, ..., q_F$ and their canonically conjugate momenta $p_1, ..., p_F$). Since the product $p_i q_i$ for any pair of canonical coordinates has the dimensions $[ML^2T^{-1}]$ of action, volumes in phase space for a system with F degrees of freedom have the dimensions $[\text{action}]^F$.

The definition (2.1) appears to make the volume of R depend on the particular choice of coordinates used to describe the dynamical state of the system, but in fact it does not. If a new set of position coordinates $q_1', ..., q_F'$ is used to describe the configurations of the system, then the old coordinates can be expressed as functions of the new:

$$q_i = f_i(q_1', ..., q_F') \quad \text{for} \quad i = 1, ..., F.$$

From these equations it follows that

$$\dot{q}_i = \sum_j (\partial f_i / \partial q_j') \, \dot{q}_j'$$

and hence, by Chap. I, eqn. (2.1), that

$$\begin{aligned} p_j' = \partial K / \partial \dot{q}_j' &= \sum_i (\partial K / \partial \dot{q}_i)(\partial \dot{q}_i / \partial \dot{q}_j') \\ &= \sum_i p_i \, \partial f_i / \partial q_j', \end{aligned}$$

the configuration being held constant in all the partial differentiations in the first line. The Jacobian of the transformation from the old to the new coordinates, expressed as the determinant of a partitioned matrix, is therefore

$$\frac{\partial(p'_1, \ldots, q'_F)}{\partial(p_1, \ldots, q_F)} = \begin{vmatrix} \left[\dfrac{\partial p'_i}{\partial p_j}\right] & \left[\dfrac{\partial p'_i}{\partial q_j}\right] \\ \left[\dfrac{\partial q'_i}{\partial p_j}\right] & \left[\dfrac{\partial q'_i}{\partial q_j}\right] \end{vmatrix} = \begin{vmatrix} \left[\dfrac{\partial f_j}{\partial q'_i}\right] & \left[\dfrac{\partial p'_i}{\partial q_j}\right] \\ [0] & \left[\dfrac{\partial f_j}{\partial q'_i}\right]^{-1} \end{vmatrix} = 1$$

so that the integral (2.1) is invariant under this *point transformation.*

On the assumption that the "dust" of phase-space points $\alpha^{(1)}, \ldots, \alpha^{(n)}$ describing the ensemble coalesces into a continuous medium as $n \to \infty$, we can now describe the density of this medium by a function $D(\alpha)$ of position in phase space, called the *phase-space density*. We define $D(\alpha)$ as the fraction of members of the ensemble per unit phase-space volume in the neighbourhood of α; or, more precisely, by

$$D(\alpha) = \lim_{\Delta\alpha \to 0} \frac{\text{fraction of ensemble having dynamical states within } \Delta\alpha}{\text{volume of } \Delta\alpha}, \tag{2.2}$$

where $\Delta\alpha$ is a phase-space region enclosing the point α. The assumption that the dust coalesces into a continuous medium means, in mathematical terms, that $D(\alpha)$ as defined by (2.2) is assumed to exist and to be independent of the way the limit $\Delta\alpha \to 0$ is approached. The double limiting process used in defining $D(\alpha)$ must be carried out in the right way: the number n of systems in the ensemble must go to infinity before, not after, the volume $\Delta\alpha$ of the elementary regions used to define the integral goes to 0. This arrangement ensures that the number of systems with dynamical states inside each region $\Delta\alpha$ remains large throughout the limit process.

In defining $D(\alpha)$ by (2.2), we assumed that $D(\alpha)$ has a well-defined finite value everywhere in phase space. This assumption places a restriction on the class of Gibbs ensembles that we can treat. In particular, pure ensembles cannot be treated by this method, since for a pure ensemble all of whose systems have dynamical state α_0, the right-hand side of (2.2) is infinite if $\alpha = \alpha_0$. It is possible to remove this restriction by generalizing our definition (2.2) of $D(\alpha)$; the more general definition makes $D(\alpha)$ a *generalized function* of α (e.g. a Dirac delta-function), which does not necessarily take a finite numerical value for each α. This generalization, however, is unnecessary for statistical mechanics.

The ensemble average of a dynamical variable G, defined in (1.1), can here be written in the form

$$\langle G \rangle = \lim_{n\to\infty} \frac{1}{n} \sum_{i=1}^{n} G(\alpha^{(i)}) \tag{2.3}$$

since in classical mechanics the value of a dynamical variable G is a function $G(\alpha)$ of the dynamical state α. Alternatively, since we assume that the dust in phase space coalesces into a continuous medium, we can rewrite the sum (2.3) as an integral over all phase space

$$\langle G\rangle = \int G(\alpha)\, D(\alpha)\, d\alpha, \tag{2.4}$$

where $d\alpha \equiv dp_1 \ldots dp_F\, dq_1 \ldots dq_F$ is the element of volume in phase space, and $D(\alpha)$ is defined by (2.2). For example, if R is a region of phase space and $G(\alpha)$ is the indicator of this region, taking the value 1 if α is in R and 0 if not, then the ensemble average of G is $\int_R D(\alpha)\, d\alpha$, and is equal to the fraction of members of the ensemble with dynamical states in R. In particular if the ensemble is a statistical ensemble and R is the dynamical image (see Chap. I, § 3) of an observational state, then $\int_R D(\alpha)\, d\alpha$ is the probability of that observational state.

The phase-space density function $D(\alpha)$ of a Gibbs ensemble satisfies two important conditions analogous to the normalization and non-negativity conditions (1.2) and (1.5) of Chap. II for probabilities. The *normalization condition* for phase-space densities may be obtained by setting $G(\alpha) \equiv 1$, so that $\langle G\rangle = 1$ by (1.1); putting this into (2.2) we obtain

$$\int D(\alpha)\, d\alpha = 1, \tag{2.5}$$

where the integral is again over all phase space. The *non-negativity* condition, which follows at once from the definition (2.2), is

$$D(\alpha) \geqq 0 \quad \text{for all } \alpha. \tag{2.6}$$

Any function of position satisfying (2.5) and (2.6) is the phase space density of a conceivable Gibbs ensemble, although there may be no statistical experiment by which this ensemble can be realized.

The name *fine-grained probability* is often used for the relative frequency of a dynamical state, or set of dynamical states, in a classical Gibbs ensemble. The mathematical properties of fine-grained probabilities are very similar to those of ordinary physical ("coarse-grained") probabilities; they differ only in that the sample space is discrete for coarse-grained and continuous for fine-grained probabilities. The conditions (2.5) and (2.6) are thus the counterparts, for fine-grained probabilities, of the normalization and non-negativity conditions (1.2) and (1.5) of Chap. II for coarse-grained; likewise, (2.3) and (2.4) for the ensemble average of a dynamical variable are the fine-grained counterparts of the "coarse-grained" eqns. (2.1) and (2.3) of Chap. II for the expectation of a random variable. It would be wrong, however, to infer from this mathematical similarity that fine- and coarse-grained probabilities are also similar in physical meaning. Coarse-grained probabilities, as explained in Chap. I, § 4, can be measured reproducibly by a well-defined experimental procedure, and can therefore be looked on as physical quantities just like mass or electric charge; but fine-grained probabilities cannot in general be measured (since the dynamical state of a large system is unobservable) nor, as will be shown in Chap. IV, § 2, are they reproducible.

2.1. Exercise

1. Verify explicitly that the point transformation induced in phase space by the transformation from Cartesian to polar position coordinates for a single particle moving in a plane has unit Jacobian.

3. The classical Liouville theorem

As the dynamical states of the systems in a Gibbs ensemble change with time, the "dust" representing these dynamical states moves through phase space in accordance with Hamilton's equations of motion, previously noted in Chap. I, eqn. (2.3),

$$\left.\begin{aligned} \dot{q}_i &= \frac{\partial H}{\partial p_i}(p_1 \ldots q_F) \\ \dot{p}_i &= -\frac{\partial H}{\partial q_i}(p_1 \ldots q_F) \end{aligned}\right\} \text{for } i = 1, \ldots, F. \tag{3.1}$$

Since these equations are of first order in t, each particle of the "dust" moves with a velocity entirely determined by its position in phase space, so that the motion of these imaginary particles is completely orderly, unlike the chaotic thermal motion of real molecules in a gas. This motion can be visualized by imagining the dust particles carried along in a fluid whose velocity at each point in phase space is given by (3.1). If the system happens to be isolated (i.e., if the Hamiltonian does not depend explicitly on time), the flow of this fluid is steady, its velocity at each point in phase space being independent of the time. The motion of this underlying fluid is called the *natural motion* in phase space.

By virtue of the canonical simplicity of the equations of motion, the flow of this $2F$-dimensional fluid satisfies a condition analogous to the condition

$$\operatorname{div} \boldsymbol{v} = 0$$

satisfied by the flow of an incompressible fluid in three dimensions; for if we treat the $2F$ canonical coordinates of the point α in phase space as a vector $\boldsymbol{\alpha}$ with components

$$\left.\begin{aligned} \alpha_i &\equiv p_i \\ \alpha_{i+F} &\equiv q_i \end{aligned}\right\} \text{for } i = 1, \ldots, N, \tag{3.2}$$

then the $2F$-dimensional velocity vector $\dot{\boldsymbol{\alpha}}$ at the point α in phase space has the components $(\dot{\alpha}_1, \ldots, \dot{\alpha}_{2F})$, and the divergence of the velocity field is therefore

$$\sum_{i=1}^{2F} \frac{\partial \dot{\alpha}_i}{\partial \alpha_i} = \sum_{i=1}^{F} \left(\frac{\partial \dot{p}_i}{\partial p_i} + \frac{\partial \dot{q}_i}{\partial q_i} \right)$$

$$= \sum_{i=1}^{F} \left(-\frac{\partial^2 H}{\partial p_i \, \partial q_i} + \frac{\partial^2 H}{\partial q_i \, \partial p_i} \right) = 0 \tag{3.3}$$

by the equations of motion (3.1). Thus we may think of the underlying fluid as incompressible and its density as a constant.

In the limit where the "dust" particles coalesce into a continuous medium, the phase-space density at any instant t, defined in accordance with (2.2) but with the time dependence made explicit, will be written $D_t(\alpha)$, or D_t for short. This continuous medium may be visualized† as a coloured substance non-uniformly dissolved in the underlying fluid, its concentration at the point α at the time t being proportional to $D_t(\alpha)$. As the underlying fluid moves through phase space, the colouring matter is carried along with it; there is no diffusion since the motion of the original dust particles was completely orderly.

If the concentration of the colouring matter happens to be a differentiable function of α and t, then by analogy with the mass conservation equation of 3-dimensional hydrodynamics,

$$\frac{\partial \varrho}{\partial t} + \operatorname{div}(\varrho \boldsymbol{v}) = 0,$$

it will satisfy the $2F$-dimensional conservation equation

$$\frac{\partial D_t}{\partial t} + \sum_{i=1}^{2F} \frac{\partial}{\partial \alpha_i} (D_t \dot{\alpha}_i) = 0. \tag{3.4}$$

Combining this with (3.3) gives

$$\frac{\partial D_t}{\partial t} + \sum_{i=1}^{2F} \dot{\alpha}_i \frac{\partial D_t}{\partial \alpha_i} = 0, \tag{3.5}$$

which we shall call *Liouville's equation*. The left-hand side of (3.5) may be interpreted as the rate of change of the value of D_t at the moving point that follows the motion of the fluid and passes through the point α at time t; the operator $\partial/\partial t + \sum_i \dot{\alpha}_i (\partial/\partial \alpha_i)$, which acts on D_t in (3.5), is the phase-space analogue of the convective time-differentiation operator $\partial/\partial t + \boldsymbol{v}.\text{grad}$ used in hydrodynamics. Accordingly Liouville's equation (3.5) has the interpretation that the phase-space density in the neighbourhood of the moving

† This method of visualizing the variation of $D_t(\alpha)$ with phase and time is due to Gibbs (*Elementary Principles in Statistical Mechanics*, p. 144).

phase point of any system in the ensemble remains constant throughout the motion. We shall refer to this fact as *Liouville's theorem.*

To give mathematical expression to Liouville's theorem, and also to extend it to non-differentiable phase-space densities, we introduce a family of vector functions $U_t(\;)$ of position in phase space so defined that if the dynamical state of a system is α at time 0, an arbitrarily chosen origin of time measurement, then its dynamical state at any time t has the position vector $U_t(\alpha)$ or $U_t\alpha$ for short. The determinism of classical mechanics guarantees that this function exists. When the abbreviated notation $U_t\alpha$ is used, it is convenient to think of U_t as an operator converting the dynamical state at time 0 into the dynamical state at time t.

We shall call these operators the *evolution operators* of the system. A special case is

$$U_0 = 1, \tag{3.6}$$

where 1 denotes the identity operator. If the Hamiltonian happens to be independent of time, then the state at time 2 is the same function of this state at time 1 as the latter is of the state at time 0, so that we have $U_2 = (U_1)^2$, and in general

$$U_t = (U_1)^t \quad \text{if } t \text{ is an integer.} \tag{3.7}$$

For general values of t, and Hamiltonians that are not necessarily independent of time, an alternative definition of U_t is as the solution of the differential equation

$$dU_t/dt = LU_t \tag{3.8}$$

that satisfies the boundary condition (3.6); here L is the operator defined by

$$L\alpha = \dot{\alpha} \quad \text{for all } \alpha.$$

To calculate the time dependence of $D_t(\alpha)$ without relying on hydrodynamical analogies, consider an arbitrary dynamical variable G and imagine each system in the ensemble permanently labelled with its own value of G at time 0. The ensemble average of the labels at time 0 is therefore

$$\int G(\alpha_0)\, D_0(\alpha_0)\, d\alpha_0 \tag{3.9}$$

by (2.4) with a change of dummy variable. Since the labels do not change with time we can equate this to the ensemble average of the labels at time t. Consider a system whose dynamical state is α at time t; then by the definition of U_t its state at time 0 is $(U_t)^{-1}\alpha$. Consequently the label on this system is $G((U_t)^{-1}\alpha)$, and the ensemble average of all labels at time t is

$$\int G((U_t)^{-1}\alpha)\, D_t(\alpha)\, d\alpha. \tag{3.10}$$

Equating the expressions (3.9) and (3.10) for the ensemble average of the labels, and at the same time transforming (3.10) to the new variable of integration

$$\alpha_0 \equiv (U_t)^{-1}\alpha, \tag{3.11}$$

we obtain

$$\int G(\alpha_0)\, D_0(\alpha_0)\, d\alpha_0 = \int G(\alpha_0)\, D_t(U_t\alpha_0) \frac{d\alpha_0}{J(U_t\alpha_0, t)}, \tag{3.12}$$

where $J(\alpha, t) \equiv \partial(\alpha_0)/\partial(\alpha)$ is the Jacobian of the transformation (3.11). Since the dynamical variable G is entirely arbitrary, it follows that

$$D_t(U_t\alpha_0) = D_0(\alpha_0)\, J(U_t\alpha_0, t) \quad \text{for all } \alpha_0. \tag{3.13}$$

To evaluate the Jacobian in (3.13) we use the divergence-free property of the flow in phase space. Defining $\alpha' \equiv U_{t+\delta t}\alpha_0$ where δt is a small time increment, we have, by the multiplication theorem for Jacobians,

$$\frac{J(\alpha, t)}{J(\alpha', t+\delta t)} = \frac{\partial(\alpha_0)/\partial(\alpha)}{\partial(\alpha_0)/\partial(\alpha')} = \frac{\partial(\alpha')}{\partial(\alpha)}. \tag{3.14}$$

To evaluate this new Jacobian $\partial(\alpha')/\partial(\alpha)$ we write the relation between α' and α in the form

$$\alpha_i' = \alpha_i + \dot{\alpha}_i\, \delta t + o(\delta t) \quad (i = 1, \ldots, 2F), \tag{3.15}$$

where $o(\delta t)$ means a quantity that becomes negligible compared with δt in the limit $\delta t \to 0$. The new Jacobian is therefore

$$\frac{\partial(\alpha')}{\partial(\alpha)} = \begin{vmatrix} 1 + \frac{\partial\dot{\alpha}_1}{\partial\alpha_1}\delta t + o(\delta t) & \ldots & \frac{\partial\dot{\alpha}_{2F}}{\partial\alpha_1}\delta t + o(\delta t) \\ \frac{\partial\dot{\alpha}_1}{\partial\alpha_2}\delta t + o(\delta t) & \ldots & \ldots \\ \ldots & \ldots & \ldots \\ \frac{\partial\dot{\alpha}_1}{\partial\alpha_{2F}}\delta t + o(\delta t) & \ldots & 1 + \frac{\partial\dot{\alpha}_{2F}}{\partial\alpha_{2F}}\delta t + o(\delta t) \end{vmatrix} \tag{3.16}$$

In the complete expansion of this determinant, all terms but the leading one involve at least two off-diagonal elements and are therefore of order $(\delta t)^2$ or smaller. Thus the expansion gives

$$\begin{aligned} \frac{\partial(\alpha')}{\partial(\alpha)} &= \prod_{i=1}^{2F}\left[1 + \frac{\partial\dot{\alpha}_i}{\partial\alpha_i}\delta t + o(\delta t)\right] + o(\delta t) \\ &= 1 + \sum_{i=1}^{2F}\frac{\partial\dot{\alpha}_i}{\partial\alpha_i}\delta t + o(\delta t) \quad \text{on multiplying out,} \\ &= 1 + o(\delta t) \end{aligned} \tag{3.17}$$

since the motion in phase space is divergence-free, by (3.3). Substituting (3.17) into (3.14) and using the definitions of α and α' we obtain

$$J(U_t\alpha_0, t) = J(U_{t+\delta t}\alpha_0, t + \delta t)\,[1 + o(\delta t)],$$

from which it follows that

$$\frac{\partial}{\partial t} J(U_t\alpha_0, t) = \lim_{\delta t \to 0} \frac{J(U_{t+\delta t}\alpha_0, t + \delta t) - J(U_t\alpha_0, t)}{\delta t}$$

$$= \lim_{\delta t = 0} \frac{J(U_{t+\delta t}\alpha_0, t + \delta t)}{\delta t} o(\delta t) = 0, \qquad (3.18)$$

the partial derivative being taken at fixed α_0. Integrating (3.18) and determining the constant of integration from the fact that $J(\alpha_0, 0)$ is, by (3.11) and (3.6), the Jacobian of the identity transformation, we find that

$$J(\alpha, t) = 1 \quad \text{for all } t, \alpha. \qquad (3.19)$$

Substituting (3.19) into (3.13) we obtain

$$D_t(U_t\alpha) = D_0(\alpha) \quad \text{for all } \alpha. \qquad (3.20)$$

That is to say, the phase-space density remains constant near the phase point of any system of the ensemble as it traces out its trajectory in phase space. Thus (3.20) provides a mathematical statement of *Liouville's theorem*, which was stated verbally after (3.5). We shall call (3.20) the *integrated Liouville equation*, since if $D_t(\alpha)$ is differentiable we can obtain Liouville's differential equation (3.5) by differentiating (3.20) with respect to t and using (3.7) and (3.8).

To illustrate the application of (3.20) let us investigate the *steady ensembles* (also called *stationary ensembles*), for which the phase-space density is independent of time. We consider only isolated systems, i.e. systems for which the Hamiltonian H is independent of time. The definition of a steady ensemble may be written

$$D_t(\alpha) = D_0(\alpha) \equiv D(\alpha) \quad \text{for all } \alpha, t, \qquad (3.21)$$

and combined with (3.20) this gives

$$D(U_t\alpha) = D(\alpha). \qquad (3.22)$$

Since α and $U_t\alpha$ are two points on the same trajectory, (3.22) has the interpretation that a steady phase-space density is constant along every trajectory that can be traced out by the system during its dynamical evolution. Thus if $G(\alpha)$ is any dynamical variable that is constant along trajectories (a *constant of the motion* or *invariant of the motion*) then we can obtain a steady ensemble by taking $D_0(\alpha)$ to be $G(\alpha)$ or any function of $G(\alpha)$. In particular, since the Hamiltonian of an isolated system is an invariant of the motion [by eqn. (2.24) of Chap. I], any phase-space density of the form

$$D(\alpha) = f(H(\alpha)) \tag{3.23}$$

is steady. Sometimes there are other simple invariants of the motion in addition to $H(\alpha)$; for example, the angular momentum for a system in a spherically symmetrical container. If $L(\alpha)$ is such an invariant, then a more general steady phase-space density than (3.23) is

$$D(\alpha) = f(H(\alpha), L(\alpha)). \tag{3.24}$$

Two important steady phase-space densities of the form (3.23) are

$$D(\alpha) \propto e^{-\beta H(\alpha)} \tag{3.25}$$

and

$$D(\alpha) \propto \begin{cases} 1 \text{ if } E - \Delta E < H(\alpha) \leq E \\ 0 \text{ otherwise.} \end{cases} \tag{3.26}$$

Here $\beta, E, \Delta E$ are given constants. The constants of proportionality can be determined from the normalization condition (2.5). An ensemble with the phase-space density (3.25) is said to be *canonical*; one with the phase-space density (3.26), *microcanonical.* In a microcanonical ensemble the phase-space density is uniform over the energy shell defined in Chap. I, eqn. (6.4); thus the right-hand side of eqn. (6.3) of Chap. I is the average of $G(\alpha)$ over a microcanonical distribution.

3.1. Exercises

1. For the canonical ensemble show that

$$\langle H \rangle = -\partial(\ln \Phi)/\partial\beta,$$

where $\Phi(\beta) \equiv \int e^{-\beta H}\, d\alpha$. Obtain an analogous formula for the ensemble average of $(H - \langle H \rangle)^2$.

2. Prove Gibbs's *principle of conservation of extension-in-phase* (i. e. of phase-space volume), which states that if a region of phase space, call it R, is transformed by the natural motion in phase space over a time interval t into a region R_t, then the phase-space volumes of R and R_t are equal.

3. For a classical harmonic oscillator with Hamiltonian $\frac{1}{2}\omega(p^2 + q^2)$, verify that

$$\begin{bmatrix} U_t p \\ U_t q \end{bmatrix} = \begin{bmatrix} \cos \omega t & -\sin \omega t \\ \sin \omega t & \cos \omega t \end{bmatrix} \begin{bmatrix} p \\ q \end{bmatrix}.$$

Draw the trajectories in phase space. If D_0 is given by

$$D_0(p, q) = \begin{cases} \text{const} & \text{if } p > 0 \text{ and } q > 0 \text{ and } p^2 + q^2 < 1 \\ 0 & \text{if not,} \end{cases}$$

evaluate the constant and find $D_t(p, q)$ for $t = \pi/2\omega$.

4. The *fine-grained entropy* of a classical Gibbs ensemble is defined to be $k \int D(\alpha) \ln [c/D(\alpha)]\, d\alpha$, where k and c are constants. Show that the fine-grained entropy is an invariant of the motion.

4. The density matrix

To develop a quantum theory of Gibbs ensembles, analogous to the classical theory in the previous two sections, we return to the basic definition of ensemble average given in (1.1),

$$\langle G \rangle = \lim_{n \to \infty} \frac{1}{n} \sum_{i=1}^{n} G^{(i)}, \tag{4.1}$$

where $G^{(i)}$ is the result of an ideal measurement of the dynamical variable G in the ith system of the ensemble. Because of the characteristic indeterminacy of quantum measurements, the value of each individual $G^{(i)}$ is not, in general, uniquely determined by the quantum state of the ith system; nevertheless, the statistical postulate of quantum mechanics does enable us to evaluate $\langle G \rangle$ in the limit of large n.

The simplest case is a pure ensemble, with all its members in the same quantum state; an example of such an ensemble is a polarized beam of light, which consists of a large number of photons all in the same quantum state. The statistical postulate of quantum mechanics† asserts that if the measurement of G on a system in the quantum state $|\alpha\rangle$, with wave function $\psi_\alpha(q_1, \ldots, q_F)$, is replicated a very large number of times, the average of the measured values is $\langle\alpha|\mathsf{G}|\alpha\rangle$, the expectation value of the operator G in the state $|\alpha\rangle$. In terms of the wave function ψ_α representing the state $|\alpha\rangle$, the expectation value of G is given by

$$\langle\alpha|\mathsf{G}|\alpha\rangle = \int \psi_\alpha^*(Q)\, \mathsf{G}\psi_\alpha(Q)\, dQ, \tag{4.2}$$

where ψ_α^* is the complex conjugate of ψ_α,

$$\left.\begin{aligned} Q &\equiv (q_1, \ldots, q_F), \\ dQ &\equiv dq_1 \ldots dq_F, \end{aligned}\right\} \tag{4.3}$$

and the integration goes over all possible configurations Q of the system. In (4.2), $\mathsf{G}\psi_\alpha(Q)$ means the wave function of the state $\mathsf{G}|\alpha\rangle$; if G has the matrix

$$G(Q; Q') \equiv \langle Q|\mathsf{G}|Q'\rangle \tag{4.4}$$

in the position representation, then $\mathsf{G}\psi_\alpha(Q)$ is given by

$$\mathsf{G}\psi_\alpha(Q) = \int G(Q; Q')\, \psi_\alpha(Q')\, dQ'. \tag{4.5}$$

Applying the statistical postulate to the pure ensemble whose members are all in the normalized quantum state $|\alpha\rangle$, we see that the definition (4.1) becomes

$$\langle G \rangle = \langle\alpha|\mathsf{G}|\alpha\rangle \quad \text{for a pure ensemble.} \tag{4.6}$$

† P. A. M. Dirac, *The Principles of Quantum Mechanics*, § 12.

To generalize (4.6) to mixed ensembles we may consider first the mixed ensemble constructed by putting together a finite number of pure sub-ensembles. If the first pure sub-ensemble comprises n_1 systems in the normalized quantum state $|\alpha_1\rangle$, the second, n_2 in the state† $|\alpha_2\rangle$, etc., then the average of G over the finite mixed ensemble comprising all these $n \equiv \sum n_k$ systems may be written in the form

$$\frac{1}{n}\sum_{i=1}^{n} G^{(i)} = \sum_k \frac{n_k}{n}\left[\frac{1}{n_k}\sum_{i(k)} G^{(i)}\right],$$

where $\sum_{i(k)}$ means a sum over the values of i that label systems belonging to the kth sub-ensemble. Taking the limit $n \to \infty$ and using (4.1) we obtain for the infinite mixed ensemble

$$\langle G\rangle = \sum_k f_k \langle G\rangle_k, \tag{4.7}$$

where $\langle G\rangle_k$ is the average of G in the kth sub-ensemble, and

$$f_k \equiv \lim_{n\to\infty} \frac{n_k}{n} \tag{4.8}$$

is the fraction of systems in this sub-ensemble. Using (4.6) to evaluate $\langle G\rangle_k$, we obtain

$$\langle G\rangle = \sum_k f_k \langle \alpha_k|G|\alpha_k\rangle. \tag{4.9}$$

The restriction to ensembles constructed by putting together a finite number of pure ensembles can be removed by a limiting process analogous to the one that makes the "dust" representing a classical Gibbs ensemble in phase space coalesce into a continuous fluid. In the quantum case, the corresponding limit process will replace the finite sum in (4.9) by an infinite sum, or possibly by an integral over Hilbert space. Thus, provided we allow the sum to represent an integral in cases where an integral is necessary, (4.9) applies to the most general type of Gibbs ensemble we shall consider.

By transforming (4.9) into the position representation, we can bring out more clearly its analogy to the corresponding classical formula for ensemble averages. In fact, substituting (4.2) and then (4.5) into (4.9) we obtain

$$\begin{aligned}\langle G\rangle &= \sum_k f_k \iint \psi_k^*(Q)\, G(Q; Q')\, \psi_k(Q')\, dQ\, dQ' \\ &= \iint G(Q; Q')\, D(Q'; Q)\, dQ'\, dQ,\end{aligned} \tag{4.10}$$

where $\psi_k(Q)$ is the wave function of the quantum state $|\alpha_k\rangle$ and $D(Q'; Q)$ is the *density matrix* of the ensemble, defined by

$$D(Q'; Q) \equiv \sum_k \psi_k(Q') f_k \psi_k^*(Q). \tag{4.11}$$

† The states $|\alpha_1\rangle$, $|\alpha_2\rangle$, ... need not be orthogonal.

Just like the classical formula $\int G(\alpha)\, D(\alpha)\, d\alpha$ [eqn. (2.4)] for ensemble averages, (4.10) expresses $\langle G \rangle$ as the integral of the product of two functions of $2F$ variables; one of these functions, $G(Q; Q')$ specifies the dynamical variable G and the other, $D(Q'; Q)$, specifies the ensemble. Accordingly we may look on (4.10) as a quantum analogue of (2.4), and the density matrix $D(Q'; Q)$ as a quantum analogue of the classical phase-space density function $D(\alpha)$.

Since the quantities f_k are real, it follows from (4.11) that the density matrix is Hermitian:

$$D(Q'; Q)^* = D(Q; Q').$$

It may be regarded as the position representative $\langle Q'|\mathsf{D}|Q\rangle$ of a certain Hermitian linear operator D; this operator is called the *density operator*, or sometimes the *statistical operator*. A simple formula for the density operator can be obtained by substituting into (4.11) this definition of D and the definition† $\psi_k(Q) = \langle Q|\alpha_k\rangle$ of a wave function. This gives

$$\langle Q'|\mathsf{D}|Q\rangle = \textstyle\sum_k \langle Q'|\alpha_k\rangle f_k \langle \alpha_k|Q\rangle, \tag{4.12}$$

and since (4.12) is true for all Q' and Q, it follows that

$$\mathsf{D} = \textstyle\sum_k |\alpha_k\rangle f_k \langle \alpha_k|. \tag{4.13}$$

That is, D is a weighted sum of the projection operators‡ $|\alpha_k\rangle\langle\alpha_k|$ of the various quantum states occurring in the systems of the ensemble. In the special case of a pure ensemble the sum in (4.13) has but one term, so that D is itself a projection operator.

To express ensemble averages in terms of the density operator instead of the density matrix, we return to the basic formula (4.9). Let $|\beta_1\rangle, |\beta_2\rangle, \ldots$ be any complete orthonormal set of quantum states (a *basic set* in Dirac's terminology). The completeness relation for these states is§

$$\textstyle\sum_j |\beta_j\rangle\langle\beta_j| = 1. \tag{4.14}$$

(An equivalent statement is that

$$\textstyle\sum_j \varphi_j(Q) \int \varphi_j^*(Q')\,\chi(Q')\, dQ' = \chi(Q)$$

holds for all wave functions $\chi(Q)$, where $\varphi_j(Q) \equiv \langle Q|\beta_j\rangle$ is the wave function of $|\beta_j\rangle$, and $\chi(Q')$ is an arbitrary wave function.) Inserting (4.14) into (4.9) and then using (4.13), we obtain

$$\begin{aligned}\langle G\rangle &= \textstyle\sum_k \sum_j f_k \langle\alpha_k|\beta_j\rangle\langle\beta_j|\mathsf{G}|\alpha_k\rangle \\ &= \textstyle\sum_j \langle\beta_j|\mathsf{GD}|\beta_j\rangle.\end{aligned} \tag{4.15}$$

† P. A. M. Dirac, *The Principles of Quantum Mechanics*, § 20.
‡ For the definition and properties of projection operators, see Chap. I, § 3.
§ P. A. M. Dirac, *op. cit.*, § 16.

The right-hand side of (4.15) is the sum of the diagonal elements of the matrix of GD in the representation with basic states $|\beta_1\rangle$, $|\beta_2\rangle$ ....

To simplify the notation in (4.15) we define the *trace* of any operator F to be

$$tr(F) = \sum_j \langle\beta_j|F|\beta_j\rangle, \tag{4.16}$$

where $|\beta_1\rangle$, $|\beta_2\rangle$, ... is an arbitrary complete orthonormal set. The definition (4.16) is independent of the complete orthonormal set chosen; for if $|\gamma_1\rangle$, $|\gamma_2\rangle$, ... is another such set, then the completeness relations for the two sets imply

$$\begin{aligned}\sum_j \langle\beta_j|F|\beta_j\rangle &= \sum_j \sum_i \langle\beta_j|\gamma_i\rangle \langle\gamma_i|F|\beta_j\rangle \\ &= \sum_i \sum_j \langle\gamma_i|F|\beta_j\rangle \langle\beta_j|\gamma_i\rangle \\ &= \sum_i \langle\gamma_i|F|\gamma_i\rangle. \end{aligned}\tag{4.17}$$

In the notation of (4.16), (4.15) takes the form

$$\langle G\rangle = tr(GD). \tag{4.18}$$

Equation (4.18) is the quantum analogue of the classical formula (2.4), $\langle G\rangle = \int GD\, d\alpha$, and the operation of taking a trace is thus the quantum analogue of an integration over classical phase space. To help in understanding and calculating with the trace, we may note a few of its basic properties.† First, if the operator F can be diagonalized, then we may take the set of basic states $|\beta_1\rangle$, $|\beta_2\rangle$, ... to be the set which diagonalizes F; then (4.16) shows that $tr(F)$ equals the sum of the eigenvalues of F. In particular, if F is the projection operator of a linear manifold L in Hilbert space, then the states $|\beta_1\rangle$, $|\beta_2\rangle$, ... may be taken to include a set spanning L; the eigenvalues corresponding to the states within L are 1 and the rest are 0, and so the trace of a projection operator equals the number of dimensions of its linear manifold. Another useful property concerns the trace of a product. If F_1 and F_2 are two operators, which need not commute, then by (4.14) and (4.16) we have

$$\begin{aligned} tr(F_1F_2) &= \sum_j \sum_i \langle\beta_j|F_1|\beta_i\rangle \langle\beta_i|F_2|\beta_j\rangle \\ &= \sum_i \sum_j \langle\beta_i|F_2|\beta_j\rangle \langle\beta_j|F_1|\beta_i\rangle \\ &= tr(F_2F_1). \end{aligned}\tag{4.19}$$

This result implies that

$$tr(F_1F_2F_3) = tr(F_2F_3F_1) = tr(F_3F_1F_2). \tag{4.20}$$

The property (4.20) is called the *cyclic invariance* of the trace.

† For a more detailed and rigorous discussion of the trace, see J. von Neumann, *Mathematical Foundations of Quantum Mechanics*, Ch. II, § 11.

The formula defining $tr(\mathsf{F})$ can easily be extended to representations labelled by continuous variables. Thus the continuous analogue of (4.16) for the position representation is

$$tr(\mathsf{F}) = \int \langle Q|\mathsf{F}|Q\rangle \, dQ, \tag{4.21}$$

and that of (4.18) is

$$\langle G\rangle = \int \langle Q|\mathsf{GD}|Q\rangle \, dQ, \tag{4.22}$$

an integral of the diagonal elements of the matrix of the operator GD. By using the law of matrix multiplication for the matrices $G(Q; Q')$ and $D(Q'; Q)$ in (4.22) we can recover the result (4.10) which was derived earlier without using the trace.

The density operator satisfies two conditions analogous to the classical normalization condition (2.5) and non-negativity condition (2.6). The quantum normalization condition is obtained, as in the corresponding classical situation, by setting $G = 1$ in (4.18); this gives

$$tr(\mathsf{D}) = 1 \tag{4.23}$$

or, in terms of the density matrix,

$$\int D(Q; Q)\, dQ = 1. \tag{4.24}$$

The normalization condition expresses the fact that the fractions f_k defined in (4.8) add up to 1. The second condition on D, corresponding to the non-negativity of the classical $D(\alpha)$, expresses the fact that the fractions f_k are all $\geqq 0$. Such a condition can be formulated by using (4.13) to calculate $\langle\beta|\mathsf{D}|\beta\rangle$ for arbitrary $|\beta\rangle$. This gives

$$\langle\beta|\mathsf{D}|\beta\rangle = \textstyle\sum_k f_k|\langle\alpha_k|\beta\rangle|^2$$

so that

$$\langle\beta|\mathsf{D}|\beta\rangle \geqq 0 \quad \text{for all } |\beta\rangle. \tag{4.25}$$

A direct physical interpretation of this inequality can be obtained by rewriting the left-hand side in the form $tr(\mathsf{D}|\beta\rangle\langle\beta|)$, that is as the ensemble average of the projection operator $|\beta\rangle\langle\beta|$: then (4.25) states that if a measurement is made on each system in the ensemble to see whether it is in the state $|\beta\rangle$ or not, the fraction found to be in the state $|\beta\rangle$ is non-negative. Mathematically (4.23) expresses the fact that the operator D is *non-negative definite*. The corresponding property of the density matrix is that

$$\iint \varphi^*(Q')\, D(Q'; Q)\, \varphi(Q)\, dQ\, dQ' \geqq 0 \tag{4.26}$$

holds for all functions $\varphi(Q)$. A further property of the operator D and the matrix $D(Q'; Q)$—that they are Hermitian—has already been mentioned in connection with (4.11); but this property can be deduced from the non-negativity condition (4.25) or (4.26) and therefore need not be listed as an independent one.

As an example of the density matrix, consider two sources of light of equal intensity capable of producing plane waves travelling along the z-axis and polarized in the x- and y-directions respectively. Let us denote the normalized quantum states of the photons produced by the two sources by $|1\rangle$ and $|2\rangle$ respectively. The beam of light produced by one of these sources, or both together, can be regarded as an ensemble of photons. If only the first source is switched on, the beam is a pure ensemble of photons all in the state $|1\rangle$; therefore, by (4.13), the density operator D is $|1\rangle\langle 1|$ and the density matrix, in the representation with basic states $|1\rangle$ and $|2\rangle$, is

$$\begin{bmatrix} \langle 1|D|1\rangle & \langle 1|D|2\rangle \\ \langle 2|D|1\rangle & \langle 2|D|2\rangle \end{bmatrix} = \begin{bmatrix} 1 & 0 \\ 0 & 0 \end{bmatrix}. \tag{4.27}$$

Likewise, if only the second source is switched on, the density operator is $|2\rangle\langle 2|$, with matrix

$$\begin{bmatrix} 0 & 0 \\ 0 & 1 \end{bmatrix}. \tag{4.28}$$

If both sources are switched on and they are incoherent, then the resulting beam may be regarded as a mixed ensemble consisting of equal proportions of photons from each source; its density operator is therefore $|1\rangle\frac{1}{2}\langle 1| + |2\rangle\frac{1}{2}\langle 2|$, by (4.13), and the density matrix is the average of (4.27) and (4.28),

$$\begin{bmatrix} \frac{1}{2} & 0 \\ 0 & \frac{1}{2} \end{bmatrix}. \tag{4.29}$$

Finally, consider the case where the two sources are coherent, with relative phase φ. Then the quantum state of a photon in the resulting beam is a superposition of those for the two beams separately, having the form $(|1\rangle + e^{i\varphi}|2\rangle)/\sqrt{2}$. Consequently the resulting beam is once again a pure ensemble, with density operator

$$D = [|1\rangle + e^{i\varphi}|2\rangle]\,[\langle 1| + e^{-i\varphi}\langle 2|]/2$$

and density matrix

$$\begin{bmatrix} \frac{1}{2} & \frac{1}{2}e^{-i\varphi} \\ \frac{1}{2}e^{i\varphi} & \frac{1}{2} \end{bmatrix}. \tag{4.30}$$

The diagonal elements are the same as in the density matrix (4.29), indicating that the probability of observing a photon in the state $|1\rangle$ or $|2\rangle$ is the same as before; but now off-diagonal elements are present as well, describing the coherent phase relationship between the states $|1\rangle$ and $|2\rangle$ in this beam. To see why these diagonal elements are present in (4.30) but not in (4.29), we may think of the incoherent sources as having a continually fluctuating relative phase. Thus, the incoherent beam can be thought of as a mixture of many coherent beams, their relative phases φ uniformly distributed in the interval $0 \leqq \varphi < 2\pi$. Replacing the sum over k in (4.13) by an integral over these values of φ, we have for the density matrix of the mixture

$$\int_0^{2\pi} \begin{bmatrix} \frac{1}{2} & \frac{1}{2}e^{-i\varphi} \\ \frac{1}{2}e^{i\varphi} & \frac{1}{2} \end{bmatrix} \frac{d\varphi}{2\pi} = \begin{bmatrix} \frac{1}{2} & 0 \\ 0 & \frac{1}{2} \end{bmatrix}. \tag{4.31}$$

Thus the two methods of thinking of the incoherent beam, either as a mixture of the two different kinds of photons produced by the two sources acting separately or as a mixture of the many different kinds of photon that the two sources can produce acting together, lead to the same density matrix, and hence to the same ensemble averages for all ideal measurements.

4.1. Exercises

1. Prove the following properties of the trace: (i) if λ_1 and λ_2 are numbers, then

$$tr(\lambda_1 F_1 + \lambda_2 F_2) = \lambda_1 tr(F_1) + \lambda_2 tr(F_2);$$

(ii) if F is Hermitian, then $tr(F)$ is real; (iii) if F_1 and F_2 are Hermitian, then $tr(F_1F_2)$ is real; (iv) if F is positive definite, then $tr(F)$ is positive.

2. Prove the following properties of the density operator D: (i) If D can be diagonalized, its eigenvalues are non-negative and their sum is 1; (ii) $D - D^2$ is non-negative definite.

3. Wigner† has shown that with any density matrix, say $D(Q'; Q)$, for a quantum system of spinless particles one can associate a function $D_w(P, Q)$ in the phase space of the corresponding classical system which behaves in many respects like a phase-space density for the ensemble whose density matrix is $D(Q'; Q)$. For a single particle on an infinite straight line the definition of this *phase-space pseudo-density* is

$$D_w(p, q) = h^{-1} \int_{-\infty}^{\infty} D(q + \tfrac{1}{2}u, q - \tfrac{1}{2}u)\, e^{-ipu/\hbar}\, du.$$

Show that $D_w(p, q)$ is real and that, for any function φ,

$$tr(\varphi(q)D) = \iint D_w(p, q)\, \varphi(q)\, dp\, dq$$

$$tr(\varphi(p)D) = \iint D_w(p, q)\, \varphi(p)\, dp\, dq.$$

5. The quantum Liouville theorem

In this section we study how quantum Gibbs ensembles change with time. As in the classical theory, we shall do this by following the change with time of the dynamical state of each system in the ensemble; this implies the use of the Schrödinger rather than the Heisenberg picture of dynamics.‡ If the dynamical state of a system is α, then α depends on time in accordance with Schrödinger's wave equation (1.2) of Chap. I, so that the Schrödinger equation of motion for $|\alpha\rangle$ is

$$\frac{d}{dt}|\alpha\rangle = -\frac{i}{\hbar} H|\alpha\rangle, \tag{5.1}$$

† E. P. Wigner, On the Quantum Correction for Thermodynamic Equilibrium, *Phys. Rev.* **40**, 749 (1932).

‡ P. A. M. Dirac, *op. cit.*, §§ 27 and 28.

where H is the Hamiltonian operator. For the present section, the detailed structure of H, studied in Chap. I, § 2, is irrelevant; all that matters is that H is Hermitian. Just as in the classical case, the equation of motion (5.1) defines a unique "velocity" (rate of change of dynamical state) at each point in dynamical space, and hence if the Gibbs ensemble is visualized as a dust or fluid in dynamical space, its velocity field is independent of time provided H is independent of time.

To obtain the quantum analogue of Liouville's theorem, we introduce a set of operators U_t so defined that if $|\alpha_0\rangle$ is the state of a system at time 0, its state at time t is $U_t|\alpha_0\rangle$. Substituting $U_t|\alpha_0\rangle$ for $|\alpha\rangle$ in (5.1) and using the fact that $|\alpha_0\rangle$ is arbitrary, we obtain

$$\frac{d}{dt}U_t = -\frac{i}{\hbar}HU_t, \tag{5.2}$$

an operator differential equation which, together with the boundary condition

$$U_0 = 1, \tag{5.3}$$

determines U_t completely. In particular, if the system happens to be isolated (i.e. if H is independent of time) then (5.2) can be integrated, and yields

$$U_t = \exp(-iHt/\hbar). \tag{5.4}$$

Since H is Hermitian, it follows from (5.2) and its Hermitian conjugate that $(d/dt)\,(U_t^\dagger U_t) = 0$, where a dagger denotes the adjoint of an operator; consequently by (5.3) we have

$$U_t^\dagger U_t = 1, \tag{5.5}$$

that is, U_t is a unitary operator.† This unitary character of U_t, indicating that the operation in Hilbert space represented by U_t is essentially a rotation, provides a quantum analogue for the volume-preserving character of the flow in classical phase space, demonstrated in § 3 [eqn. (3.19)]. As we saw in Chap. I, § 3, the quantum analogue of a region R in phase space with indicator variable $J_R(\alpha)$ is a linear manifold L in Hilbert space with projection operator J_L. The volume of R, as defined in (2.1), may be written $\int J_R(\alpha)\,d\alpha$. Since the quantum analogue of an integration over all phase space is a trace, the quantum analogue of volume for L is $tr(J_L)$ which, as shown in connection with (4.16), is the number of dimensions of L. Thus the analogue of the preservation of volumes under the flow in classical phase space is the preservation of the number of dimensions of a linear manifold under rotations in Hilbert space.

† P. A. M. Dirac, *op. cit.*, § 26.

Using the operator U_t we can express the density operator at any time t in terms of the density operator D_0 at time 0. Using (4.13) for the time 0 we have

$$D_0 = \sum_k |\alpha_k\rangle f_k \langle\alpha_k|,$$

where f_k is the fraction of members of the ensemble that are in state $|\alpha_k\rangle$ at time 0; then all the systems constituting this fraction will be in the state $U_t|\alpha_k\rangle$ at time t, and so the density operator at time t is

$$\begin{aligned} D_t &= \sum_k U_t|\alpha_k\rangle f_k \langle\alpha_k| U_t^\dagger \\ &= U_t D_0 U_t^\dagger. \end{aligned} \tag{5.6}$$

Equation (5.6), by expressing D_t in terms of D_0, performs for the quantum density operator the same service that the integrated Liouville equation (3.20) performs for the classical phase space density. Accordingly we call (5.6) the *integrated quantum Liouville equation.* For an isolated system, eqn. (5.4) may be used in (5.6) and gives

$$D_t = \exp(-iHt/\hbar)\, D_0 \exp(iHt/\hbar). \tag{5.7}$$

This operator equation may alternatively be written as a matrix equation; the most convenient representation is the energy representation

$$\langle m|D_t|n\rangle = \exp[i(E_n - E_m)t/\hbar]\, \langle m|D_0|n\rangle, \tag{5.8}$$

where $|1\rangle$, $|2\rangle$, ... are the stationary states of the system, with energies $E_1, E_2, \ldots$, so that

$$H|m\rangle = E_m|m\rangle \quad \text{for} \quad m = 1, 2, \ldots. \tag{5.9}$$

To find the quantum version of Liouville's differential equation (3.5), we differentiate (5.6) with respect to t and use (5.2) and its adjoint, obtaining

$$\frac{dD_t}{dt} = -\frac{i}{\hbar}(HD_t - D_t H). \tag{5.10}$$

This may be called the *quantum Liouville equation.* It can† be put into a form closely analogous to the classical Liouville equation (3.5). To see this, we write the latter in the form

$$\frac{\partial D}{\partial t} = \sum_{i=1}^{fN} \left(\frac{\partial H}{\partial q_i}\frac{\partial D}{\partial p_i} - \frac{\partial H}{\partial p_i}\frac{\partial D}{\partial q_i} \right) \tag{5.11}$$

obtained with the help of the definition (3.2) of α_i and the classical equations of motion (3.1). The right-hand side of (5.11) is called the *Poisson bracket* of the classical dynamical variables H and D, often written $\{H, D\}$. Now if H and D are two quantum dynamical variables instead, then the analogue

† P. A. M. Dirac, *op. cit.*, § 33.

of the Poisson bracket is† $(HD - DH)/i\hbar$ and therefore the quantum analogue of (5.11) is indeed the quantum Liouville equation (5.10).

As in the classical theory, a simple application of Liouville's theorem is to find the steady ensembles, whose density operators are independent of time, for an isolated system. The matrix form (5.8) of Liouville's theorem shows that, in a steady ensemble, the matrix element $\langle m|D_0|n\rangle$ must vanish unless $E_m = E_n$. In the simple case where there are no degenerate energy levels, therefore, the density matrix of a steady ensemble must be diagonal in the energy representation; the diagonal element $\langle n|D|n\rangle$ of such a density matrix is the probability that the system is in the energy level E_n, and the vanishing of the off-diagonal elements indicates that there are no phase relationships between different energy levels. Since the energy levels are non-degenerate in this case it is possible to find a function $f(\,)$ such that $\langle n|D|n\rangle = f(E_n)$; this function has the property

$$D = f(H) \tag{5.12}$$

which may be verified by expressing both sides as matrices in the energy representation. In the general case, where H does have degenerate eigenvalues, it is still possible to diagonalize D and H simultaneously, but (5.12) generalizes to

$$D = f(H, L), \tag{5.13}$$

where L is some invariant of the motion (an operator commuting with H, such as the angular momentum in spherically symmetrical systems) not expressible as a function of H. Equations (5.12 and 13) are the quantum analogues of (3.23) and (3.24).

The most important steady ensembles are again the *canonical ensemble*, whose density matrix is

$$D = \text{const.} \exp(-\beta H), \tag{5.14}$$

i.e. $\langle m|D|n\rangle = \text{const.}\ e^{-\beta E_m}\,\delta_{mn}$, where δ_{mn} is defined in Chap. II, eqn. (4.22), and the *microcanonical ensemble*, whose density matrix is

$$D = \begin{Bmatrix} \text{const. if } E - \Delta E < H \leqq E \\ 0 \text{ if not,} \end{Bmatrix} \tag{5.15}$$

i.e.
$$\langle m|D|n\rangle = \begin{Bmatrix} \text{const. } \delta_{mn} \text{ if } E - \Delta E < E_m \leqq E \\ 0 \text{ if not.} \end{Bmatrix} \tag{5.16}$$

As in the classical case, the numbers β, E, ΔE, are arbitrary positive constants, and the constants of proportionality are fixed by the normalization condition (4.23) (which in the energy representation reads $\sum_m \langle m|D|m\rangle = 1$).

† P. A. M. Dirac, *op.cit.*, § 21.

5.1. *Exercises*

1. For a harmonic oscillator with angular frequency ω, show that the density operator D_t is periodic in t with period $2\pi/\omega$.

2. For a single particle of mass m in a cubical box of side L, whose wave functions satisfy the periodic boundary conditions

$$\psi(0, y, z) = \psi(L, y, z) \quad \text{and} \quad \frac{\partial \psi}{\partial x}(0, y, z) = \frac{\partial \psi}{\partial x}(L, y, z)$$

$$\psi(x, 0, z) = \psi(x, L, z) \quad \text{and} \quad \frac{\partial \psi}{\partial y}(x, 0, z) = \frac{\partial \psi}{\partial y}(x, L, z)$$

$$\psi(x, y, 0) = \psi(x, y, L) \quad \text{and} \quad \frac{\partial \psi}{\partial z}(x, y, 0) = \frac{\partial \psi}{\partial z}(x, y, L),$$

show that the density matrix for a canonical ensemble is given approximately by

$$D(\mathbf{q}, \mathbf{q}') \cong L^{-3} \exp\left(-m|\mathbf{q} - \mathbf{q}'|^2/2\beta\hbar^2\right),$$

where $\mathbf{q}$ stands for (x, y, z) and $\mathbf{q}'$ for (x', y', z'). Assume that $\hbar(\beta/2m)^{\frac{1}{2}} \ll L$ and $|\mathbf{q} - \mathbf{q}'| \ll L$.

3. The *fine-grained entropy* of a quantum Gibbs ensemble is defined to be $-k\, tr\, [D \ln (c^{-1}D)]$, where k and c are constants. Show that the fine-grained entropy is an invariant of the motion.

CHAPTER IV

Probabilities from Dynamics

1. Dynamical images of events

In the previous two chapters we followed two distinct streams of argument. In Chapter II the arguments were statistical, arising from the Markovian postulate; but in Chapter III they were dynamical, arising from the laws of dynamics applied to arbitrary Gibbs ensembles. In the present chapter the two streams meet, and their confluence marks the beginning of true statistical mechanics. The central result is a set of formulae expressing statistical quantities (probabilities) in terms of dynamical quantities (volumes of phase-space regions in classical mechanics, dimension numbers of linear manifolds in quantum mechanics).

The method we shall use is to apply the general theory of Gibbs ensembles from Chapter III to realizable Gibbs ensembles of isolated systems. In practice, if a statistical experiment was conducted by replicating a trial $\mathfrak{T}$ simultaneously at many different places, the limitations on experimental procedure would make it impossible to ensure that every system in the resulting ensemble had exactly the same Hamiltonian; consequently the ensemble would only approximately satisfy the definition of a Gibbs ensemble, given at the beginning of Chap. III, § 1. In accordance with the discussion leading up to eqn. (3.11) of Chap. I, the resulting errors could be allowed for by assuming an error of order t/τ in applying Liouville's theorem over a time interval of duration t. We shall need this refinement eventually (in Chap. IV, § 7), but until then we shall work with the simplifying assumption that the theory of Gibbs ensembles from Chapter III holds exactly (i.e. that $\tau = \infty$) for realizable ensembles.

The first stage in the argument is to express the probabilities for the various possible outcomes of $\mathfrak{T}$ in terms of D_0, the phase-space density or density matrix describing the Gibbs ensemble at the initial time $t = 0$. In classical mechanics this is easily done using indicator variables such as $J_A(\alpha)$, defined as in eqn. (3.1) of Chap. I to take the value 1 if the dynamical state α implies the observational state A and the value 0 otherwise. In the simplest case where the observation stage of the trial $\mathfrak{T}$ consists of a single observation made at a time $t > 0$, the probability that this observation gives the outcome A equals the fraction of systems in the ensemble whose dynamical states at time t, denoted by α_t, satisfy $J_A(\alpha_t) = 1$. In view of the dynamical relation $\alpha_t = U_t\alpha_0$ where U_t is defined in Chap. III, § 3, these systems also satisfy $J_A(U_t\alpha_0) = 1$, and therefore, by eqn. (2.8) of Chap. II and eqn. (2.4) of

Chap. III, we have

$$\begin{aligned} Pr(A|\mathfrak{T}) &= \langle J_A(\alpha_t)\rangle = \langle J_A(U_t\alpha_0)\rangle \\ &= \int J_A(U_t\alpha_0)\, D_0(\alpha_0)\, d\alpha_0 \\ &= \int J_{A,t}(\alpha)\, D_0(\alpha)\, d\alpha, \end{aligned} \tag{1.1}$$

where

$$J_{A,t}(\alpha) \equiv J_A(U_t\alpha). \tag{1.2}$$

The dynamical variable $J_{A,t}(\alpha)$ may be thought of as giving a prediction, based on the laws of dynamics, of the value that the observable $J_A(\alpha)$ will take t units of time later. Unless the observational states are deterministic, the observational evidence available at any given time is insufficient to give a certain prediction of the observational state t units of time later, and so $J_{A,t}(\alpha)$ is not itself an observable.

The formalism is easily extended to the case where the outcome of the trial $\mathfrak{T}$ depends on the result of a compound observation. If this compound observation consists of n instantaneous observations made at the instants $t^1, t^2, \ldots t^n$, where $0 < t^1 < t^2 < \ldots < t^n$, then a typical elementary event for $\mathfrak{T}$ is $A_{t^1}B_{t^2} \ldots G_{t^n}$, defined to occur if and only if the system is in observational state A at time t^1, then B at time t^2, ..., and finally G at time t^n. An equivalent statement is that this elementary event occurs if and only if all n of the relations $J_A(\alpha_{t^1}) = 1$, $J_B(\alpha_{t^2}) = 1$, ..., $J_G(\alpha_{t^n}) = 1$ hold. By an extension of the argument used in (1.1), it follows that

$$\begin{aligned} Pr(G_{t^n} \ldots B_{t^2}A_{t^1}|\mathfrak{T}) &= \langle J_G(U_{t^n}\alpha_0) \ldots J_B(U_{t^2}\alpha_0)\, J_A(U_{t^1}\alpha_0)\rangle \\ &= \int J_{G,t^n}(\alpha) \ldots J_{B,t^2}(\alpha)\, J_{A,t^1}(\alpha)\, D_0(\alpha)\, d\alpha \end{aligned} \tag{1.3}$$

using the notation for probabilities introduced in Chap. II, § 4. That is, the probability of the event $E \equiv G_{t^n} \ldots B_{t^2}A_{t^1}$ equals the ensemble average, evaluated at time 0, of a dynamical variable

$$J_E(\alpha) \equiv J_{G,t^n}(\alpha) \ldots J_{B,t^2}(\alpha)J_{A,t^1}(\alpha) \equiv J_E(\alpha), \tag{1.4}$$

which is an indicator variable since it takes the values 0 and 1 only. The region of phase space where $J_E(\alpha)$ takes the value 1 may be denoted by $\omega(E)$; we shall call it the *dynamical image* of the event E, since E occurs if and only if the dynamical state at time 0 is in this region. With this notation, (1.3) can be written more simply as

$$\begin{aligned} Pr(E|\mathfrak{T}) &= \int J_E(\alpha)\, D_0(\alpha)\, d\alpha \\ &= \int_{\omega(E)} D_0(\alpha)\, d\alpha \end{aligned} \tag{1.5}$$

expressing the fact that $Pr(E|\mathfrak{T})$ is the fraction of systems in the ensemble whose dynamical states are in $\omega(E)$ at time 0. The formula (1.5) also applies when the event E is not elementary, the indicator variable for a non-elementary event being constructed from those given by (1.4) for the elementary events composing it by means of the rules given in Chap. II, eqn. (2.9).

The extension of these ideas to quantum mechanics is straightforward, provided the outcome of the trial depends on just one observation. Just as in the classical case, the quantum indicator variable of an observational state A is (see Chap. I, § 3) a dynamical variable J_A, taking the value 1 if and only if the system is in the observational state A, and the value 0 if and only if the system is in some other observational state. If the observation completing $\mathfrak{T}$ is carried out at time t, the ensemble average of the observed values of J_A is by definition equal to the probability of A, and therefore by eqn. (4.18) of Chap. III we have

$$\begin{aligned} Pr(A|\mathfrak{T}) &= \langle J_A \rangle \text{ evaluated at time } t \\ &= tr(J_A D_t), \end{aligned} \tag{1.6}$$

where D_t is the density operator at the time t. Applying Liouville's theorem [Chap. III, eqn. (5.6)] and the cyclic invariance of the trace [Chap. III, eqn. (4.20)] we can put (1.6) into a form that brings out the analogy with the classical formula (1.1):

$$\begin{aligned} Pr(A|\mathfrak{T}) &= tr(J_A U_t D_0 U_t^\dagger) \\ &= tr(J_{A,t} D_0), \end{aligned} \tag{1.7}$$

where

$$J_{A,t} \equiv U_t^\dagger J_A U_t \tag{1.8}$$

is the quantum analogue of the classical dynamical variable $J_{A,t}$ defined in (1.2). In going from (1.6) to (1.7) we have in effect gone from the Schrödinger picture,† with fixed dynamical variables J_A and time-dependent density operator D_t, to the Heisenberg picture with fixed density operator and time-dependent dynamical variables $J_{A,t}$.

In order to extend (1.6) to a trial whose outcome depends on a compound observation we shall use the *postulate of compatibility*, stated verbally in Chap. I, § 3. This postulate asserts that instantaneous observations made on a large system at different times are compatible. Since compatible measurements correspond to commuting operators, the postulate can be expressed mathematically as a requirement that the Heisenberg operators describing observations made at different times must commute,‡ and in particular that

$$J_{A,t^1} J_{B,t^2} = J_{B,t^2} J_{A,t^1} \tag{1.9}$$

† P. A. M. Dirac, *op. cit.*, § 28.

‡ To keep the theory as simple as possible, we have adopted an unnecessarily strong statement as our formulation of the postulate of compatibility. It seems unlikely that any realistic model would satisfy the operator equation (1.9). It should be possible, however, to base the theory instead on a statement referring only to observable quantities—that is, to expectation values rather than to operators. A likely candidate is the assumption that for a large system all quantities of the form $tr(J_E)$ and $tr(J_E D_0)$, with J_E defined in (1.11) and D_0 the density operator of any realizable ensemble, are very nearly independent of the order of the factors in the operator product (1.11).

for all A, B, t^1, t^2. Suppose now that the observation stage of the trial $\mathfrak{T}$ comprises n observations made at the instants $t^1, t^2, \ldots, t^n$, where $0 \leqq t^1 < t^2 < \ldots < t^n$. Then the probability that a system will have the observational state A at time t^1, then B at time t^2, ... and finally G at time t^n, equals the probability that simultaneous ideal measurements of the n operators $J_{A,t^1}, J_{B,t^2}, \ldots, J_{G,t^n}$ made on the system at time 0 will all yield the value 1. Since these operators all commute, by (1.9), the simultaneous measurements do not interfere, and may be regarded as a single measurement whose result is the set of values taken simultaneously by the n operators. The n operators all take the value 1 if and only if their product is 1, and hence we can express the required probability as the expectation of this product:

$$Pr(G_{t^n} \ldots B_{t^2}A_{t^1}|\mathfrak{T}) = tr(J_{G,t^n} \ldots J_{B,t^2}J_{A,t^1}D_0). \tag{1.10}$$

Like the corresponding classical formula (1.3), this formula can be interpreted as the ensemble average, evaluated at time 0, of a dynamical variable

$$J_E \equiv J_{G,t^n} \ldots J_{B,t^2}J_{A,t^1} \tag{1.11}$$

whose only eigenvalues are 0 and 1. This dynamical variable is therefore a projection operator, and the linear manifold of quantum states for which its eigenvalue is 1 may be called the *dynamical image* $\omega(E)$ of the event $E \equiv G_{t^n} \ldots B_{t^2}A_{t^1}$, since this event is certain to occur in a given system if and only if its quantum state at time 0 is in this linear manifold. Equation (1.10), abbreviated to

$$Pr(E|\mathfrak{T}) = tr(J_E D_0), \tag{1.12}$$

is the quantum analogue of (1.5); it may be interpreted as a statement that the probability of observing the event E equals the fraction of systems in the ensemble for which an ideal measurement of J_E at time 0 would give the value 1 (indicating a quantum state in $\omega(E)$) rather than 0 (indicating a quantum state orthogonal to $\omega(E)$). Like its classical analogue, (1.12) also applies when the event E is not elementary and J_E is constructed using (1.11) and the rules given in Chap. II, eqn. (2.9).

1.1. Exercise

1. Show that the statement (1.9) of the postulate of compatibility is equivalent to the statement that there exists a complete set of quantum states $|\beta_1\rangle, |\beta_2\rangle, \ldots$ with the property that, for every $t \geqq 0$ and $n > 0$ and every A, $U_t|\beta_n\rangle$ is an eigenstate of J_A. (These states $|\beta_1\rangle, |\beta_2\rangle, \ldots$ are the closest quantum analogues of phase-space points, so that the $|\beta\rangle$ representation is helpful in making the transition from classical to quantum statistical mechanics—see § 5 below.)

2. Observational equivalence

The formulae arrived at in the previous section show how all the probabilities associated with a statistically regular trial $\mathfrak{T}$ can be calculated from the density D_0 of a Gibbs ensemble created by replicating this trial in a large number of places at the same time. Since $\mathfrak{T}$ is statistically regular, these probabilities are completely determined by the instructions specifying $\mathfrak{T}$. It does not follow, however, that D_0 itself is completely determined by $\mathfrak{T}$; in fact, as we shall now show by examples, D_0 is in general only partially determined by $\mathfrak{T}$. These examples refer to classical systems, though the corresponding quantum situations can be treated nearly as easily.

An artificially simple example arises when the energy of the system is the only observable, so that each observational state comprises all the dynamical states within a specified range of energies:

$$J_A(\alpha) = \begin{cases} 1 \text{ if } E_{A-1} < H(\alpha) \leqq E_A \\ 0 \text{ if not} \end{cases} \quad \text{for } A = 1, 2, \dots, \tag{2.1}$$

where $E_0 < E_1 < E_2 < \dots$ are constants and $H(\alpha)$ is the Hamiltonian. This model rather trivially satisfies the Markovian postulate: the observational state at any time is determined, through conservation of energy, by the initial observational state and, as shown in Chap. I, § 5, observational states that are deterministic are also Markovian. If K is any observational state and $\mathfrak{T}$ is a trial whose preparation stage is defined solely by the condition that at time 0 the observational state is K, then D_0 is determined by $\mathfrak{T}$ only to the extent that

$$[1 - J_K(\alpha)]D_0(\alpha) = 0; \tag{2.2}$$

that is, $D_0(\alpha)$ must vanish outside $\omega(K)$, the dynamical image of K. Provided $\omega(K)$ contains more than one dynamical state, there are many different functions $D_0(\alpha)$ satisfying (2.2) together with the normalization and non-negativity conditions (2.5) and (2.6) of Chap. III.

As a less artificial example illustrating the same point, let us once again consider a classical system with a Markovian observational description, but drop the assumption (2.1). A statistical experiment to measure the probability $Pr(A_t|K)$ can be conducted in many different ways; one of them is to bring a large collection of replicas of the system into some state L at time -1, then isolate all of them and use the ones that have the observational state K at time 0 as the statistical ensemble. By the Markovian postulate, the probabilities measured in this statistical ensemble are independent of the choice of L. The dynamical properties of the ensemble, on the other hand, depend strongly on L; for the phase-space density must satisfy the condition

$$[1 - J_L(U_{-1}\alpha)]D_0(\alpha) = 0,$$

so that the phase-space densities for different values of L are completely distinct, being confined to non-overlapping regions in phase space. This construction illustrates how two ensembles can be *observationally equivalent* for $t \geqq 0$—that is they can give identical probabilities for all events observable at or after $t = 0$—and yet have entirely different phase-space densities.

These examples show that statistical regularity of a trial $\mathfrak{T}$ is not enough to ensure the complete reproducibility of D_0, even though it does ensure the reproducibility of the probabilities of all the observable events belonging to $\mathfrak{T}$. In the language of Chap. III, § 2, the coarse-grained probabilities are reproducible but the fine-grained probabilities are not; but since the fine-grained probabilities are unobservable, their lack of reproducibility has no observable consequences.

To see in more detail how two different ensembles can be observationally equivalent, let us consider a specific dynamical model. The model is far from realistic, but is simple enough to permit the mathematical problems which are so complicated for real dynamical systems to be solved exactly. This model is a classical system whose phase space is the square

$$0 \leqq p < 1, \quad 0 \leqq q < 1, \tag{2.3}$$

and whose dynamics is given not by the usual continuous motion but by a discrete transformation, known as the *baker's transformation*† since it recalls the kneading of a piece of dough, repeated at every instant of observation. The baker's transformation can be carried out in two steps: in the first step the square is squeezed into a $2 \times \frac{1}{2}$ rectangle, and in the second this

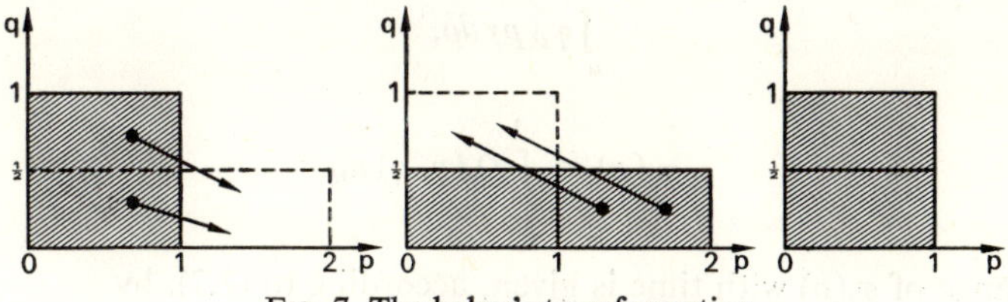

FIG. 7. The baker's transformation.

rectangle is cut in half and the two pieces fitted together to make a new 1×1 square (Fig. 7). Mathematically, the first step is

$$(p, q) \to (2p, \tfrac{1}{2}q)$$

and the second is

$$(2p, \tfrac{1}{2}q) \to \begin{cases} (2p, \tfrac{1}{2}q) \text{ if } 0 \leqq 2p < 1 \\ (2p - 1, \tfrac{1}{2}q + \tfrac{1}{2}) \text{ if } 1 \leqq 2p < 2, \end{cases}$$

† P. R. Halmos, *Lectures on Ergodic Theory* (Mathematical Society of Japan, 1965), p.9.

so that the complete transformation is

$$U_1(p, q) = \begin{cases} (2p, \tfrac{1}{2}q) \text{ if } 0 \leqq p < \tfrac{1}{2} \\ (2p - 1, \tfrac{1}{2}q + \tfrac{1}{2}) \text{ if } \tfrac{1}{2} \leqq p < 1. \end{cases} \tag{2.4}$$

The inverse transformation is

$$(U_1)^{-1}(p, q) = \begin{cases} (\tfrac{1}{2}p, 2q) \text{ if } 0 \leqq q < \tfrac{1}{2} \\ (\tfrac{1}{2}p + \tfrac{1}{2}, 2q - 1) \text{ if } \tfrac{1}{2} \leqq q < 1. \end{cases} \tag{2.5}$$

The Jacobian of the transformations U_1 and U_1^{-1} is 1, and hence Liouville's theorem still holds in its integrated form eqn. (3.20) of Chap. III,

$$D_t(\alpha) = D_0((U_t)^{-1}\alpha) \tag{2.6}$$

(although not in its differential form, since there are no differential equations of motion here). From (2.6) we obtain the recurrence formula $D_{t+1}(\alpha) = D_t(U_1^{-1}\alpha)$, that is,

$$D_{t+1}(p, q) = \begin{cases} D_t(\tfrac{1}{2}p, 2q) \text{ if } 0 \leqq q < \tfrac{1}{2} \\ D_t(\tfrac{1}{2}p + \tfrac{1}{2}, 2q - 1) \text{ if } \tfrac{1}{2} \leqq q < 1. \end{cases} \tag{2.7}$$

In addition to defining the dynamics of our model, we must also define the observational states. We take these to be strips parallel to the q-axis, specifiable by inequalities of the form $a \leqq p < b$, where a and b are constants. At the instant t, such an observational state has the probability

$$\int_a^b \varphi_t(p)\, dp,$$

where

$$\varphi_t(p) \equiv \int_0^1 D_t(p, q)\, dq. \tag{2.8}$$

The change of $\varphi_t(p)$ with time is given, according to (2.7), by

$$\begin{aligned} \varphi_{t+1}(p) &= \int_0^{\frac{1}{2}} D_{t+1}(p, q)\, dq + \int_{\frac{1}{2}}^1 D_{t+1}(p, q)\, dq \\ &= \int_0^{\frac{1}{2}} D_t(\tfrac{1}{2}p, 2q)\, dq + \int_{\frac{1}{2}}^1 D_t(\tfrac{1}{2}p + \tfrac{1}{2}, 2q - 1)\, dq \\ &= \tfrac{1}{2}[\varphi_t(\tfrac{1}{2}p) + \varphi_t(\tfrac{1}{2}p + \tfrac{1}{2})]. \end{aligned} \tag{2.9}$$

It follows by induction on t that if two different ensembles have the same $\varphi_t(p)$ at any one time, say $t = 0$, they also have the same $\varphi_t(p)$ for all times $t > 0$ and are therefore observationally equivalent for $t \geqq 0$.

2.1. Exercise

1. Show that the baker's transformation (2.4) can be described in a very simple way using the binary representations of the numbers p, q, etc. What is the corresponding description for the iterated transformation U_t?

3. The classical accessibility postulate

Since all the probabilities associated with a trial can be expressed in terms of the initial ensemble density D_0 by the method of § 1, the conditions imposed by the Markovian postulate on the values of these probabilities restrict the class of realizable ensemble densities. The more observations are included in the observation stage of $\mathfrak{T}$, the more information about D_0 is provided by these conditions. Even in the limit of a trial $\mathfrak{T}$ with an infinitely long observation stage, however, the information provided by the Markovian postulate does not, on its own, tell us enough about D_0 to fix the numerical values of any probabilities. To complete the theory, we shall introduce one further postulate, the *postulate of accessibility*;† the additional information about D_0 supplied by this new postulate is sufficient to yield expressions for probabilities which (unlike those in § 1) are independent of the particular form of D_0. In the present section the accessibility postulate for classical mechanics will be explained; its extension to quantum mechanics will be treated in § 4, and its application to the calculation of probabilities in § 5.

An important property of trials with very long observation stages is the law of large numbers, proved in Chap. II, § 9. This law refers to trials in which the system is isolated throughout the observation stage, so that the successive observational states of the system after preparation form a stationary Markov chain. The initial state K is taken to be persistent and the observation stage is taken to consist of observations made at the times $1, 2, \ldots, s$. If A is any observational state reachable from K, then the fraction of times the system is in this state, during the observation stage of the trial, is a random variable

$$\bar{J}_{A,s} \equiv \frac{1}{s} \sum_{t=1}^{s} J_{A,t} \tag{3.1}$$

where $J_{A,t}$ is 1 if the observational state is A at time t and is 0 if not. In accordance with the point of view taken in § 1, the value of this random variable can be expressed as a function of the dynamical state of the system at the initial time:

$$\bar{J}_{A,s} = \bar{J}_{A,s}(\alpha_0) \equiv \frac{1}{s} \sum_{t=1}^{s} J_{A,t}(\alpha_0), \tag{3.2}$$

where $J_{A,t}(\alpha_0)$ is defined in (1.2).

† The concept of accessibility used here is related to, but not identical with, the one used by R. Fowler and E. A. Guggenheim in §§ 105 and 106 of *Statistical Thermodynamics* (Cambridge University Press, London, 1960); on the other hand, it is completely unrelated to the concept of accessibility that is central to Carathéodory's treatment of thermodynamics (see M. Born, Kritische Betrachtungen zur traditionellen Darstellung der Thermodynamik, *Phys. Z.* **22**, 218, 249, 282 (1921)).

The law of large numbers [eqn. (9.11) of Chap. II] states that in the limit where s becomes large, the probability distribution of the random variable $\bar{J}_{A,s}$ becomes sharply concentrated near the value π_A, the equilibrium probability of the state A; that is,

$$\lim_{s\to\infty} Pr(E_{s,\delta}|K) = 0 \text{ for any } \delta > 0, \tag{3.3}$$

where $E_{s,\delta}$ is the event defined by

$$E_{s,\delta} \text{ occurs if and only if } |\bar{J}_{A,s} - \pi_A| \geqq \delta. \tag{3.4}$$

Using (1.5) we can express the law of large numbers (3.3) as a condition on the initial phase-space density D_0:

$$\lim_{s\to\infty} \int_{\omega(E_s)} D_0(\alpha)\, d\alpha = 0 \text{ for any } \delta > 0, \tag{3.5}$$

where $\omega(E_s)$ is the dynamical image of the event $E_s \equiv E_{s,\delta}$, defined by the property that E_s occurs if and only if the initial state is in $\omega(E_s)$. By (3.4) and (3.2) this definition is equivalent to the statement

$$\alpha \text{ is in } \omega(E_s) \text{ if and only if } |\bar{J}_{A,s}(\alpha) - \pi_A| \geqq \delta. \tag{3.6}$$

One would like to be able to infer from (3.5) that the phase-space volume of the region $\omega(E_s)$ tends to 0 as s tends to ∞. To make such inferences, however, one must be able to rule out the alternative possibility that the average value of $D_0(\alpha)$ over the region $\omega(E_s)$ tends to 0. This can be done only by making some further assumption about the class of realizable ensemble densities such as D_0. The purpose of the accessibility postulate is to provide just such an assumption. It is necessary to formulate the postulate carefully, however, to allow for the cases where the probability is small (in fact, zero) for other reasons than a small phase-space volume for $\omega(E_s)$.

A primitive version of the accessibility postulate is already present in the non-statistical mechanics used for small systems: it is assumed that any dynamical state whatever may be used as the initial condition for the motion of such a system, or in other words that there is no restriction (apart from the equations of motion) on the part of phase space accessible to the system. That such a restriction, consistent with the laws of dynamics but not derivable from them, is at least a logical possibility is shown by the example of Bohr's theory of the hydrogen atom, in which the classical equations of motion are supplemented by the restriction that the only accessible dynamical states are the ones lying on trajectories satisfying the quantum condition $h^{-1} \oint p\, dq =$ integer or ∞. Thus for small systems the "accessibility postulate" is merely the negative statement that there is no physical principle restricting the dynamical state of the system at the initial time $t = 0$. Since observational states and dynamical states coincide for a small system, the hypothesis embodied in this "postulate" can be tested by direct observation: if a set of dynamical states were discovered with the property that the system could never be prepared in one of these states, the hypothesis would be falsified. The fact that, for classical system, no such set of states has ever been discovered is our justification for regarding the hypothesis as true.

For large systems, too, we shall make the assumption that there is no physical law, apart from the laws of dynamics themselves, restricting the part of phase space accessible to the system. Since each observational state now embraces many dynamical states, the experimental testing of the hypothesis embodied in this assumption is more difficult than for a small system; but the hypothesis can still be falsified if it is possible to demonstrate experimentally the existence of a region in phase space, say ω_0, having positive volume and the property that in any realizable Gibbs ensemble the fraction of systems with dynamical states in ω_0 vanishes:

$$\int_{\omega_0} d\alpha > 0 \text{ and } \int_{\omega_0} D_0(\alpha)\, d\alpha = 0. \tag{3.7}$$

The condition on the volume of ω_0 is needed to prevent the second integral from vanishing identically.

In order to demonstrate experimentally the existence of a region ω_0 satisfying (3.7) it would be sufficient to find an observable event, say E, depending only on the results of observations made after time 0, and an observational state, say K, with the properties

$$\left.\begin{aligned} \Omega(EK) \equiv \int_{\omega(EK)} d\alpha > 0, \\ Pr(E|K) = 0, \end{aligned}\right\} \tag{3.8}$$

where $\omega(EK)$ denotes the intersection of the image regions $\omega(E)$ and $\omega(K)$. To see this, we show that if (3.8) holds then we can satisfy (3.7) by taking $\omega_0 = \omega(EK)$. The first half of (3.7) then follows immediately from the first half of (3.8). To derive the second half of (3.7), we use the second half of (3.8), which asserts that if we select from a realizable Gibbs ensemble the sub-ensemble consisting of those systems that are in the state K at time 0, then a vanishingly small fraction of the sub-ensemble is in the image region $\omega(E)$ at time 0. If the phase-space density of the original ensemble is $D_0(\alpha)$, the phase-space density of the sub-ensemble is proportional to $J_K D_0(\alpha)$, and (1.5) therefore gives

$$Pr(E|K) = \int_{\omega(E)} J_K D_0(\alpha)\, d\alpha \Big/ \int J_K D_0(\alpha)\, d\alpha\,.$$

If the second half of (3.8) holds, then the numerator on the right must vanish, and this numerator is precisely equal to the second integral in (3.7). Thus we see that if (3.8) can be satisfied, then so can (3.7), enabling us to falsify the assumption that all parts of phase space are accessible. If we wish to affirm that all parts of phase space are accessible, we must therefore deny (3.8). The negation of (3.8) can be written

$$\Omega(EK) > 0 \text{ implies } Pr(E|K) > 0 \tag{3.9}$$

or equivalently

$$Pr(E|K) = 0 \text{ implies } \Omega(EK) = 0. \tag{3.10}$$

Either (3.9) or (3.10) may be taken as the mathematical formulation of the (weak) *accessibility postulate* for a classical system.†

To deal with the limit process $s \to \infty$ used in the law of large numbers, a stronger assumption than (3.10) is necessary: we must assume not only that zero probability implies zero phase-space volume but also that very small probability implies very small phase-space volume. Here a "small" volume means one much less than the volume occupied by the entire set of possible initial dynamical states for the trial $\mathfrak{T}$. This set of possible initial dynamical states is simply the dynamical image $\omega(K)$ of the initial observational state K, and accordingly its phase-space volume is

$$\Omega(K) \equiv \int_{\omega(K)} d\alpha. \tag{3.11}$$

We take $\Omega(K)$ to be non-zero, since no real observation is so accurate that it can locate the dynamical state of the system within a region of zero volume. Consequently our strengthened version of (3.10) can be written

$$\text{small } Pr(E|K) \text{ implies small } \Omega(EK)/\Omega(K). \tag{3.12}$$

A mathematically precise statement of the assumption (3.12) is:

there exists a function $\nu(x)$ such that

(i) $$\Omega(EK)/\Omega(K) \leqq \nu\{Pr(E|K)\}$$

for any event E in a trial whose initial observational state is K,

(ii) $$\nu(0) = 0 \text{ and } \lim_{x\to 0} \nu(x) = 0. \tag{3.13}$$

We shall call this statement‡ the *strong accessibility postulate*. The form of the function $\nu(x)$ is unimportant, apart from its limiting behaviour near $x = 0$, but this form must not be affected by any other limiting processes, such as the limit $N \to \infty$, that may be used in the later stages of the theory.

† When classical statistical mechanics is based on the ergodic theorem of general dynamics instead of on the Markovian postulate (see Chap. I, § 6) it is assumed that regions of phase space having zero volume (or measure) can be ignored. This is tantamount to assuming that $\Omega(E) = 0$ implies $Pr(E) = 0$. This "converse" of (3.10) will not be required here.

‡ In the language of measure theory, the statement is that volume (measure) in phase space is absolutely continuous with respect to probability.

3.1. Exercises

1. Prove the following corollary of the accessibility postulate:

$$1 - \Omega(FK)/\Omega(K) \leqq \nu\{1 - Pr(F|K)\}, \tag{3.14}$$

where F is any event in a trial whose initial state is K.

2. Show that the image region of an ergodic set Z is almost an invariant of the motion (i.e. that the set of phase-space points that are in $U_t\omega(Z)$ but not in $\omega(Z)$ has zero volume).

4. The quantum accessibility postulates

The notion of accessibility, formulated for classical systems in the preceding section, can be extended to quantum systems, but it becomes a little more complicated because of the symmetry requirements on the wave functions of many-particle systems. Corresponding to the various symmetry requirements that can be imposed on the wave functions, there are various accessibility postulates that can be used for a quantum system, and each of them leads to a different form of statistical mechanics. In order to specify a quantum system completely for statistical mechanics, therefore, it is necessary to specify not only its Hamiltonian but also which type of *statistics*—that is, which of the various possible forms of the accessibility postulate—it satisfies.

The simplest quantum accessibility postulate is obtained when no symmetry requirement at all is imposed on the wave function—that is, when it is postulated, in close analogy with the classical argument, that there is no physical principle restricting the dynamical state (wave function) of the system at time 0. The form of statistical mechanics that follows from the adoption of this postulate is called *Boltzmann statistics*. Although Boltzmann statistics does not exactly apply to any real system, it has the advantage of being simpler than the more realistic types of statistics that we shall formulate later in this section, and provides a convenient standard of comparison against which to appraise the effect of the symmetry requirements in determining the physical properties of real systems.

Since the physical content of the accessibility postulate of Boltzmann statistics—that there is no physical principle restricting the dynamical state of the system at time 0—is the same as that of the classical accessibility postulate, it can be formulated mathematically by finding the quantum analogues of the classical formulae (3.9), (3.10), and (3.13). To do this we seek the quantum analogues of the phase-space volumes $\Omega(K)$ and $\Omega(EK)$ appearing in these formulae. These volumes are defined, in (3.8) and (3.11), as integrals over regions of phase space, but they may equally well be written, using the indicator variables of these regions, as integrals over the whole of phase space,

$$\Omega(K) = \int J_K(\alpha)\, d\alpha, \quad \Omega(EK) = \int J_E(\alpha) J_K(\alpha)\, d\alpha, \tag{4.1}$$

where $J_E(\alpha)$ is the indicator of $\omega(E)$, the phase-space region with the property that E occurs if and only if the dynamical state at time 0 belongs to $\omega(E)$. The method of writing down an explicit formula for $J_E(\alpha)$, both for elementary and non-elementary events E, is explained in the paragraph containing (1.4).

To find the quantum analogues of the classical definitions (4.1) we use the fact, noted in connection with eqn. (4.18) of Chap. III, that the quantum analogue of an integral over phase space is a trace. The analogues of the phase-space volumes defined in (4.1) are therefore the quantities defined by

$$\Omega(K) \equiv tr(J_K), \tag{4.2}$$

$$\Omega(EK) \equiv tr(J_E J_K). \tag{4.3}$$

Here J_K is the quantum analogue of $J_K(\alpha)$ (that is, the projection operator of $\omega(K)$, the linear manifold forming the dynamical image of the observational state K) and J_E is the projection operator of the linear manifold $\omega(E)$, defined as in § 1 to comprise all states such that the event E is certain to occur if and only if the quantum state at time 0 is a member of $\omega(E)$. In the discussion of the formula (4.16) of Chap. III defining the trace we showed that the trace of a projection operator equals the dimension of its linear manifold; thus (4.2) and (4.3) show that the quantum analogue of the volume of a phase space region such as $\omega(E)$ or $\omega(EK)$ is the dimension of the corresponding linear manifold.

With these new definitions for the symbols $\Omega(K)$ and $\Omega(EK)$, we can now state the accessibility postulate for Boltzmann statistics by means of precisely the same formulae as we used for classical statistics—

weak Boltzmann accessibility postulate:

$$\Omega(EK) > 0 \text{ implies } Pr(E|K) > 0 \tag{4.4}$$

or equivalently

$$Pr(E|K) = 0 \text{ implies } \Omega(EK) = 0; \tag{4.5}$$

strong Boltzmann accessibility postulate: (4.6)

there exists a function $v(x)$ such that

(i) $$\Omega(EK)/\Omega(K) \leq v\{Pr(E|K)\}$$

for any event E in a trial whose initial observational state is K, and

(ii) $$v(0) = 0 \text{ and } \lim_{x\to 0} v(x) = 0.$$

The physical assumption expressed by the accessibility postulate of Boltzmann statistics— that there is no physical principle restricting the wave function of the system at time 0—is not true for real systems containing identical particles. For any such system, there is a symmetry condition † restricting the

† P. A. M. Dirac, *op. cit.*, § 54.

wave function of the system to a subspace of the dynamical space of all wave functions.

As a simple example, let us consider a system of just two identical spinless particles, with position vectors $\boldsymbol{q}_1, \boldsymbol{q}_2$. Then, if the particles are *bosons*, the wave functions of the system are restricted to the *symmetric* subspace, consisting of all wave functions satisfying the condition

$$\psi(\boldsymbol{q}_1, \boldsymbol{q}_2) = \psi(\boldsymbol{q}_2, \boldsymbol{q}_1) \text{ for all } \boldsymbol{q}_1, \boldsymbol{q}_2; \tag{4.7}$$

but if the particles are *fermions*, then the wave functions are restricted to the *antisymmetric* subspace, consisting of all wave functions satisfying

$$\psi(\boldsymbol{q}_1, \boldsymbol{q}_2) = -\psi(\boldsymbol{q}_2, \boldsymbol{q}_1) \text{ for all } \boldsymbol{q}_1, \boldsymbol{q}_2. \tag{4.8}$$

Alternatively, these symmetry restrictions may be written

$$\left.\begin{aligned} J_+\psi &= \psi \text{ (bosons)},\\ J_-\psi &= \psi \text{ (fermions)}, \end{aligned}\right\} \tag{4.9}$$

where, for the two-particle system,

$$J_\pm\psi(\boldsymbol{q}_1, \boldsymbol{q}_2) \equiv \tfrac{1}{2}\psi(\boldsymbol{q}_1, \boldsymbol{q}_2) \pm \tfrac{1}{2}\psi(\boldsymbol{q}_2, \boldsymbol{q}_1). \tag{4.10}$$

The operators J_+ and J_- are linear; we call J_+ the *symmetrizing operator*, and J_- the *antisymmetrizing operator*. It may be verified, using the definition given in Chap. I, § 3, that J_+ and J_- are projection operators; the corresponding linear manifolds are the manifolds consisting of all symmetric and all antisymmetric wave functions respectively.

For a system of N identical particles the symmetrizing and antisymmetrizing operators are given by

$$J_\pm \equiv (1/N!) \textstyle\sum_P (\pm)^P P, \tag{4.11}$$

where the sum goes over the $N!$ permutations P of N objects, and $(+)^P$ means $+1$ whilst $(-)^P$ means $+1$ for even permutations and -1 for odd permutations. The permutation P acting on a wave function $\psi(\boldsymbol{q}_1, \ldots, \boldsymbol{q}_N)$ means the linear operator whose effect is to make the appropriate permutation of the suffixes; for example, if $N = 4$ and P is the permutation turning 1 into 2, 2 into 4, 3 into 3, and 4 into 1, then

$$P\psi(\boldsymbol{q}_1, \boldsymbol{q}_2, \boldsymbol{q}_3, \boldsymbol{q}_4) = \psi(\boldsymbol{q}_2, \boldsymbol{q}_4, \boldsymbol{q}_3, \boldsymbol{q}_1).$$

Just as in the case of 2 particles, the operators J_+ and J_- are projection operators, and the corresponding linear manifolds are the manifolds consisting of all symmetric and all antisymmetric wave functions respectively. If the N identical particles are bosons, then (4.9) shows that the wave function is restricted to the linear manifold of J_+; similarly, if they are fermions, it is restricted to that of J_-.

The fact that the particles are indistinguishable implies that all the observables associated with the system, and also its Hamiltonian, are symmetrical under permutations of the particles. Thus if G is any observable, or the Hamiltonian, and ψ is any dynamical state (not necessarily symmetrical or antisymmetrical), then a permutation P applied to the product $G\psi$ affects only the factor ψ, so that

$$PG\psi = GP\psi. \tag{4.12}$$

It follows, by (4.11), that $J_\pm G\psi = GJ_\pm\psi$ and hence, since ψ is arbitrary, that

$$J_\pm G = GJ_\pm. \tag{4.13}$$

In particular, since the projection operator J_K is an observable, we have

$$J_\pm J_K = J_K J_\pm. \tag{4.14}$$

Since J_K and $J_\pm$ are projection operators it follows that $J_\pm J_K$ also satisfies the definition of a projection operator; its linear manifold, which we denote by $\omega_\pm(K)$, comprises the symmetric (or antisymmetric) wave functions in $\omega(K)$, and has the dimension number

$$\Omega_\pm(K) \equiv tr(J_\pm J_K) = tr_\pm(J_K), \tag{4.15}$$

where the upper sign refers to Bose statistics (the statistics of bosons) and the lower to Fermi statistics (that of fermions), and we have defined

$$tr_\pm(F) \equiv tr(J_\pm F) \tag{4.16}$$

for any linear operator F.

Since the Hamiltonian commutes with $J_\pm$ [eqn. (4.13)], the time-shift operator U_t defined in eqns. (5.2) and (5.3) of Chap. III does so too; therefore, by using first (4.13) with $G = J_A$, then the definition (1.8) of $J_{A,t}$, and, finally, the definition of J_E given in § 1, we can show that J_E commutes with $J_\pm$. Accordingly we have

$$J_\pm J_E J_K = J_E J_K J_\pm, \tag{4.17}$$

and so $J_\pm J_E J_K$, like $J_\pm J_K$, satisfies the definition of a projection operator. Its linear manifold $\omega_\pm(EK)$ has dimension

$$\Omega_\pm(EK) \equiv tr_\pm(J_E J_K). \tag{4.18}$$

The natural way to extend the accessibility postulate to a system of N identical bosons or fermions is to assume that there is no physical principle, except the symmetry condition, restricting the dynamical state of the system at time 0. That is, we now assume that all dynamical states with symmetric (or antisymmetric) wave functions are accessible in the same sense that all states within the entire Hilbert space were accessible for Boltzmann statistics. Thus the statements (4.4), (4.5), and (4.6) of the weak and strong accessibility postulates need to be modified only by the replacement of $\Omega(K)$

by $\Omega_\pm(K)$, the dimension of the symmetric (or antisymmetric) part of $\omega(K)$, and likewise $\Omega(EK)$ by $\Omega_\pm(EK)$. In particular, the *strong accessibility postulate* for a system of N identical bosons or fermions reads as follows:

$$\left.\begin{array}{l}\text{there exists a function } \nu(x) \text{ such that}\\ \text{(i)} \qquad \Omega_\pm(EK)/\Omega_\pm(K) \leqq \nu\{Pr(E|K)\}\\ \text{for any event } E \text{ in a trial whose initial observational state}\\ \text{is } K\text{, and}\\ \text{(ii)} \qquad \nu(0) = 0 \quad \text{and} \lim_{x\to 0} \nu(x) = 0.\end{array}\right\} \tag{4.19}$$

The upper signs refer to Bose statistics and the lower to Fermi statistics.

The postulate (4.19) can be extended to systems that contain particles of more than one type, either in the form of a mixture of free particles or combined into atoms or compound molecules. The only modification necessary is to replace the purely Bose or Fermi dimension numbers $\Omega_\pm$, defined in (4.15) and (4.18), by dimension numbers Ω_* defined by

$$\Omega_*(K) \equiv tr(J_* J_K), \tag{4.20}$$

where J_* is the projection operator of the linear manifold of wave functions with the symmetry appropriate to the system. For example, if the system is a fully ionized deuterium plasma, consisting of N electrons and N deuterons, the appropriate projection operator is

$$J_* = J_{\text{elec}-} J_{\text{deut}+}, \tag{4.21}$$

where $J_{\text{elec}-}$ antisymmetrizes with respect to the N electrons, and $J_{\text{deut}+}$ symmetrizes with respect to the N deuterons (which obey Bose statistics since each deuteron comprises two elementary particles, a proton and a neutron). The formula (4.21) is easily generalized to a system with any number fo different particle types, each of which may obey either Bose or Fermi statistics.

If the electrons and nuclei composing the system are permanently bound into atoms and molecules, a useful approximation is obtained by treating each atom as a particle and each molecule as a rigid body. If the atoms forming the molecule are all different, as in the HCl molecule, then the appropriate projection operator is once again of the form $J_\pm$, the upper sign being used if the total number of electrons, protons, and neutrons in the molecule is even and the lower if it is odd. If the atoms are not all different, as in the H_2O molecule, then the projection operator to use is

$$J_* = J_\pm \prod_i \left(\frac{1}{\sigma} \sum_R (\pm')^R R\right), \tag{4.22}$$

where $J_\pm$ symmetrizes or antisymmetrizes over permutations of the molecules, the sum in brackets goes over all the (finite) rotations R of the ith molecule whose net effect is the same as a permutation of atoms within the molecule, and the symbol $(\pm')^R$ takes the value $+1$ if the permutation of fermions brought about by the rotation R is even and -1 if this permutation is odd. The number of such rotations is called the *symmetry number* of the molecule, and denoted by σ. For the H_2O molecule the only such rotations are the identity and the one that interchanges the two H atoms, so that $\sigma = 2$. Other molecules may have higher symmetry numbers; for example, the carbon tetrachloride molecule CCl_4, in which the four chlorine atoms are arranged at the vertices of a regular tetrahedron, has the symmetry number 12.

4.1. *Exercises*

1. What are the symmetry numbers of the five regular solids? What is the symmetry number of a benzene molecule?

2. Show that $tr_-(F) = \sum_i \langle\alpha_i|F|\alpha_i\rangle$ where $|\alpha_1\rangle, |\alpha_2\rangle, \ldots$ are a set of states with the property $\sum_i|\alpha_i\rangle\langle\alpha_i| = J_-$, and F is any operator. Compare this formula for $tr_-(F)$ with the corresponding formula, eqn. (4.16) of Chap. III, for $tr(F)$.

5. The equilibrium ensemble

We can now return to the discussion of the law of large numbers, left unfinished in § 3. As stated in (3.3), this law, which applies to isolated systems, is

$$\lim_{s\to\infty} Pr(E_{s,\delta}|K) = 0 \text{ for any } \delta > 0, \tag{5.1}$$

where

$$E_{s,\delta} \text{ occurs if and only if } |\bar{J}_{A,s} - \pi_A| \geqq \delta \tag{5.2}$$

and

$$\bar{J}_{A,s} \equiv \frac{1}{s}\sum_{t=1}^{s} J_{A,t} \tag{5.3}$$

with K and A any persistent states in the same ergodic set Z. Using the accessibility postulate we can infer from (5.1) that the phase-space volume (or, in quantum mechanics, the dimension number) of the image region $\omega(E_{s,\delta})$ becomes negligible as $s \to \infty$, so that the values of $J_{A,s}$ and π_A are close together in nearly all the important parts of dynamical space. This leads to a method for calculating π_A in terms of purely dynamical quantities.

To avoid complicating the proof with inessentials, we consider first the case of a classical system. The first step is to combine (5.1) with the classical accessibility postulate (3.13). Since $\lim_{x\to 0} \nu(x) = 0$, eqn. (5.1) implies

$$\lim_{s\to\infty} \nu\{Pr(E_{s,\delta}|K)\} = 0, \tag{5.4}$$

and so (3.13) gives

$$\lim_{s\to\infty} \frac{\Omega(E_{s,\delta} K)}{\Omega(K)} = 0 \tag{5.5}$$

since all phase volumes are non-negative. Using the definition (5.2) of $E_{s,\delta}$, we can now estimate the integral

$$\int_{\omega(K)} [\bar{J}_{A,s}(\alpha) - \pi_A]\, d\alpha .$$

In the part of $\omega(K)$ that is complementary to $\omega(E_{s,\delta} K)$, the absolute value of the integrand is at most δ, by (5.2); the volume of this part of $\omega(K)$ is $\Omega(K) - \Omega(E_{s,\delta} K)$. In $\omega(E_{s,\delta} K)$ itself, the absolute value of the integrand is at most 1 since $\bar{J}_{A,s}$ is by definition a proper fraction and π_A is a probability. Thus the absolute value of the entire integral satisfies the inequality

$$\left| \int_{\omega(K)} [\bar{J}_{A,s}(\alpha) - \pi_A]\, d\alpha \right| \leq \int_{\omega(K)} |\bar{J}_{A,s}(\alpha) - \pi_A|\, d\alpha$$
$$\leq [\Omega(K) - \Omega(E_{s,\delta} K)]\,\delta + \Omega(E_{s,\delta} K). \tag{5.6}$$

Dividing by $\Omega(K)$ and taking the limit $s \to \infty$ we obtain, using (5.5), the result

$$\limsup_{s\to\infty} \frac{1}{\Omega(K)} \left| \int_{\omega(K)} [\bar{J}_{A,s}(\alpha) - \pi_A]\, d\alpha \right| \leq \delta . \tag{5.7}$$

Since the value of the positive number δ can be arbitrarily small and the left-hand side of (5.7) is intrinsically non-negative, this left-hand side must vanish; it follows, by the definition (3.11) of $\Omega(K)$, that

$$\pi_A = \lim_{s\to\infty} \frac{1}{\Omega(K)} \int_{\omega(K)} \bar{J}_{A,s}(\alpha)\, d\alpha . \tag{5.8}$$

Equation (5.8) expresses π_A in terms of purely dynamical quantities. To obtain a simpler expression we multiply both sides of (5.8) by $\Omega(K)/\Omega(Z)$, where

$$\Omega(Z) \equiv \sum_{K\in Z} \Omega(K) \tag{5.9}$$

is the volume of $\omega(Z)$, the dynamical image of the ergodic set Z containing A and K. Summing the resulting equation over all K in Z, we obtain

$$\pi_A = \lim_{s\to\infty} \frac{1}{\Omega(Z)} \int_{\omega(Z)} \bar{J}_{A,s}(\alpha)\, d\alpha . \tag{5.10}$$

By the definitions (5.3) and (1.2), this is equivalent to

$$\pi_A = \lim_{s\to\infty} \frac{1}{\Omega(Z)\, s} \sum_{t=1}^{s} \int_{\omega(Z)} J_A(U_t \alpha)\, d\alpha . \tag{5.11}$$

To calculate the integral we transform to the new variable of integration $\alpha' \equiv U_t\alpha$ and use the fact, basic to the proof of Liouville's theorem, that the Jacobian of this transformation is 1 [eqn. (3.19) of Chap. III]. This gives the inequality

$$\int_{\omega(Z)} J_A(U_t\alpha)\, d\alpha \leqq \int J_A(\alpha')\, d\alpha' = \Omega(A), \tag{5.12}$$

the second integral going over all phase space. In this way, (5.11) reduces to

$$\pi_A \leqq \Omega(A)/\Omega(Z). \tag{5.13}$$

By eqn. (6.4) of Chap. II the sum of the left-hand side of (5.13), taken over all A in the ergodic set Z, is 1; and by (5.9) the sum of the right-hand side over all A in Z is also 1. Consequently, the two sides of (5.13) must be equal for all A in Z:

$$\pi_A = \Omega(A)/\Omega(Z). \tag{5.14}$$

That is to say, the equilibrium probabilities of the various observational states are proportional to the volumes of their image regions in phase space. This theorem is fundamental in equilibrium statistical mechanics.

An alternative statement of the result (5.14) can be obtained by using the first formula of (4.1), with K replaced by A, to express $\Omega(A)$ as the integral of $J_A(\alpha)$ over all phase space. We obtain

$$\pi_A = \int J_A(\alpha)\, D_Z(\alpha)\, d\alpha, \tag{5.15}$$

where

$$D_Z(\alpha) \equiv J_Z(\alpha)/\Omega(Z) \tag{5.16}$$

and

$$J_Z(\alpha) \equiv \sum_{K \in Z} J_K(\alpha) \tag{5.17}$$

is the indicator of the region $\omega(Z)$. This new phase-space density $D_Z(\alpha)$ is normalized, is uniform inside $\omega(Z)$, and vanishes outside. Equation (5.15) shows that, in an ensemble with phase-space density $D_Z(\alpha)$, the fraction of systems having the observational state A [which, by eqn. (2.8) of Chap. II, is the same as the ensemble average of $J_A(\alpha)$] is equal to π_A, the equilibrium probability of the state A. Since this is true of every state A in Z, we call $D_Z(\alpha)$ the *equilibrium phase-space density* associated with the ergodic set Z, and an ensemble having this phase-space density we call an *equilibrium ensemble.*

To clarify the physical meaning of the equilibrium phase-space density $D_Z(\alpha)$, we may apply the result (7.23) of Chap. II which states that, for any states A and K belonging to an aperiodic ergodic set Z, π_A is the limiting value of $Pr(A_t|K)$ for very large t. Using this result in (5.15), and noting that $Pr(A_t|K)$ can be thought of as the ensemble average of the indicator variable $J_A(\alpha)$ evaluated at time t, we obtain the formula

$$\lim_{t\to\infty} \int J_A(\alpha)\, D_t(\alpha)\, d\alpha = \int J_A(\alpha)\, D_Z(\alpha)\, d\alpha, \tag{5.18}$$

where $D_t(\alpha)$ stands for the phase-space density at time t in an ensemble constructed by putting a large number of systems into the state K at time 0. Equation (5.18) shows that if Z is aperiodic then the true phase-space density $D_t(\alpha)$ and the equilibrium phase-space density $D_Z(\alpha)$ give the same probabilities for any single observation made sufficiently long after the initial time 0. In the language of § 2, we may say that in the limit of large t the true ensemble and the equilibrium ensemble become observationally equivalent with respect to single observations. It will be shown in § 6 that this equivalence extends to compound observations as well, so that the equilibrium ensemble can be used in place of the true ensemble for all calculations of equilibrium properties. Owing to the particularly simple form of the equilibrium ensemble, this equivalence property is extremely valuable.

An analogy with the mixing of fluids, pointed out by Gibbs,† helps in understanding the nature of the equivalence between the equilibrium and the true ensemble. As explained in Chap. III, § 3, Gibbs suggested visualizing $D_t(\alpha)$ as the density of a coloured material non-uniformly dissolved in an incompressible fluid in phase space. The motion of this fluid, described quantitatively by Hamilton's equations of motion, may be visualized as a stirring motion. Initially, all the colouring matter is inside $\omega(K)$, the image region of the initial observational state. As the stirring proceeds, colouring matter will be carried into the image regions of all the observational states in the ergodic set Z, but (by definition of an ergodic set) no significant amount of colouring matter will be carried outside $\omega(Z)$. One would expect the colouring matter eventually to become thoroughly mixed with the fluid throughout $\omega(Z)$, so that the amount of colouring matter in any part of $\omega(Z)$ tends in the limit to the value it would have if the colouring matter were uniformly distributed over $\omega(Z)$. Equation (5.18) shows that, provided Z is aperiodic, this expectation is indeed fulfilled for regions of the form $\omega(A)$. We shall show in § 6 that it is also fulfilled for the image regions $\omega(E)$ of events defined in terms of the results of compound observations. The expectation is not, however, fulfilled in general for *all* regions: for example, if $\omega(Z)$ is the region between two energy surfaces $H(\alpha) = E_0$ and $H(\alpha) = E_1$, and the colouring matter is initially confined to the phase-space region in which $E_0 < H(\alpha) \leqq \frac{1}{2}(E_0 + E_1)$, then by conservation of energy it will never spread to the other part of the image region $\omega(Z)$, in which $\frac{1}{2}(E_0 + E_1) < H(\alpha) \leqq E_1$.

Although the actual distribution of colouring matter becomes equivalent to the uniform equilibrium distribution in this coarse-grained sense, the two distributions remain quite distinct from a fine-grained point of view. Since the carrier fluid is incompressible and there is no diffusion, the region actually occupied by colouring matter preserves its initial volume, which is at most $\Omega(K)$ and therefore less, in general, than the volume of $\omega(Z)$; on

† J. W. Gibbs, *Elementary Principles in Statistical Mechanics*, pp. 144–8.

the other hand, the shape of this region becomes very complicated as t increases, so that eventually it can "fill" the region $\omega(Z)$ of larger volume in the same way that the shavings obtained by planing a piece of wood can fill a box much larger than the original piece of wood.

The result (5.14) may be extended to quantum mechanics. For simplicity we consider in detail only Boltzmann statistics, but the further extension to Bose and Fermi statistics is trivial. The quantum analogue of the classical calculation leading to (5.6) is an estimation of $tr([\bar{J}_{A,s} - \pi_A]J_K)$, where $\bar{J}_{A,s}$ is still related to $J_{A,t}$ as in (5.3) but $J_{A,t}$ is now defined as $U_t^\dagger J_A U_t$ in accordance with (1.8). To estimate this trace we choose a set of orthonormal quantum states $|\alpha_1\rangle, |\alpha_2\rangle, \ldots$, spanning the entire Hilbert space of the system and so chosen that the states $|\alpha_1\rangle \ldots |\alpha_{\Omega(K)}\rangle$ span the subspace $\omega(K)$ of Hilbert space and the states $|\alpha_1\rangle \ldots |\alpha_{\Omega(EK)}\rangle$ span $\omega(EK)$. This is possible because $\omega(EK)$ is a linear subspace of $\omega(K)$. Using these states in the definition [(4.16) of Chap. III] of a trace we obtain

$$tr([\bar{J}_{A,s} - \pi_A]J_K) = \sum_{i=1}^{\Omega(K)} \langle\alpha_i|(\bar{J}_{A,s} - \pi_A)|\alpha_i\rangle, \tag{5.19}$$

since $J_K|\alpha_j\rangle = 1$ if $j \leqq \Omega(K)$ and $= 0$ if not. The first $\Omega(EK)$ terms in the sum have magnitudes not exceeding 1, and the remainder have magnitudes less than δ, by (5.2); it follows that

$$|tr([\bar{J}_{A,s} - \pi_A]J_K)| \leqq \Omega(EK) + [\Omega(K) - \Omega(EK)]\,\delta. \tag{5.20}$$

The same steps that, in the classical case, led from (5.6) to (5.11) now bring us to the formula

$$\pi_A = \lim_{s\to\infty} \frac{1}{\Omega(Z)\,s} \sum_{t=1}^{s} tr(J_Z J_{A,t}), \tag{5.21}$$

where

$$J_Z \equiv \sum_{K\in Z} J_K, \tag{5.22}$$

and

$$\Omega(Z) \equiv \sum_{K\in Z} \Omega(K) = tr(J_Z) \tag{5.23}$$

is the number of dimensions of $\omega(Z)$, the dynamical image of the ergodic set Z containing A.

To estimate the traces in (5.21) we use a complete orthonormal set $|\beta_1\rangle$, $|\beta_2\rangle, \ldots$, chosen to include a subset $|\beta_1\rangle, \ldots, |\beta_{\Omega(Z)}\rangle$ spanning the linear manifold $\omega(Z)$ associated with J_Z; this gives

$$\begin{aligned} tr(J_Z J_{A,t}) &= \sum_{i=1}^{\Omega(Z)} \langle\beta_i|J_{A,t}|\beta_i\rangle \\ &\leqq \sum_{i=1}^{\infty} \langle\beta_i|J_{A,t}|\beta_i\rangle = tr(J_{A,t}) \\ &= \Omega(A). \end{aligned} \tag{5.24}$$

The inequality is a consequence of the fact that any projection operator, such as $J_{A,t}$, is non-negative definite; the final step is a consequence of the definition (1.8), which makes $J_{A,t}$ and J_A unitary transforms of each other so that they have the same eigenvalues and the same trace. Substituting (5.24) in (5.21) and proceeding as in the discussion of (5.13) we conclude that

$$\pi_A = \Omega(A)/\Omega(Z) \tag{5.25}$$

for Boltzmann statistics, as for classical statistics. The result (5.25) generalizes to Bose and Fermi statistics in a very natural way: one need only replace the dimension numbers $\Omega(A)$ and $\Omega(Z)$ of the subspaces $\omega(A)$ and $\omega(Z)$ by the dimension numbers $\Omega_\pm(A)$ and $\Omega_\pm(Z)$ of the symmetric or anti-symmetric parts of these subspaces, as given in (4.15). This generalization is, therefore,

$$\left.\begin{aligned} \pi_A &= \Omega_+(A)/\Omega_+(Z) \quad \text{for Bose statistics,} \\ \pi_A &= \Omega_-(A)/\Omega_-(Z) \quad \text{for Fermi statistics.} \end{aligned}\right\} \tag{5.26}$$

The proofs of these two formulae are closely analogous to that of (5.25), being based on the estimation of $tr(J_\pm[\bar{J}_{A,s} - \pi_A]J_K)$. We shall therefore not give the details of these proofs.

The quantum analogue of the equilibrium phase-space density of classical mechanics is a normalized *equilibrium density operator*, which we denote by D_Z. It is defined by

$$D_Z \equiv \begin{cases} J_Z/\Omega(Z) & \text{(Boltzmann statistics),} \\ J_+J_Z/\Omega_+(Z) & \text{(Bose statistics),} \\ J_-J_Z/\Omega_-(Z) & \text{(Fermi statistics).} \end{cases} \tag{5.27}$$

In terms of the equilibrium density operator, (5.25) or (5.26) can be written analogously to (5.15) or (5.18):

$$\pi_A = tr(J_A D_Z) \tag{5.28}$$

or

$$\lim_{t\to\infty} tr(J_A D_t) = tr(J_A D_Z), \tag{5.29}$$

where D_t denotes the density operator of the true ensemble at time t. Equation (5.29) shows that, just as in the classical case, the true ensemble and the equilibrium ensemble become observationally equivalent for large t, at least with respect to single observations.

The simplest and most important equilibrium ensemble density, both for classical and quantum mechanics, is the *microcanonical ensemble*. This applies to the case when Z comprises all the observational states with a given value for the observable energy. Its dynamical image therefore comprises all the dynamical states whose dynamical energies lie within the corresponding tolerance interval, say $E - \Delta E$ to E. Explicit formulae for the microcanonical ensemble density are given in eqn. (3.26) of Chap. III for classical statistics and in eqn. (5.15) of Chap. III for Boltzmann statistics. For Bose or Fermi statistics they can be obtained, as in (5.27), by multiplying the Boltzmann microcanonical density operator by J_+ or J_- and making the appropriate adjustment of the normalization factor.

5.1. Exercises

1. Show, using classical mechanics, that the equilibrium ensemble is almost stationary, i.e. that if $D_0(\alpha) = D_Z(\alpha)$ holds for all α then $D_t(\alpha) = D_Z(\alpha)$ also holds for all α except, possibly, in a region with zero phase-space volume. (*Hint*: Use the result of exercise 2 of Chap. IV, § 3.1.)

2. Generalize (5.18) to periodic ergodic sets.

6. Coarse-grained ensembles

The deductions made in the previous section from the law of large numbers and the accessibility postulate do not exhaust the information to be drawn from this type of argument. As noted earlier, in Chap. II, § 9, the law of large numbers can be generalized to give information about the rates of occurrence of successive pairs, triplets, etc., of observational states. For example, the generalized law of large numbers applying to successive pairs of observational states is [see Chap. II, eqn. (9.20)]

$$\lim_{s\to\infty} Pr(E'_{s,\delta}|K) = 0 \quad \text{for any} \quad \delta > 0, \tag{6.1}$$

where, by definition,

$$E'_{s,\delta} \text{ occurs if and only if} |\bar{J}_{BA,s+1} - w_{BA}\pi_A| \geqq \delta, \tag{6.2}$$

and

$$\bar{J}_{BA,s+1} \equiv \frac{1}{S}\sum_{t=1}^{s} J_{B,t+1}J_{A,t}, \tag{6.3}$$

and K, A, B are any observational states in the same ergodic set. From this generalized law of large numbers, and the further generalizations to triplets, quadruplets, and so on, we can supplement (5.14) and its quantum analogues (5.25) and (5.26) with further purely dynamical expressions for probabilities.

The simplest of these is an expression for the transition probabilities, obtained from (6.1). In classical mechanics an argument corresponding to the one which led from (5.1) to (5.8) now brings us to

$$w_{BA}\pi_A = \lim_{s\to\infty} \frac{1}{\Omega(K)} \int_{\omega(K)} \bar{J}_{BA,s}(\alpha)\, d\alpha. \tag{6.4}$$

Continuing as in the derivation of (5.11) and (5.13), we obtain

$$w_{BA}\pi_A = \lim_{s\to\infty} \frac{1}{\Omega(Z)s} \sum_{t=1}^{s} \int_{\omega(Z)} J_B(U_{t+1}\alpha)\, J_A(U_t\alpha)\, d\alpha. \tag{6.5}$$

The transformation $\alpha' = U_t\alpha$ now gives

$$\begin{aligned}\int_{\omega(Z)} J_B(U_{t+1}\alpha)\, J_A(U_t\alpha)\, d\alpha &\leqq \int J_B(U_{t+1}(U_t)^{-1}\alpha')\, J_A(\alpha')\, d\alpha' \\ &= \int J_B(U_1\alpha')\, J_A(\alpha')\, d\alpha' \end{aligned} \tag{6.6}$$

since, by eqn. (3.7) of Chap. III, $U_n = (U_1)^n$ for an isolated system. The last two integrations go over all phase space. In the notation of (3.8), the last integral in (6.6) is $\Omega(B_1A)$, where B_1 stands for the event that the observational state at time 1 is B. Equations (6.5) and (6.6) combine to give

$$w_{BA}\pi_A \leqq \Omega(B_1A)/\Omega(Z). \tag{6.7}$$

By eqn. (4.15) of Chap. II the sum of the left-hand side of (6.7), taken over all B in Z, is π_A; and by the expression (6.6) for $\Omega(B_1A)$, combined with (5.17), the sum of the right-hand side over all B in Z is

$$\int J_Z(U_1\alpha)\, J_A(\alpha)\, d\alpha/\Omega(Z) \leqq \Omega(A)/\Omega(Z) = \pi_A \tag{6.8}$$

since $J_Z \leqq 1$, the integration covering all phase space and the last equality being merely (5.14). Consequently, the two sides of (6.7) must be equal for all B in Z:

$$w_{BA}\pi_A = \Omega(B_1A)/\Omega(Z). \tag{6.9}$$

Eliminating π_A with the help of (5.14), we obtain a purely dynamical expression for the transition probabilities of a classical system:

$$w_{BA} = \Omega(B_1A)/\Omega(A), \tag{6.10}$$

analogous to (5.14) for the equilibrium probabilities. Using the definition [(4.9) of Chap. II] of w_{BA}, eqn. (6.10) may be written

$$Pr(B_1|A) = \Omega(B_1A)/\Omega(A). \tag{6.11}$$

The argument leading to (6.11) can be generalized further by starting not from (6.1) but from the generalized law of large numbers referring to triplets [eqn. (9.21) of Chap. II]. The result obtained in place of (6.11) is then

$$Pr(C_2B_1|A) = \Omega(C_2B_1A)/\Omega(A). \tag{6.12}$$

Likewise, by applying the method to $(n+1)$-tuplets of states, we can obtain the result for the general case:

$$Pr(G_nF_{n-1} \dots C_2B_1|A) = \Omega(G_nF_{n-1} \dots B_1A)/\Omega(A) \tag{6.13}$$

where $G, F, \ldots, A$ are persistent states all in the same ergodic set, and

$$\Omega(G_n \ldots B_1 A) \equiv \int J_{G,n}(\alpha) \ldots J_{B,1}(\alpha)\, J_A(\alpha)\, d\alpha, \tag{6.14}$$

the integration covering all phase space. The quantum analogues of (6.11), (6.12), and (6.13) have the same form, but with the classical phase-space volumes $\Omega(\ldots)$ replaced by the dimension numbers

$$\left.\begin{aligned} \Omega(G_n \ldots B_1 A) &\equiv tr(J_{G,n} \ldots J_{B,1} J_A) \text{ (Boltzmann)},\\ \Omega_+(G_n \ldots B_1 A) &\equiv tr_+(J_{G,n} \ldots J_{B,1} J_A) \text{ (Bose)},\\ \Omega_-(G_n \ldots B_1 A) &\equiv tr_-(J_{G,n} \ldots J_{B,1} J_A) \text{ (Fermi)}. \end{aligned}\right\} \tag{6.15}$$

They can be proved by adapting the proofs of (6.11), (6.12), and (6.13) in the same way that the classical proof of (5.14) was adapted to quantum mechanics to give proofs of (5.25) and (5.26).

The proof we have indicated for the basic result (6.13) applies only in the case where the observational states $G, F, \ldots, A$ are all persistent and in the same ergodic set. The validity of this result can easily be extended, however, to cover all cases where A is persistent; for if any of the states $G, F, \ldots, B$ are either transient or else in a different ergodic set from A, then the probability on the left of (6.13) vanishes—the only states reachable with non-vanishing probability from A being the persistent states in its own ergodic set (see Chap. II, § 5)—so that, by the accessibility postulate (3.10) or (3.13), the expression on the right-hand side of (6.13) also vanishes. Thus (6.13) is valid for all cases where A is persistent. The remaining case, where A is transient, will be considered in § 8 of this chapter.

The result (6.13) may be compared with a formula obtainable from the hypothesis of equal *a priori* probabilities, considered in Chap. I, § 6, as a possible alternative to the Markovian postulate as a basic postulate of statistical mechanics. To make the comparison, we write E for the event $G_n F_{n-1} \ldots B_1$ and write $J_E(\alpha)$ for its indicator variable $J_{G,n}(\alpha) \ldots J_{B,1}(\alpha)$; then (6.13) takes the form

$$Pr(E|A) = \int_{\omega(A)} J_E(\alpha)\, d\alpha \Big/ \int_{\omega(A)} d\alpha. \tag{6.16}$$

A similar formula can be obtained by applying the hypothesis of equal *a priori* probabilities, eqn. (6.5) of Chap. I, to the dynamical variable $J_E(\alpha)$ and the phase-space region $\omega(A)$; this formula is

$$\langle J_E(\alpha) \rangle = \int_{\omega(A)} J_E(\alpha)\, d\alpha \Big/ \int_{\omega(A)} d\alpha. \tag{6.17}$$

The left-hand side of (6.17) means the expectation value of $J_E(\alpha)$ evaluated at a moment when the only information available about the system is that its dynamical state lies within $\omega(A)$; since $J_E(\alpha)$ takes the value 1 at time 0 if E will occur and the value 0 if not, this expectation is just the probability of the event E. Thus the results (6.16) and (6.17), one obtained from the Markovian postulate and the other from the hypothesis of equal *a priori* probabilities, are essentially equivalent. It does not follow, however, that the Markovian postulate and the hypothesis of equal *a priori* probabilities always lead to equivalent results: the Markovian postulate has consequences (the consistency conditions on the observational states, derived in the next section) which do not follow from the hypothesis of equal *a priori* probabilities and is, therefore, in a sense, the stronger assumption; on the other hand, the hypothesis of equal *a priori* probabilities is stronger than the Markovian postulate in the sense that it assigns definite values to the expectations of all dynamical variables, whereas the theory based on the Markovian postulate only assigns them to the expectation values of observables.

Another useful way of looking at the result (6.13) is to convert it into a statement that two ensembles are observationally equivalent, generalizing the equivalence property of the equilibrium ensemble treated in the previous section. The simplest such equivalence is obtained by using (1.5) to express the left-hand side of (6.16) as an ensemble average; the resulting formula can be written

$$\int J_E(\alpha)\, D_0(\alpha)\, d\alpha = \int J_E(\alpha)\, D_A(\alpha)\, d\alpha, \tag{6.18}$$

where

$$D_A(\alpha) \equiv J_A(\alpha)/\Omega(A) \tag{6.19}$$

and $D_0(\alpha)$ is the initial phase-space density of some ensemble constructed by replicating a trial with a given persistent initial state A. The result (6.18) shows that the two ensembles whose phase-space densities at time 0 are $D_0(\alpha)$ and $D_A(\alpha)$ are observationally equivalent with respect to any sequence of observations beginning at time 1; they are also observationally equivalent at time 0 since all systems of both ensembles are in the state A at this instant. It follows that the two ensembles are observationally equivalent for all times from 0 onwards.

More generally, let us consider the case where all the observational states are persistent, and the phase-space density $D_0(\alpha)$ is that of an arbitrary ensemble that is realizable at time 0 but is not necessarily confined to any one observational state.† Let the probability distribution over observational states in this ensemble at time 0 be denoted by $p_0(A)$, so that

$$p_0(A) = \int J_A(\alpha)\, D_0(\alpha)\, d\alpha. \tag{6.20}$$

There are two ways to write the probability of an event of the form $G_n F_{n-1} \ldots B_1 A_0$ for this ensemble. One, following from the formalism of (1.5), is

$$Pr(G_n \ldots B_1 A_0) = \int J_E(\alpha)\, J_A(\alpha)\, D_0(\alpha)\, d\alpha \tag{6.21}$$

† For example, the instant denoted here by 0 might be not the initial instant of the trial but the instant, denoted in eqn. (4.17) of Chap. II by t^*, at which the system first becomes isolated.

since, in the notation of this section, the indicator of $G_n \ldots A_0$ is $J_E(\alpha)J_A(\alpha)$. The other, obtained by using the definition [(1.13) of Chap. II] of conditional probability and then (6.13), is

$$Pr(G_n \ldots B_1 A_0) = Pr(G_n \ldots B_1|A_0)\,p_0(A)$$
$$= \int J_E(\alpha)\,D_A(\alpha)\,d\alpha\,p_0(A). \quad (6.22)$$

Combining (6.21) and (6.22), and using the properties of indicators given in eqn. (3.2) of Chap II and eqn. (3.3) of Chap. I, we obtain

$$\int J_E(\alpha)\,J_A(\alpha)\,D_0(\alpha)\,d\alpha = \int J_E(\alpha)\,J_A(\alpha)\,\tilde{D}_0(\alpha)\,d\alpha, \quad (6.23)$$

where $$\tilde{D}_0(\alpha) \equiv \textstyle\sum_A p_0(A)\,D_A(\alpha) = \sum_A p_0(A)\,J_A(\alpha)/\Omega(A). \quad (6.24)$$

The result (6.23) shows that the ensembles with phase-space densities $D_0(\alpha)$ and $\tilde{D}_0(\alpha)$ at time 0 are observationally equivalent for all times from 0 onwards.

In the argument leading to the equivalence property (6.23), there is no special significance in the choice of the time 0 at which the probabilities in (6.20) are evaluated; any arbitrary time t could have been used instead. That is, if $D_t(\alpha)$ is any phase-space density realizable at time t, and $p_t(A)$ is the corresponding probability distribution,† given by

$$p_t(A) = \int J_A(\alpha)\,D_t(\alpha)\,d\alpha \quad (6.25)$$

then by equating two expressions for $Pr(EA_t) \equiv Pr(G_{t+n}F_{t+n-1} \ldots B_{t+1}A_t)$ we obtain the relation, analogous to (6.23),

$$\int J_E(\alpha)\,J_A(\alpha)\,D_t(\alpha)\,d\alpha = \int J_E(\alpha)\,J_A(\alpha)\,\tilde{D}_t(\alpha)\,d\alpha, \quad (6.26)$$

where $$\tilde{D}_t(\alpha) \equiv \textstyle\sum_A p_t(A)\,D_A(\alpha) = \sum_A p_t(A)\,J_A(\alpha)/\Omega(A). \quad (6.27)$$

The result (6.26) shows that the two ensembles with phase-space densities $D_t(\alpha)$ and $\tilde{D}_t(\alpha)$ at time t give the same probabilities for the event $G_{t+n}F_{t+n-1} \ldots B_{t+1}A_t$, and therefore for any event depending only on the results of observations made at or after the time t; the two ensembles are therefore observationally equivalent for all times from t onwards.

The phase-space density $\tilde{D}_t(\alpha)$ defined by (6.27) is called the *coarse-grained* phase-space density corresponding to the (fine-grained) phase-space density $D_t(\alpha)$. The definition shows that $\tilde{D}_t(\alpha)$ is uniform throughout each of the image regions $\omega(A)$ of observational states, the value in each image region being the average value of $D_t(\alpha)$ over that region. The equivalence property (6.27) shows that this averaging or smoothing procedure (analogous to the procedure used by a statistician when he draws a histogram) does not affect the probabilities of events that can be observed in the systems of the ensemble at times from t onwards.

† In Chapter II this same quantity was denoted by $p_A(t)$.

The equivalence properties derived before we reached the general result (6.26) can now be seen to be special cases of it. Thus, in § 5 we showed that any realizable ensemble (with a persistent initial state belonging to an aperiodic ergodic set Z) becomes equivalent, in the limit $t \to \infty$, to the equilibrium ensemble of Z, defined in (5.16) and denoted there by $D_Z(\alpha)$. This result can equally well be deduced from (6.26), for the definition (6.27) of $\tilde{D}_t(\alpha)$ implies that the coarse-grained ensemble becomes identical with the equilibrium ensemble in the limit $t \to \infty$:

$$\begin{aligned} \lim_{t\to\infty} \tilde{D}_t(\alpha) &= \sum_{A\in Z} \pi_A J_A(\alpha)/\Omega(A) \\ &= \sum_{A\in Z} J_A(\alpha)/\Omega(Z) \\ &= J_Z(\alpha)/\Omega(Z) = D_Z(\alpha) \end{aligned} \tag{6.28}$$

by eqns. (7.23) of Chap. II, (5.14), (5.17), and (5.16). At the other end of the possible range of variation of t, we have the result (6.18), showing that if initially (at time 0) all the systems of the ensemble are in a single persistent observational state, say K, then the ensemble is equivalent to one with a phase-space density $D_K(\alpha)$ that is uniform over $\omega(K)$ and vanishes outside. This result is again a special case of (6.26), since if $p_0(A) = \delta_{AK}$, then (6.27) reduces to

$$\tilde{D}_0(\alpha) = J_K(\alpha)/\Omega(K) = D_K(\alpha). \tag{6.29}$$

The quantum analogue of a coarse-grained phase-space density is a *coarse-grained density operator*, defined, in analogy with (6.27), by

$$\tilde{\mathsf{D}}_t \equiv \begin{cases} \sum_A p_t(A) \mathsf{J}_A/\Omega(A) & \text{(Boltzmann)}, \\ \sum_A p_t(A) \mathsf{J}_+ \mathsf{J}_A/\Omega_+(A) & \text{(Bose)}, \\ \sum_A p_t(A) \mathsf{J}_- \mathsf{J}_A/\Omega_-(A) & \text{(Fermi)}, \end{cases} \tag{6.30}$$

where

$$p_t(A) = tr(\mathsf{J}_A \mathsf{D}_t) \tag{6.31}$$

and D_t is a density operator that is realizable at time t. As in the classical case, the fundamental property of the coarse-grained density operator $\tilde{\mathsf{D}}_t$ is that its ensemble is observationally equivalent to the true ensemble (whose density is D_t) for all times from t onwards. The proof and discussion of this property follow closely their classical counterparts and therefore need not be given here.

6.1. *Exercises*

1. Express $Pr(G_n|F_{n-1} \ldots B_1 A)$ in terms of dynamical quantities without explicitly using the Markov postulate.

2. Show that the function $\nu(x)$ appearing in the strong accessibility postulate (3.13) may be taken to be x itself.

3. The results (6.10) and (6.13) were derived on the assumption that the system is isolated. To what extent do they apply to non-isolated systems?

7. The consistency condition

In § 6 we saw how the Markovian and accessibility postulates lead to expressions for all the probabilities referring to an isolated system with a persistent initial state. The Markovian postulate also leads, however, to relations connecting the numerical values of these probabilities, and therefore to relations connecting the corresponding dynamical expressions. The general form of the Markovian relation connecting probabilities is given in Chap. II, eqn. (4.6); for a stationary Markov chain it can be written (taking $t^1 = 1$, $t^2 = 2$, etc., for simplicity)

$$Pr(G_nF_{n-1} \dots C_2B_1|A) = Pr(G_n|F_{n-1}) \dots Pr(C_2|B_1)\, Pr(B_1|A)$$
$$= w_{GF} \dots w_{CB}w_{BA}. \tag{7.1}$$

Combining (7.1) with (6.10) and (6.13), we obtain a condition

$$\frac{\Omega(G_nF_{n-1} \dots B_1A)}{\Omega(A)} = \frac{\Omega(G_1F)}{\Omega(F)} \dots \frac{\Omega(C_1B)}{\Omega(B)}\,\frac{\Omega(B_1A)}{\Omega(A)} \tag{7.2}$$

which must be satisfied for every positive n and every set of $n+1$ persistent states $G, F, \dots, A$ if the system obeys classical or Boltzmann statistics. For Bose or Fermi statistics, the corresponding condition is obtained by writing Ω_+ or Ω_- in place of Ω throughout (7.2).

We shall call (7.2), or its analogue for Bose or Fermi statistics, the *consistency condition*.† It is a statement about the dynamics of the system and the choice of observational states, and by itself gives no information about probabilities; but it is a necessary condition for the applicability of our basic probability hypothesis, the Markovian postulate, for if the system does not satisfy (7.2) then the adoption of the Markovian postulate would lead to a contradiction [that the probabilities (6.13), derived from the Markovian postulate, do not themselves satisfy this postulate]. In classical mechanics, the consistency condition is not only necessary but also sufficient to ensure that the Markovian postulate can be adopted without leading to a contradiction: to show this we need only exhibit a possible ensemble whose probabilities obey this postulate. Such an ensemble is the one whose initial phase space density, denoted in (6.19) by $D_A(\alpha)$, is uniform over the dynamical image of the initial observational state A; for the consistency condition (7.2) is equivalent to the statement that the probabilities in this ensemble do obey the Markovian postulate. The consistency condition is not, however, a guarantee that the actual probabilities are equal to the ones in this hypo-

† The need for a consistency condition in this type of theory was pointed out by N. G. Van Kampen on p. 180 of his article in the volume *Fundamental Problems in Statistical Mechanics*, edited by E. G. D. Cohen (North-Holland, Amsterdam, 1962). His consistency condition is not as strong as ours, however, being based on the Chapman–Kolmogorov equation (4.4) of Chap. II rather than on the full Markov condition (4.6) of Chap. II.

thetical ensemble, and therefore does not relieve us of the need to include in the theory some postulate, such as the Markovian postulate, referring to the values of measurable physical probabilities.

In quantum mechanics, the consistency condition by itself is not sufficient to ensure the self-consistency of the theory, but the consistency condition and the postulate of compatibility (1.9) taken together are. The proof of sufficiency is closely parallel to the classical proof.

A simple but unrealistic example where the consistency condition can quite easily be verified may be constructed using the model described in Chap. IV, § 2, whose dynamics are given by the baker's transformation. Let us define the observational states of the system by dividing the phase space $0 \leqq p < 1, 0 \leqq q < 1$ into four equal strips parallel to the q-axis and taking these strips to be the dynamical images of four observational states labelled 0, 1, 2, 3 (Fig. 8). The indicator variables of these four regions are

$$J_A(\alpha) \equiv J_A(p, q) = \begin{cases} 1 \text{ if } \tfrac{1}{4}A \leqq p < \tfrac{1}{4}(A+1) \\ 0 \text{ if not,} \end{cases} \tag{7.3}$$

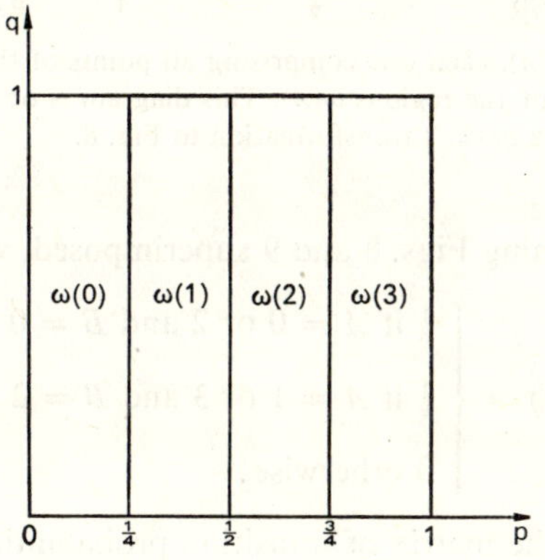

FIG. 8. Image regions $\omega(A)$ of the four observational states of a model based on the baker's transformation.

where A can take the values 0, 1, 2, or 3; their "volumes" (areas) are

$$\Omega(A) = \int_0^1 \int_0^1 dp\, dq\, J_A(p, q) = \tfrac{1}{4}. \tag{7.4}$$

To evaluate $\Omega(B_1 A)$ we use the definitions (6.14) and (1.2), obtaining

$$\begin{aligned} \Omega(B_1 A) &= \int d\alpha\, J_A(\alpha)\, J_B(U_1 \alpha) \\ &= \int d\alpha_1\, J_A(U_1{}^{-1} \alpha_1)\, J_B(\alpha_1), \end{aligned} \tag{7.5}$$

since the Jacobian of the transformation $\alpha \to \alpha_1 \equiv U_1\alpha$ is 1. The function $J_A(U_1^{-1}\alpha_1)$ is the indicator of a region $\omega'(A)$ comprising the points α_1 that are of the form $U_1\alpha$ with α in the strip $\omega(A)$. By the dynamical law (2.4) for this model, the regions $\omega'(A)$ are the four squares shown in Fig. 9. According to (7.5), the value of $\Omega(B_1A)$ is the area of overlap of the strip $\omega(B)$ and the

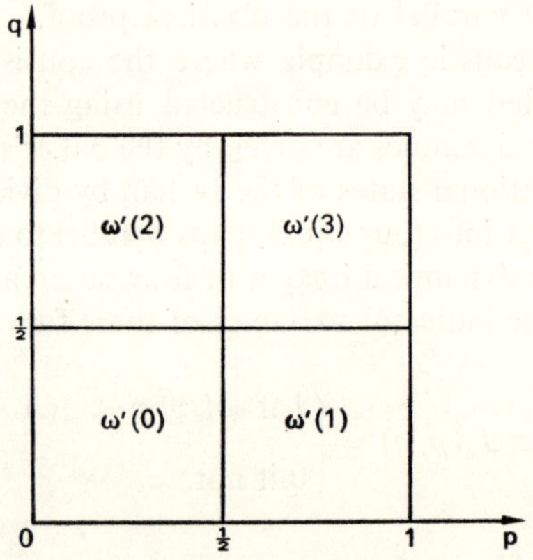

FIG. 9. Regions $\omega'(A)$, each one comprising all points of the form $U\alpha$ with α ranging over one of the regions $\omega(A)$. This diagram is obtained by applying the baker's transformation to Fig. 8.

square $\omega'(A)$. Imagining Figs. 8 and 9 superimposed, we find that

$$\Omega(B_1A) = \begin{cases} \frac{1}{8} \text{ if } A = 0 \text{ or } 2 \text{ and } B = 0 \text{ or } 1, \\ \frac{1}{8} \text{ if } A = 1 \text{ or } 3 \text{ and } B = 2 \text{ or } 3, \\ 0 \text{ otherwise}. \end{cases} \tag{7.6}$$

By (6.10) and (7.4) the matrix of transition probabilities is therefore

$$[w_{BA}] = \begin{bmatrix} \frac{1}{2} & 0 & \frac{1}{2} & 0 \\ \frac{1}{2} & 0 & \frac{1}{2} & 0 \\ 0 & \frac{1}{2} & 0 & \frac{1}{2} \\ 0 & \frac{1}{2} & 0 & \frac{1}{2} \end{bmatrix}. \tag{7.7}$$

In a similar way we can evaluate $\Omega(C_2B_1A)$: it is the area of overlap of the regions $\omega(C)$, $\omega'(B)$, and a region $\omega''(A)$ comprising all the points reachable in exactly 2 steps from a point in $\omega(A)$. The regions $\omega''(A)$ are shown in Fig. 10. The area of overlap is either $^1/_{16}$ or 0, depending on the values of A, B, and C. Since $\omega''(A)$ overlaps $\omega(C)$ for all A and C, the three regions fail to overlap only if either B cannot be reached in 1 step from A

or C cannot be reached in 1 step from B; that is, only if $w_{CB}w_{BA} = 0$. Accordingly we obtain

$$\Omega(C_2B_1A) = \begin{cases} {}^1/_{16} \text{ if } w_{CB}w_{BA} \neq 0, \\ 0 \text{ if not.} \end{cases} \tag{7.8}$$

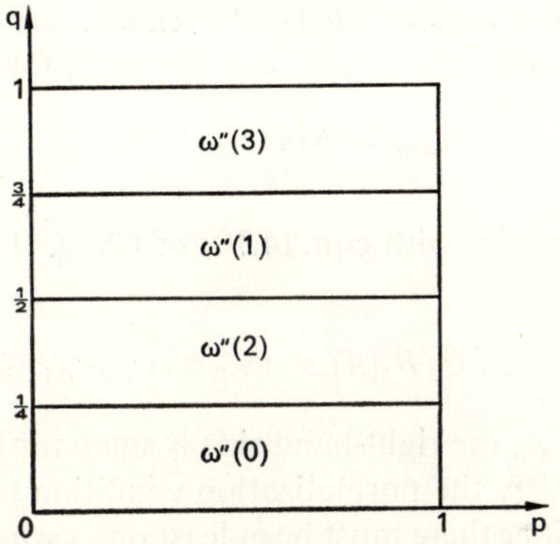

FIG. 10. The regions $\omega''(A)$. This diagram bears the same relation to Fig. 9 as Fig. 9 does to Fig. 8.

Using (7.4) and noting that, by (7.7), the product $w_{CB}w_{BA}$ is either $\frac{1}{4}$ or 0, we find that

$$\frac{\Omega(C_2B_1A)}{\Omega(A)} = w_{CB}w_{BA} \tag{7.9}$$

so that the consistency condition (7.2) is satisfied with $n = 2$ for all A, B, C. This argument can be extended to general values of n.

As an illustration of a situation where the consistency condition is not satisfied, let us consider a model with the same dynamics as in the previous example, but with only two observational states, defined by the indicator variables

$$\left.\begin{aligned} J_0(p, q) &= \begin{cases} 1 \text{ if } 0 \leqq p < \frac{1}{4}, \\ 0 \text{ if not,} \end{cases} \\ J_1(p, q) &= \begin{cases} 1 \text{ if } \frac{1}{4} \leqq p < 1, \\ 0 \text{ if not.} \end{cases} \end{aligned}\right\} \tag{7.10}$$

That is, the observational state 1 of the present model is the union of the states 1, 2, and 3 of the previous model. It may be verified by a calculation similar to the previous one that

$$\begin{aligned} \Omega(1) &= \tfrac{3}{4}, \\ \Omega(1_1 1) &= \tfrac{5}{8}, \\ \Omega(1_2 1_1 1) &= \tfrac{1}{2}, \end{aligned} \tag{7.11}$$

and so (7.2) (with $n = 2$, $A = B = C = 1$) is not satisfied for this model.

For systems more realistic than those based on the baker's transformation, it is a difficult mathematical problem to investigate the consistency condition (7.2) in full. One important conclusion can be drawn quite easily, however: it is that no finite quantum system can exactly satisfy the postulates of the theory in a non-trivial way. This disconcerting result is a consequence of the fact that the dimension numbers $\Omega(A)$, $\Omega_{\pm}(A)$, etc., are integers. Its proof is simplest in the case where there is an ergodic set Z for which

$$w_{\max} \equiv \operatorname*{Max}_{B,A \in Z} w_{BA} < 1. \tag{7.12}$$

Combining this inequality with eqn. (4.10) of Chap. II from Markov chain theory, we obtain

$$Pr(G_n F_{n-1} \ldots C_2 B_1 | A) = w_{GF} \ldots w_{CB} w_{BA} \leqq (w_{\max})^n \tag{7.13}$$

for all $G, F, \ldots, A$ in Z; the right-hand side is small for large n. On the other hand, in order to satisfy the normalization condition (1.17) of Chap. II for conditional probabilities there must be at least one sequence of observational states $B, C, \ldots, F, G$, in the ergodic set of A, such that $Pr(G_n \ldots B_1|A) > 0$; and since this probability is, by (6.13), a fraction with denominator $\Omega(A)$ or $\Omega_{\pm}(A)$ and integral numerator, we have

$$Pr(G_n \ldots B_1|A) \geqq 1/\Omega(A) \quad \text{or} \quad 1/\Omega_{\pm}(A) \tag{7.14}$$

for at least one sequence $B \ldots G$. The two statements (7.13) and (7.14) are clearly incompatible when

$$n \geqq \frac{\ln \Omega(A)}{\ln (1/w_{\max})} \text{ or } \frac{\ln \Omega_{\pm}(A)}{\ln (1/w_{\max})}, \tag{7.15}$$

and it must be concluded that the postulates we have adopted are not exactly self-consistent for a finite quantum system if (7.12) holds. This result can be extended to cases where (7.12) does not hold—in fact to all cases where the persistent states are not all deterministic (that is, where the system has a pair of persistent states A, B for which $0 < w_{BA} < 1$).

Another way of arriving at the same result, that a finite quantum system cannot exactly satisfy the postulates of our theory, is to use the recurrence theorem of Ono and Percival.† This theorem states that for a finite quantum system the expectation $\langle G \rangle_t$ of any dynamical variable G at the time t in an arbitrary Gibbs ensemble is an almost periodic function of t: that is to say, for every positive number ε there exist arbitrarily large numbers u such that the inequality

$$|\langle G \rangle_t - \langle G \rangle_{t+u}| < \varepsilon$$

† S. Ono, *Mem. Fac. Kyushu Univ.* **11**, 125 (1949); I. C. Percival, Almost Periodicity and the Quantum *H* Theorem, *J. Math. Phys.* **2**, 235 (1961) and **3**, 386 (1962).

holds for all t. Consequently, taking G to be the indicator of an observational state A, we find that the probability $p_t(A)$ must come arbitrarily close to its initial value at arbitrarily large times, which is difficult to reconcile with the implication of the Markovian postulate that this probability in general tends to a limit $\pi(A)$ that is different from its initial value.

This difficulty has its origin in the fact, already discussed in Chap. I, § 3, that in principle one can completely determine the dynamical state of a finite quantum system whose Hamiltonian is known exactly, by making a sufficiently large number of compatible measurements on it. In Chap. I, eqn. (3.10), the number of measurements required was estimated as $\ln W/\ln M$, or $\ln \Omega(Z)/\ln M$ in the present notation; the estimate $\ln \Omega(A)/\ln (1/w_{\max})$ provided by (7.15) is essentially the same, though somewhat larger in general because of the correlation between successive observational states. Just as in Chap. I, § 3, we can escape from the paradox by allowing for the fact that the Hamiltonian of a real system is never known precisely. As indicated at the start of Chap. IV, § 1, this refinement replaces the Gibbs ensemble of identical systems, with which we have worked so far, by an ensemble in which the systems have slightly different Hamiltonians. For this ensemble there is a small error, say of order t/τ with τ a constant, in the integrated Liouville equation for a time interval t; and the order of magnitude of τ is restricted by an upper bound proportional to $\ln \Omega(Z)$, as in Chap. I, eqn. (3.11). Before using the law of large numbers, which involves the limit process $t \to \infty$, we must therefore remove this upper bound on the time interval τ over which Liouville's equation may be usefully integrated, by means of a limit process under which $\Omega(Z)$ tends to infinity. Such a limit process normally involves taking the number of particles in the system to infinity, as in the bulk limit defined in Chap. I, § 1, and we shall therefore denote it by $N \to \infty$.

This limit process is easily allowed for in the theory described in §§ 5 and 6, where the basic expressions for probabilities in terms of dynamical quantities are obtained. It is necessary first to assume that a correspondence between the observational states for different values of N can be set up, so that it makes sense to take the limit $N \to \infty$ while keeping fixed the label A of the observational state under consideration. Having done this, we need only replace the single limit operation $\lim_{t\to\infty}$, wherever it occurs in the derivations and results, by the double limit $\lim_{t\to\infty} \lim_{N\to\infty}$. Thus the final results of these calculations become, for Boltzmann (or classical) statistics,

$$\pi_A = \lim_{N\to\infty} \Omega(A)/\Omega(Z), \tag{7.16}$$

$$w_{BA} = \lim_{N\to\infty} \Omega(B_1 A)/\Omega(A), \tag{7.17}$$

$$Pr(G_n \dots B_1|A) = \lim_{N\to\infty} \Omega(G_n \dots B_1 A)/\Omega(A). \tag{7.18}$$

The corresponding formulae for Bose or Fermi statistics can be written down, as usual, by substituting Ω_+ or Ω_- for Ω everywhere. The modified consistency condition is (again for Boltzmann statistics)

$$\lim_{N\to\infty} \frac{\Omega(G_n F_{n-1} \ldots B_1 A)}{\Omega(A)} = \lim_{N\to\infty} \frac{\Omega(G_1 F)}{\Omega(F)} \cdots \frac{\Omega(B_1 A)}{\Omega(A)} \tag{7.19}$$

corresponding to the requirement that the probabilities given by (7.18) must satisfy the Markovian condition. The finite-system inequality (7.14), which we found to be incompatible with the Markovian assumption, is now replaced by the innocuous

$$Pr(G_n \ldots B_1|A) \geqq \lim_{N\to\infty} 1/\Omega(A) = 0 \tag{7.20}$$

since $\Omega(A) \to \infty$ as $N \to \infty$.

The dynamical problem of verifying that the consistency condition (7.2) or (7.19) can be satisfied, by suitable choice of observational states, in a realistic dynamical model has not yet been attacked directly. There has, however, been progress in the simpler problem of showing that the basic probability formulae of statistical mechanics, eqns. (6.10) and (6.13), are consistent with the master equation (4.21) of Chap. II,

$$p_A(t+1) - p_A(t) = \sum_B [w_{AB}p_B(t) - w_{BA}p_A(t)] \tag{7.21}$$

where

$$p_A(t) \equiv Pr(A_t|K)$$
$$= \Omega(A_t K)/\Omega(K) \quad \text{here.} \tag{7.22}$$

The master equation is implied by the basic Markovian condition (4.5) of Chap. II, but does not imply it; it is therefore easier to prove (7.21) than the full consistency relation (7.2). A number of derivations leading to master equations essentially the same as (7.21) have been published recently,† but since these derivations are complicated and do not in the end verify the consistency condition,‡ they will not be summarized here. It is worth noting, however, that all these derivations involve the bulk limit, whether the systems they treat obey classical or quantum mechanics. If, therefore, the methods used in these derivations can be extended to the problem of verifying the consistency condition, they will verify it only in the limit of infinite Ω as in

† L. van Hove, Quantum-mechanical Perturbations Giving Rise to a Statistical Transport Equation, *Physica* **21**, 517 (1955); R. Brout, Statistical Mechanics of Irreversible Processes, *Physica* **22**, 509 (1956); I. Prigogine, *Non-equilibrium Statistical Mechanics* (Interscience, New York, 1962), Chaps. II § 7, IV § 2, VI § 6, XIII § 4, and Appendix II; G. Sewell, Quantum-statistical Theory of Irreversible Processes, *Physica* **31**, 1520 (1965) and **34**, 493 (1967).

‡ The fact that a derivation of the master equation does not suffice to demonstrate Markovian behaviour is pointed out by I. Oppenheim and K. E. Shuler in Master Equations and Markov Processes, *Phys. Rev.* **138**, B 1007 (1965).

(7.19) and not for finite Ω as in (7.2). Thus the weakening of (7.2) to (7.19), which we have already seen to be necessary in quantum mechanics, is perhaps also necessary in classical mechanics once we pass beyond the highly artificial model based on the baker's transformation.

7.1. Exercises

1. Apply the classification method of Chap. II, § 5, to the Markov chain with transition probability matrix (7.7). Also find its equilibrium probabilities, both from (7.7) and from (7.4).

2. Show that the consistency condition is equivalent to the following statement: for all observational states A the ensemble whose density is $D_A \equiv J_A/\Omega(A)$ at time 0 is observationally equivalent, for all times not earlier than time 1, to the ensemble whose density is $\sum_B D_B w_{BA}$ at time 1.

3. Use the result of the preceding exercise, together with the last result obtained in § 2, to give a more concise proof that the model defined by (7.3) and the baker's transformation satisfies the consistency condition.

4. A quantum system has just 2^N orthogonal dynamical states, which we denote by $|a_1\, a_2 \dots a_N\rangle$ where each of the variables a_i can take the values 0 and 1 only. The law of observation is that only the first two numbers in the string $a_1, a_2, \dots, a_N$ can be measured, so that there are just four observational states. The law of motion is $U_1|x_1\, x_2 \dots x_N\rangle = |x_2 \dots x_N x_1\rangle$. Verify that the compatibility condition is satisfied, and that the consistency condition is satisfied in the form (7.19) [but not in the form (7.2) if $n \geqq N-2$]. This model is the quantum analogue of the classical model based on the baker's transformation.

8. Transient states

The argument used in § 6 to derive the main formula of this chapter,

$$w_{BA} = \Omega(B_1 A)/\Omega(A), \qquad (8.1) = (6.10)$$

and its generalization (6.13), applies only when the initial state A is persistent. In the present section we consider the alternative possibility that the initial state may be transient. As in the previous section it is necessary to consider not only the case of a finite system but also the limit where the system becomes infinite. We shall find that transient states can arise only in the latter case.

To show that transient states cannot arise in a finite system, we argue by *reductio ad absurdum*. Suppose that a finite system, exactly satisfying the postulates of the theory, does have a transient state A. The set of observational states that are irreversibly reachable from A is non-empty and may be denoted, as in Chap. II, § 5, by Q. It was shown in Chap. II, § 5, that every state reachable from a given state C in Q is itself in Q, so that

$$Pr(Q_t|C) = 1 \quad \text{for all } C \text{ in } Q \text{ and all } t > 0, \qquad (8.2)$$

where Q_t is the event that the observational state is in Q at time t. It was also shown in Chap. II, § 5, that the probability of the state A tends to 0 as $t \to \infty$; it follows (by interchanging A and B in eqn. (4.17) of Chap. II, and taking the limit $t \to \infty$ at fixed $u - t$) that all the states from which A can be reached also have vanishingly small probabilities in this limit. Consequently, as $t \to \infty$ the probability approaches 1 that the system will be in a state from which A cannot be reached. Since all the states of this type reachable from A are in Q, it follows that

$$\lim_{t\to\infty} Pr(Q_t|A) = 1. \tag{8.3}$$

Applying the corollary (3.14) of the accessibility postulate, with $F \equiv Q_t$, we obtain from (8.2) and (8.3) the results

$$\frac{\Omega(Q_tQ)}{\Omega(Q)} = 1 \quad \text{for all } t > 0 \tag{8.4}$$

and

$$\lim_{t\to\infty} \frac{\Omega(Q_tA)}{\Omega(A)} = 1. \tag{8.5}$$

In classical mechanics, the definitions (3.8) and (3.11) for weights have the consequence that

$$\begin{aligned} \Omega(Q_tA) + \Omega(Q_tQ) &= \int J_Q(U_t\alpha)\,[J_Q(\alpha) + J_A(\alpha)]\,d\alpha \\ &\leqq \int J_Q(U_t\alpha)\,d\alpha = \int J_Q(\alpha_t)\,d\alpha_t = \Omega(Q). \end{aligned} \tag{8.6}$$

The step leading to the second line depends on the fact that A is not a member of Q, so that $J_A + J_Q \leqq 1$. The step leading to the last line depends on the fact that the Jacobian of the transformation $\alpha \to \alpha_t \equiv U_t\alpha$ is 1, as shown in the proof of Liouville's theorem [eqn. (3.19) of Chap. III]. Combining (8.4), (8.5), and (8.6) we obtain

$$\frac{\Omega(A)}{\Omega(Q)} = \lim_{t\to\infty} \frac{\Omega(Q_tA)}{\Omega(Q_tQ)} \leqq \lim_{t\to\infty} \frac{\Omega(Q)}{\Omega(Q_tQ)} - 1 = 0. \tag{8.7}$$

For a finite system $\Omega(Q)$ is finite, and therefore (8.7) implies that $\Omega(A) = 0$. As we noted in connection with (3.11), however, no real observation can be accurate enough to locate the phase point within a region of zero volume, and therefore $\Omega(A)$ must be positive; thus our initial hypothesis, that the state A is transient, leads to a contradiction and is therefore false. Consequently, transient states cannot occur in a finite system to which the postulates apply exactly. A parallel argument can be applied in quantum mechanics, but it carries less weight there, since we have already shown that the theory as a whole cannot apply exactly to a finite quantum system.

In the alternative case, where the limit of a very large system is considered, this objection to transient states no longer applies, for the derivation of (8.7) must now be modified, as in § 7, replacing all quotients of weights by the values they take in the limit $N \to \infty$. In consequence, (8.7) is now replaced by the weaker result

$$\lim_{N\to\infty} \Omega(A)/\Omega(Q) = 0 \tag{8.8}$$

which, unlike (8.7), does not contradict the requirement $\Omega(A) > 0$.

In order to calculate the transition probabilities for the transient states that can arise in the limit of an infinite system it would be desirable to generalize the formula

$$w_{BA} = \lim_{N\to\infty} \frac{\Omega(B_1 A)}{\Omega(A)}, \qquad (8.9) = (7.17)$$

obtained in § 7 for the case where A is persistent, to the case of transient A. Unfortunately, the proof outlined in § 7 for this formula does not generalize to the new case. There is, however, an important special case where (8.9) can still be justified using another argument. This is the case where the transient states are deterministic, as, for example, in the theory of heat conduction, or any other branch of macroscopic physics using too coarse an observational description to admit the possibility of fluctuations. We shall discuss this case using the formulae appropriate to classical or Boltzmann statistics; to convert these formulae to Bose or Fermi statistics the symbol Ω should be replaced everywhere by Ω_+ or Ω_-.

If the state A is deterministic, then it has a unique successor state, denoted here as in Chap. II, § 5, by UA, with the property that

$$w_{BA} = \begin{cases} 1 \text{ if } B = UA \\ 0 \text{ if not.} \end{cases} \qquad (8.10) = [(5.1) \text{ of Chap. II}]$$

For classical or Boltzmann statistics it follows, by the accessibility postulate (3.13) or (4.6), that

$$\lim_{N\to\infty} \frac{\Omega(B_1 A)}{\Omega(A)} = 0 \quad \text{if} \quad B \neq UA. \tag{8.11}$$

Since $\sum_B J_B = 1$ [eqn. (3.4) of Chap. I], we have $\sum_B J_{B,1} J_A = J_A$ by (1.2) or (1.8); consequently, by (6.14) or (6.15), we find that $\sum_B \Omega(B_1 A) = \Omega(A)$. Combining this formula with (8.11) we obtain

$$\lim_{N\to\infty} \frac{\Omega(B_1 A)}{\Omega(A)} = 1 \quad \text{if} \quad B = UA. \tag{8.12}$$

Equations (8.10), (8.11) and (8.12) show that (8.9) holds for all B whenever A is deterministic. In a similar way we can prove the more general result, analogous to (7.18), that

$$Pr(G_n F_{n-1} \ldots B_1 | A) = \lim_{N\to\infty} \frac{\Omega(G_n F_{n-1} \ldots B_1 A)}{\Omega(A)} \tag{8.13}$$

if $F, \ldots, B, A$ are deterministic.

The result (8.13) has a corollary which will be used in Chap. V, §§ 1 and 4, where the non-decrease property of entropy is discussed. According to the definitions given in Chap. II, § 5, a state B is not reachable from a state A (written $B \nleftarrow A$) if and only if $Pr(B_t|A) = 0$ for all $t > 0$. Using (8.13), we obtain, by steps analogous to those used in (8.6), the inequality

$$Pr(B_t|A) = \lim_{N\to\infty} \frac{\Omega(B_t A)}{\Omega(A)}$$

$$\leq \lim_{N\to\infty} \frac{\Omega(B_t)}{\Omega(A)} = \lim_{N\to\infty} \frac{\Omega(B)}{\Omega(A)}. \tag{8.14}$$

Consequently, if the observational states are deterministic, then

$$\lim_{N\to\infty} \frac{\Omega(B)}{\Omega(A)} = 0 \quad \text{implies} \quad B \nleftarrow A. \tag{8.15}$$

This corollary may be compared with an earlier result, the inequality (9.32) of Chap. II, which shows, on interchanging the symbols B and A, that if A and B are persistent states in the same ergodic set then a small value of π_B/π_A [i.e. of $\Omega(B)/\Omega(A)$] implies a small probability of reaching the state B from A. The new result (8.15) shows that a zero value of $\Omega(B)/\Omega(A)$ (in the limit $N \to \infty$) implies a zero probability of reaching B from A; it thus extends the earlier result, which concerns only quasi-transient states, to genuinely transient states.

The remaining case, where some of the observational states are transient but not deterministic,† is not covered by either of the arguments used in this chapter to justify (8.9) and its generalization. It appears, in fact, that for this case, (8.9) cannot be justified in general (although *ad hoc* justifications may be possible in particular cases) without going beyond the limited system of postulates adopted in this book. This is a technical defect in the theory, but as it does not affect the basic applications we shall not try to remedy it here.

8.1. Exercise

1. Discuss the consistency condition for deterministic observational states.

† An example of a Markov chain of this type is the random walk with absorbing boundaries.

CHAPTER V

Boltzmann Entropy

1. Two fundamental properties of entropy

One of Boltzmann's greatest contributions to physics was to show how entropy, the most characteristic concept in the apparently non-mechanical science of thermodynamics, can in fact be linked to mechanics, the link being provided by the statistical approach.† This connection gives to entropy a more direct physical interpretation than thermodynamics alone can provide, and lies at the heart of the main application of statistical mechanics, the calculation of equilibrium thermodynamic properties of bulk matter from the mechanical properties of its constituent molecules. The connection is also valuable in furnishing generalizations taking the notion of entropy beyond its original domain, the thermodynamics of equilibrium; these generalizations enable one to apply the notion of entropy in divers fields, including the kinetic theory of gases, the theory of fluctuations, and the theory of communication. The purpose of this and the next chapter is to formulate the concepts of entropy that are used in statistical mechanics and to indicate how they operate in these applications.

In thermodynamics, the two fundamental properties of entropy are:

(i) its non-decrease: if no heat enters or leaves a system, its entropy cannot decrease; and (1.1)

(ii) its additivity: the entropy of two systems taken together is the sum of their separate entropies. (1.2)

When these postulated properties of thermodynamic entropy are adjoined to the first law of thermodynamics it is possible to derive all the rest of equilibrium thermodynamics‡ with the exception of the parts depending on the third law.

† L. Boltzmann, *Vorlesungen über Gastheorie* (J. A. Barth, Leipzig, 1896), § 8. The English translation, by S. G. Brush, is *Lectures on Gas Theory* (University of California Press, Berkeley and Los Angeles, 1964).

‡ H. B. Callen, *Thermodynamics* (Wiley, New York, 1960). Callen postulates only that the equilibrium state of an isolated system maximizes its entropy. E. A. Guggenheim The Definition of Entropy in Classical Thermodynamics, *Proc. Phys. Soc.* **78**, 1079 (1962), pointed out that Callen's system of postulates is incomplete but can be completed by adding the postulate that the entropy is constant in a quasi-static adiabatic process. Our statement (1.1) implies both Callen's and Guggenheim's postulates, but cannot be derived from them since it does not refer only to equilibrium states. Another treatment based on postulates similar to (1.1) and (1.2) is that of P. Fong in *Foundations of Thermodynamics* (Oxford University Press, London, 1963).

As an example of this type of derivation, let us consider the derivation of the "zeroth law of thermodynamics", which asserts that two or more bodies in thermal equilibrium must have equal temperatures. Consider an isolated system $\mathfrak{S}$ comprising a number of subsystems $\mathfrak{S}', \mathfrak{S}'', \ldots$, which are able to exchange heat energy. As time proceeds the energies $E', E'', \ldots$ of the individual subsystems change, whilst the total energy

$$E = E' + E'' + \cdots \tag{1.3}$$

remains fixed. Eventually the system achieves an equilibrium distribution of energy over the subsystems; we assume this equilibrium distribution to be uniquely determined by the total initial energy. The non-decrease condition implies that this equilibrium distribution has at least as much entropy as any possible initial distribution with the same value of E; consequently it may be found by maximizing the entropy S with respect to $E', E'', \ldots$ subject to a fixed E. The additivity condition implies that

$$S = S'(E') + S''(E'') + \ldots, \tag{1.4}$$

where $S', S'', \ldots$ are the entropies of the subsystems, which we assume for simplicity to depend on their energies only. A necessary condition for S to be a maximum at fixed E can be obtained by the method of Lagrange multipliers; it is

$$\frac{dS'}{dE'} = \frac{dS''}{dE''} = \cdots. \tag{1.5}$$

Accordingly, if we define the temperature of $\mathfrak{S}'$ to be an arbitrary single-valued function of dS'/dE', that of $\mathfrak{S}''$ to be the same function of dS''/dE'' and so on, the derivation of the zeroth law is complete. By convention, this single-valued function is taken to be the reciprocal, so that the thermodynamic temperature of any system is given by

$$T \equiv \left(\frac{dS}{dE}\right)^{-1}. \tag{1.6}$$

In order to extend the thermodynamic concept of entropy to statistical mechanics, we shall look for a quantity in statistical mechanics that shares with thermodynamic entropy the properties (1.1) and (1.2) of non-decrease and additivity, though not the property of being defined only at equilibrium. We shall find that in the statistical mechanics of finite systems it is impossible to satisfy both (1.1) and (1.2) exactly; instead, statistical mechanics offers two distinct definitions of entropy, one satisfying (1.1) exactly but (1.2) only approximately, the other statisfying (1.2) exactly but (1.1) only approximately.† By taking the limit $N \to \infty$ in a suitable way, we can render the ap-

† The two definitions of entropy are contrasted by P. and T. Ehrenfest in § 24 of *The Conceptual Foundations of the Statistical Approach in Mechanics* (Cornell University Press, Ithaca, 1959). A more concise discussion is given by M. J. Klein, Entropy and the Ehrenfest Urn Model, *Physica* **22**, 569 (1956).

proximations negligible so that the two definitions become equivalent; once the limit has been taken, therefore, either definition may be used to establish the connection with thermodynamics, which then becomes a consequence of statistical mechanics instead of an independent theory.

It is perhaps not surprising that there are two distinct definitions for entropy in statistical mechanics, for the non-decrease property (1.1) of entropy is closely linked with the tendency of dissipative systems to approach equilibrium, and we have already seen in Chap. II, §§ 5 and 6, that there are two distinct attitudes we can take to the concept of equilibrium. The simpler of the two attitudes, which we may call the Boltzmann attitude, is to look on equilibrium as an observable property of individual systems: a system is said to be in equilibrium whenever its observational state is an equilibrium state – one from which the probability of a transition to any other state is zero. The alternative attitude, which we may call the Gibbs attitude, is to look on equilibrium as an observable property of ensembles rather than of individual systems: an ensemble may be said to be in equilibrium whenever the probabilities for all observational states are independent of time,† regardless of how the systems in the ensemble behave as individuals. The definitions of entropy that arise from these two attitudes will be considered in this chapter and the next respectively.

According to the Boltzmann point of view, we regard not only equilibrium but also entropy as an observable property of an individual system. The Boltzmann entropy of a system is therefore a function of its observational state; we shall write this function $S(A)$, where A is the observational state. The form of this function is essentially determined by the non-decrease and additivity conditions (1.1) and (1.2). We noted at the beginning of Chap. II, § 6, that the Boltzmann concept of equilibrium, since it presupposes that the Markov chain of observational states includes some deterministic states—the equilibrium states—applies most naturally when the entire Markov chain is deterministic. We may therefore anticipate that the Boltzmann concept of entropy, too, will operate best for deterministic observational states, and in particular that it will not satisfy the non-decrease condition exactly unless the observational states are deterministic. Accordingly, we consider only deterministic observational states in the present introductory section; non-deterministic observational states will be discussed in § 4.

To see how the function $S(A)$ should be chosen so that the non-decrease condition (1.1) is satisfied for deterministic observational states, we first consider the case where the system is isolated. Being isolated, the system can neither lose nor gain heat, and so (1.1) implies that its entropy cannot decrease. Since the transition probabilities for a deterministic

† That is, the coarse-grained probabilities must be independent of time: the fine-grained probabilities need not be. An ensemble is in equilibrium if it is observationally equivalent to (but not necessarily identical with) an equilibrium ensemble of the type discussed in Chap. IV, § 5.

Markov chain take the values 0 or 1 only, this non-decrease condition can be written

$$S(B) \geqq S(A) \quad \text{if} \quad w_{BA} = 1. \tag{1.7}$$

To link this condition with mechanics it is necessary to use the limit process $N \to \infty$, since all non-trivial deterministic Markov chains have transient states and we showed in Chap. IV, § 8, that transient states cannot occur for finite systems. Accordingly, the link with mechanics is provided by eqn. (8.14) of Chap. IV which implies, for classical or Boltzmann statistics, that

$$\lim_{N\to\infty} [\Omega(B)/\Omega(A)] \geqq 1 \quad \text{if} \quad w_{BA} = 1. \tag{1.8}$$

This formula may also be written in the form

$$\Omega(B) \geqq (1-\delta)\,\Omega(A) \quad \text{if} \quad w_{BA} = 1, \tag{1.9}$$

where

$$\lim_{N\to\infty} \delta = 0. \tag{1.10}$$

Comparison of (1.9) with (1.7) suggests that the connection between entropy and mechanics be taken, for classical and Boltzmann statistics, to be

$$S(A) = f\{\Omega(A)\}, \tag{1.11}$$

where $f(\ )$ is an increasing function. If the small quantity δ in (1.9) is neglected, then (1.11) and (1.9) together imply that $S(A)$ has the non-decrease property (1.7). More rigorously (that is, not neglecting δ), (1.9) and (1.11) imply, with the help of Taylor's theorem, that

$$\begin{aligned} S(B) - S(A) &\geqq f\{(1-\delta)\,\Omega(A)\} - f\{\Omega(A)\} \\ &= -\delta\,\Omega(A)\,g\{(1-\theta\delta)\,\Omega(A)\} \\ &= -\frac{\delta}{1-\theta\delta}\,xg(x), \end{aligned} \tag{1.12}$$

where $g(x) \equiv df(x)/dx$, $0 \leqq \theta \leqq 1$, and $x \equiv (1-\theta\delta)\,\Omega(A)$. Taking the limit $N \to \infty$ in (1.12) with the help of (1.10), we see that a sufficient condition for (1.7) to hold in this limit is that the function $f(\ )$, in addition to being non-decreasing, should satisfy

$$x\,df/dx < \text{const.} \quad \text{as} \quad x \to \infty. \tag{1.13}$$

These considerations can be extended to the case where the system is not isolated yet no heat enters or leaves it. Since the Hamiltonian of a non-isolated system depends on the time, some care is necessary to decide whether the consequent changes in its energy are to be classified as heat or mechanical work. In practice the distinction is this: when two systems transfer energy by mechanical work, only a few observable degrees of freedom are involved, so that the amount of energy transferred can be measured by direct observation of forces and displacements; but when the energy is transferred as heat many unobservable molecular degrees of freedom are involved, so that the amount of energy transferred must be determined indirectly, through temperature measurements and a knowledge of physical constants including the mechanical equivalent of heat. The distinction between heat and mechanical work thus depends on whether or not the external parameters $x_1, \ldots, x_n$, used in Chap. I, § 2, to describe the time variation of the Hamiltonian, can be measured experimentally.† If these parameters and their associated generalized forces $X_1, \ldots, X_n$ (where $X_i \equiv \partial H/\partial x_i$) can all be measured experimentally, then the change in energy resulting from a change in the parameters can, in principle, be calculated directly from measured quantities by integration, using the formula

$$\frac{dH}{dt} = \sum_{i=1}^{n} X_i \frac{dx_i}{dt}; \qquad (1.14) = [(2.26) \text{ of Chap. I}]$$

this energy change would therefore be classified as mechanical work. Only if the right-hand side of (1.14) contains quantities that are not measurable (e.g., if x_i is the position of an individual molecule outside the system) do we need to classify part or all of the energy change as heat. Thus a necessary condition for thermal isolation of the system is that the Hamiltonian should depend on time only through external parameters $x_1, \ldots, x_n$ whose time dependence may be regarded as known to the observer.

Changes in the Hamiltonian resulting from changes in such parameters were considered in the discussion of active observations in Chap. I, § 3. It was shown there that consistency with the model of observation used in this book could best be achieved by postulating, first, that the only possible changes in the external parameters (and hence in the Hamiltonian) are discontinuous jumps occurring at the equally spaced instants when observations are possible and, secondly, that the indicator variables J_A ($A = 1, 2, \ldots$) describing the observational states do not explicitly depend on the parameters. From these assumptions it follows, as shown in Chap. I, § 3, that a jump in the values of the parameters $x_1, \ldots, x_n$ has no immediate effect on the dynamical or observational state of the system. Moreover, the value of Ω for each individual observational state is unaffected by a change in H, since $\Omega(A)$ can be expressed [using eqns. (3.11) or (4.2) of Chap. IV] in terms of J_A which is independent of H. Thus a jump in the external parameters has no immediate effect on the current value of $\Omega(A)$ and, therefore, provided the function $f(\,)$ appearing in (1.11) is independent of $x_1, \ldots, x_n$, none on the current value of $S(A)$ either. It is true that such a jump can, and in general will, influence the values taken by $\Omega(A)$ and $S(A)$ at later times because it alters the dynamical laws of the system and, through them, the transition probabilities w_{BA}; nevertheless, as long as the transitions remain

† Even when they can be measured experimentally, the external parameters are not "observables" in the sense defined in Chap. I, § 3, since they are not dynamical variables. Their associated generalized forces, on the other hand, are dynamical variables and may therefore legitimately be classified as observables.

deterministic, (1.8) holds for every time step, and it follows by the argument used before that the non-decrease condition is still satisfied.

As an example we may consider a thermodynamic system consisting of two subsystems which may either be kept completely isolated from one another or else brought into thermal contact. Such a system may be represented by a Hamiltonian of the form

$$H = H' + H'' + xH_{\text{int}}, \tag{1.15}$$

where H' and H'' are the Hamiltonians of the two subsystems, x is a parameter capable of the values 0 and 1 only, and H_{int} is an interaction term large enough to transfer energy at an appreciable rate from one sub-system to the other, yet small enough for its contribution to the total energy at any moment to be negligible. If x is set at 0, the energies E' and E'' of the two sub-systems are both invariants of the motion, so that there is an equilibrium state for each pair of values (E', E''). On the other hand, if x is set at 1, the only constant of the motion is the total energy, so that there is one equilibrium state for each value of $E' + E''$, and the equilibrium values of the individual energies E' and E'' are related by the condition of equal temperatures, eqn. (1.5). Thus a jump in the value of x from 0 to 1, corresponding physically to the establishment of thermal contact between two previously isolated systems, alters the transition probabilities in such a way that states of the form (E', E'') with $T' \neq T''$ are no longer equilibrium states. As a result, if before the jump both systems were in equilibrium with $T' \neq T''$, the entropy will begin to increase as soon as x jumps to the value 1, and will continue to do so until either a new equilibrium state with $T' = T''$ is reached or x is restored to the value 0. Jumps of this type in the Hamiltonian, altering the number of constants of the motion and hence the nature of the equilibrium behaviour of the system, are given prominence in some presentations of thermodynamics,† where the jump from $x = 0$ to $x = 1$ is described as *removing a constraint*. In our example, the constraint is the invariance of the energies of the individual subsystems.

The Boltzmann entropy of a system may be interpreted as a measure of the *degree of disorder* in the arrangement of its molecules. Order and disorder can be defined in various ways, but the ordered arrangements are necessarily much less numerous than the disordered ones; for example, the letters of the alphabet can be arranged in 26! ways, of which only one is the perfectly ordered arrangement *ABC ... XYZ*, all the rest having varying degrees of disorder. Consequently, an ordered observational state comprises many fewer arrangements than a disordered one, and so if each arrangement contributes roughly the same amount to Ω an ordered state has a much smaller Ω, and consequently less entropy, than a disordered one. For example, a crystal, with its ordered array of atoms, has less entropy than a fluid

† See H. Callen, *Thermodynamics*, § 1.8.

made from the same atoms and occupying the same volume. Anyone who has worked with papers at a desk or table will have experienced the natural tendency of inanimate objects towards a state of maximal disorder. If we interpret the disorder of the papers as a form of entropy, this tendency may be taken as an illustration of the non-decrease property of entropy as defined in (1.7). Such interpretations should be used with care, however, since they can lead to misleading results if carried too far.†

2. Composite systems

In addition to the condition of non-decrease (1.1) used in the preceding section, entropy must also satisfy the additivity condition (1.2). This condition states that if a system $\mathfrak{S}$ consists of two spatially separated subsystems $\mathfrak{S}'$ and $\mathfrak{S}''$, then the entropy of the composite system equals the sum of the entropies of the two subsystems. In order to apply this condition, it is first necessary to find how the observational states of $\mathfrak{S}$ and their dynamical images are related to the corresponding states and images for $\mathfrak{S}'$ and $\mathfrak{S}''$.

For the observational states themselves, this relationship is very simple. If an observation of $\mathfrak{S}'$ shows it to be in an observational state A' and simultaneously another observation shows $\mathfrak{S}''$ to be in a state A'', then the two observations taken together may be regarded as a single observation of the composite system $\mathfrak{S}$, showing it to be in an observational state which may be denoted by $A'A''$. Moreover, if $J_{A'}$ denotes the indicator variable that takes the value 1 if $\mathfrak{S}'$ is in the observational state A' and 0 otherwise, and $J_{A''}$ the corresponding indicator for the state A'' of $\mathfrak{S}''$, then the indicator of the observational state $A'A''$ of $\mathfrak{S}$ is given by

$$J_{A'A''} = J_{A'}J_{A''}. \tag{2.1}$$

To relate the dynamical images of the observational state $A'A''$ in system $\mathfrak{S}$ to those of A' and A'' in systems $\mathfrak{S}'$ and $\mathfrak{S}''$, we must first show how the dynamical space of $\mathfrak{S}$ is related to the dynamical spaces of $\mathfrak{S}'$ and $\mathfrak{S}''$. For simplicity, we shall only treat the case where $\mathfrak{S}'$ consists of N' identical particles confined by a container to a region R' in physical space, and $\mathfrak{S}''$ consists of N'' identical particles confined to a region R''. The particles of $\mathfrak{S}'$ may either be identical with the particles of $\mathfrak{S}''$ or different—the two cases will be treated separately.

As in Chap. I, eqn. (2.6), we shall represent the f position coordinates of the ith particle by a vector $\boldsymbol{q}_i$. The set of all possible configurations of a system will be called its *configuration space*. The configuration space C' of the system $\mathfrak{S}'$ consists of all configurations Q' having the form

$$Q' \equiv (\boldsymbol{q}_1, \ldots, \boldsymbol{q}_{N'}) \tag{2.2}$$

with

$$\boldsymbol{q}_i \in R' \quad \text{for} \quad i = 1, \ldots, N'. \tag{2.3}$$

† See M. J. Klein, Order, Organization and Entropy, *Brit. J. Philos. Sci.* **4**, 158 (1953–4), and J. M. Burgers, Entropy and Disorder, *Brit. J. Philos. Sci.* **5**, 70 (1954–5).

If the particles of $\mathfrak{S}'$ have f degrees of freedom each, then C' has fN' dimensions. Likewise, the configuration space C'' of $\mathfrak{S}''$ is an fN''-dimensional space consisting of all configurations Q'' of the form

$$Q'' \equiv (q_{N'+1}, \ldots, q_N) \tag{2.4}$$

with

$$q_i \in R'' \quad \text{for} \quad i = N'+1, \ldots, N, \tag{2.5}$$

where $N \equiv N' + N''$.

The nature of the configuration space of $\mathfrak{S}$ depends on whether the particles of $\mathfrak{S}'$ and $\mathfrak{S}''$ are different or the same. If they are different, then $\mathfrak{S}$ is a system comprising N' particles of one type in the region R' and N'' of another type in R''. If we use the labels $1, \ldots, N'$ for the particles of the first type and $N'+1, \ldots, N$ for the rest, the configuration space of this system is an N-dimensional space consisting of all configurations Q having the form

$$Q \equiv (q_1, \ldots, q_N) \equiv Q'Q'', \tag{2.6}$$

where

$$\left.\begin{aligned} &q_i \in R' \quad \text{for} \quad i = 1, \ldots, N', \\ &q_i \in R'' \quad \text{for} \quad i = N'+1, \ldots, N. \end{aligned}\right\} \tag{2.7}$$

This space is called the *product space* of the configuration spaces C' and C''; we denote it by $C'C''$.

On the other hand, if the particles in $\mathfrak{S}'$ and $\mathfrak{S}''$ are identical, then $\mathfrak{S}$ is a system comprising N identical particles of which N' are in R' and N'' are in R''. The configuration space of this system is invariant under all permutations P of the N identical particles; that is, if Q is a configuration of $\mathfrak{S}$ then so is $\mathsf{P}Q$ for any permutation P. The product space $C'C''$ defined by (2.6) and (2.7) is only a part of the configuration space of $\mathfrak{S}$, since it does not include states of the form $\mathsf{P}Q$, where Q is in $C'C''$ and P is a permutation that interchanges particles between R' and R''. To include these states we must replace (2.7) by

$$\left.\begin{aligned} &q_i \in R' \quad \text{for } N' \text{ different integers } i \text{ from the set } 1, \ldots, N; \\ &q_i \in R'' \quad \text{for the remaining values of } i \text{ from this set.} \end{aligned}\right\} \tag{2.8}$$

The configuration space C of $\mathfrak{S}$, specified by (2.6) and (2.8), consists of $N!/N'!N''!$ parts, one for each method of choosing N' integers from the set $1, \ldots, N$. One of these parts is $C'C''$, and the rest are congruent to $C'C''$, being obtainable from it by permutations of the particle labels $1, \ldots, N$ (Fig. 11).

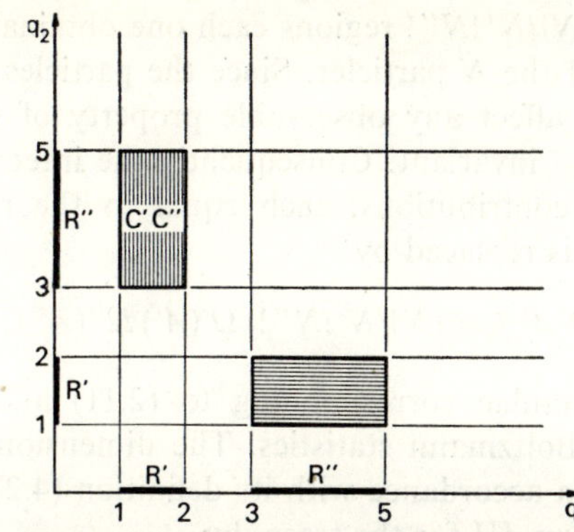

FIG. 11. The configuration space C for a one-dimensional system defined to consist of two identical particles, one of which is confined to the segment R' of one-dimensional space where $1 \leqq q \leqq 2$, and the other to the segment R'' where $3 \leqq q \leqq 5$. The space C consists of two disjoint parts, one (vertical shading) being the product space $C'C''$ defined in (2.7), and the other part (horizontal shading) being congruent to it under an interchange of the particle labels 1 and 2.

In classical mechanics, the phase-space volume of the observational state A' for the system $\mathfrak{S}'$ is given, in accordance with eqn. (3.11) of Chap. IV, by the integral

$$\Omega'(A') = \int dP' \int_{C'} dQ' J_A(P', Q'), \tag{2.9}$$

where P' means $(p_1, \ldots, p_{fN'})$, dP' signifies an integration in which each of the fN' momentum variables goes from $-\infty$ to $+\infty$, and dQ' stands for $dq_1 \ldots dq_{fN'}$. There is a similar expression for $\Omega''(A'')$. In the case where the particles in R' are different from those in R'', the phase-space volume of the observational state $A'A''$ for system $\mathfrak{S}$ is similarly given by

$$\Omega(A'A'') = \int dP \int_{C'C''} dQ J_{A'A''}(P, Q). \tag{2.10}$$

Using the definition given by (2.6) and (2.7), we can reduce the integral to

$$\Omega(A'A'') = \int dP' \int dP'' \int_{C'} dQ' \int_{C''} dQ'' J_{A'}(P', Q')\, J_{A''}(P'', Q'')$$

$$= \Omega'(A')\, \Omega''(A''). \tag{2.11}$$

The factorization of $J_{A'A''}$ is justified by (2.1).

In the alternative case where the particles in R' and R'' are identical, (2.10) is replaced by

$$\Omega(A'A'') = \int dP \int_C dQ\, J_{A'A''}(P, Q), \tag{2.12}$$

where C consists of $N!/N'!N''!$ regions each one obtainable from $C'C''$ by some permutation of the N particles. Since the particles are identical, this permutation cannot affect any observable property of the system and it therefore leaves $J_{A'A''}$ invariant. Consequently the integral in (2.12) is the sum of $N!/N'!N''!$ contributions, each equal to the right-hand side of (2.10), and so (2.11) is replaced by

$$\Omega(A'A'') = (N!/N'!N''!)\,\Omega'(A')\,\Omega''(A''). \tag{2.13}$$

The quantum formulae corresponding to (2.11) and (2.13) are most easily obtained for Boltzmann statistics. The dimension of $\omega(A')$ for the system $\mathfrak{S}'$ is given, in accordance with its definition (4.2) of Chap. IV and formula (4.21) of Chap. III for the trace, by

$$\Omega'(A') = \int_{C'} dQ' \,\langle Q'|J_A|Q'\rangle. \tag{2.14}$$

The formula for $\Omega''(A'')$ is similar. In the case where the particles in R' are different from those in R'', the dimension of $\omega(A'A'')$ simplifies as follows, with the help of (2.1):

$$\begin{aligned}\Omega(A'A'') &= \int_{C'C''} dQ\langle Q|J_{A'A''}|Q\rangle \\ &= \int_{C'} dQ' \int_{C''} dQ''\langle Q'Q''|J_{A'}J_{A''}|Q'Q''\rangle \\ &= \int_{C'} dQ' \int_{C''} dQ''\langle Q'|J_{A'}|Q'\rangle\,\langle Q''|J_{A''}|Q''\rangle \\ &= \Omega'(A')\,\Omega''(A''),\end{aligned} \tag{2.15}$$

where $Q'Q''$ is, according to (2.2), (2.4), and (2.6), simply another way of writing Q, and the factorization of the matrix $\langle|J_{A'}J_{A''}|\rangle$ is justified† by the fact that $J_{A'}$ affects only the subsystem $\mathfrak{S}'$ and $J_{A''}$ affects only the subsystem $\mathfrak{S}''$. In the case where the particles in R' and R'' are identical, the calculation is similar to that in (2.15), but once again there is an extra factor $N!/N'!N''!$ due to the larger configuration space C, and the resulting formula is once again

$$\Omega(A'A'') = (N!/N'!N''!)\,\Omega'(A')\,\Omega''(A''). \tag{2.16}$$

Thus in each case the result for Boltzmann statistics is formally identical with its classical counterpart, differing only in the meaning attached to the symbols Ω, Ω', Ω''.

Finally, we consider a quantum system obeying Bose or Fermi statistics. The dimension of the symmetric or antisymmetric part of $\omega(A')$ for system $\mathfrak{S}'$ may be written, using eqn. (4.15) of Chap. IV and eqn. (4.21) of Chap. III, in the form

$$\Omega'_{\pm}(A') = \int_{C'} dQ'\langle Q'|J'_{\pm}J_{A'}|Q'\rangle, \tag{2.17}$$

† See P. A. M. Dirac, *The Principles of Quantum Mechanics*, § 20.

where $J'_\pm$ is the symmetrizing or antisymmetrizing operator for N' particles, defined as in eqn. (4.11) of Chap. IV by

$$J'_\pm \equiv \frac{1}{N'!}\sum_{P'}(\pm)^{P'}P' \tag{2.18}$$

with the sum ranging over all the permutations P' of N' particles. If the particles in R' and R'' are not identical, then $\mathfrak{S}$ is a system with two kinds of particle, and formula (4.15) of Chap. IV for $\Omega_\pm(A)$ must be generalized. To avoid having to invent new symbols, we assume that the particles in $\mathfrak{S}'$ obey the same kind of statistics (Bose or Fermi) as those in $\mathfrak{S}''$; then by forming the appropriate generalization of eqn. (4.15) of Chap. IV and then using eqn. (4.21) of Chap. III again we obtain

$$\begin{aligned}\Omega_\pm(A'A'') &= tr(J'_\pm J''_\pm J_{A'A''})\\ &= \int_{C'C''} dQ\langle Q|J'_\pm J''_\pm J_{A'A''}|Q\rangle,\end{aligned} \tag{2.19}$$

where $J'_\pm$ symmetrizes or antisymmetrizes over permutations P' of particles labelled $1 \ldots N'$ in the formula (2.6) for Q, and $J''_\pm$ symmetrizes or antisymmetrizes for the particles labelled $N'+1, \ldots, N$. Proceeding as in (2.15), we can simplify (2.19) to

$$\begin{aligned}\Omega_\pm(A'A'') &= \int_{C'} dQ' \int_{C''} dQ''\langle Q'|J'_\pm J_{A'}|Q'\rangle\langle Q''|J''_\pm J_{A''}|Q''\rangle\\ &= \Omega'_\pm(A')\,\Omega''_\pm(A''),\end{aligned} \tag{2.20}$$

the factorization of the matrix $\langle|J'_\pm J''_\pm J_{A'}J_{A''}|\rangle$ being justified, as before, by the fact that $J'_\pm J_{A'}$ acts only on $\mathfrak{S}'$ and $J''_\pm J_{A''}$ only on $\mathfrak{S}''$.

If the particles in R' and R'' are identical, the calculation of $\Omega_\pm(A'A'')$ is a little more complicated than previously, but leads to a remarkably simple result. Since all N particles are now identical we must start not from (2.19) but from

$$\begin{aligned}\Omega_\pm(A'A'') &= \int_C dQ\langle Q|J_\pm J_{A'A''}|Q\rangle\\ &= \frac{N!}{N'!\,N''!}\int_{C'} dQ' \int_{C''} dQ''\langle Q'Q''|J_\pm J_{A'A''}|Q'Q''\rangle,\end{aligned} \tag{2.21}$$

where

$$J_\pm \equiv \frac{1}{N!}\sum_P (\pm)^P P \qquad (2.22) = [(4.11) \text{ of Chap. IV}]$$

is the symmetrizing or antisymmetrizing operator for the N particles in the entire system.

To obtain a factorized form for (2.21) we must express $J_\pm$ in terms of $J'_\pm$ and $J''_\pm$. This is done by factorizing every permutation in the sum (2.22) into the form

$$P = P_0 P' P'', \tag{2.23}$$

where P' permutes the particles 1, ..., N', P'' permutes the particles $N' + 1$, ..., N, and P_0 intermingles the two subsets of particles 1, ..., N' and $N' + 1$, ..., N without changing the order within either subset. For example suppose that $N = 4$, $N' = 2$, and that P is the permutation that changes (1234) into (4231). Then the factorization (2.23) is given by the permutations P_0, P', P'', that effect the changes

$$P''(1234) = (1243),$$

$$P'(1243) = (2143),$$

$$P_0(2143) = (4231),$$

so that, as required, $P_0P'P''$ (1234) = (4231). The permutation P'' interchanges 3 and 4, then P' interchanges 1 and 2, and, finally, P_0 suitably intermingles the two subsets (12) and (34) without affecting their internal order. If the general factorization (2.23) is substituted into (2.22), the result is

$$J_\pm = \frac{1}{N!} \sum_{P_0} \sum_{P'} \sum_{P''} (\pm)^{P_0} (\pm)^{P'} (\pm)^{P''} P_0 P' P''$$

$$= \frac{N'!\,N''!}{N!} \sum_{P_0} (\pm)^{P_0} P_0 J'_\pm J''_\pm , \tag{2.24}$$

where the sum goes over all the $N!/N'!N''!$ permutations that intermingle the two subsets of particles 1, ..., N' and $N' + 1$, ..., $N' + N''$ without changing the order within either subset.

We can now simplify (2.21) by substituting (2.24) into it: the substitution gives

$$\Omega_\pm(A'A'') = \sum_{P_0} (\pm)^{P_0} \int_{C'} dQ' \int_{C''} dQ'' \langle Q'Q''|P_0 J'_\pm J''_\pm J_{A'} J_{A''}|Q'Q''\rangle \tag{2.25}$$

and we now show that only the term where P_0 is the unit permutation gives a non-vanishing contribution to the sum. The integrand in (2.25) can be written $\langle Q'Q''|P_0|\alpha\rangle$, where $Q' \in C'$, $Q'' \in C''$, and

$$|\alpha\rangle \equiv J'_\pm J''_\pm J_{A'} J_{A''}|Q'Q''\rangle. \tag{2.26}$$

Each of the four operators $J'_\pm$, $J''_\pm$, $J_{A'}$ and $J_{A''}$ acts either on system $\mathfrak{S}'$ alone or on system $\mathfrak{S}''$ alone; consequently, since $|Q'Q''\rangle$ is a state for which particles 1, ..., N' make up $\mathfrak{S}'$ and particles $N' + 1$, ..., N make up $\mathfrak{S}''$, the same is true of the state $|\alpha\rangle$ formed from it in accordance with (2.26). Any intermingling permutation P_0 applied to this state will, unless it is the unit permutation, transfer at least one of the first N' particles from $\mathfrak{S}'$ to $\mathfrak{S}''$. If j is the label of one of the transferred particles, it follows that

$\langle q_1 \ldots q_j \ldots q_N | P_0 | \alpha \rangle$ must vanish, since q_j lies in C' but the amplitude for the jth particle to lie in C' is zero for the state $P_0|\alpha\rangle$. All terms with $P_0 \neq 1$ in (2.25) therefore vanish, and it reduces to

$$\Omega_\pm(A'A'') = \int_{C'} dQ' \int_{C''} dQ'' \langle Q'Q''|J'_\pm J''_\pm J_{A'} J_{A''}|Q'Q''\rangle$$
$$= \Omega_\pm(A')\,\Omega_\pm(A'') \qquad (2.27)$$

by the same method that gave (2.20). The result (2.27) is analogous to (2.13) for classical and (2.16) for Boltzmann statistics, but simpler because the combinatorial factor no longer appears.

2.1. Exercise

1. Generalize the derivation of (2.13) to the case where there are two types of particle, system $\mathfrak{S}'$ containing N'_1 of type 1 and N'_2 of type 2, and $\mathfrak{S}''$ containing N''_1 of type 1 and N''_2 of type 2. Check that the resulting formula for $\Omega(A'A'')$ reduces to (2.11) and to (2.13) in the appropriate special cases where some of the particle numbers vanish.

3. The additivity of entropy

The additivity condition (1.2) for Boltzmann entropy can be written in the form

$$S(A'A'') = S'(A') + S''(A''), \qquad (3.1)$$

where S is the entropy of a composite system $\mathfrak{S}$ consisting, as in § 2, of two spatially separated subsystems $\mathfrak{S}'$ and $\mathfrak{S}''$ whose observational states are A' and A'' and whose entropies are S' and S''. In the present section it will be shown how the condition (3.1) is sufficient to determine the form of the expression for Boltzmann entropy almost completely.

We have already seen (1.11) how the condition of non-decrease indicates that the Boltzmann entropy of a system in the observational state A, obeying classical or Boltzmann statistics, should have the form $f\{\Omega(A)\}$, where f is an increasing function which may depend on the nature of the system but not on the values of any external parameters. Substituting this formula for the entropy into (3.1) we obtain

$$f\{\Omega(A'A'')\} = f'\{\Omega'(A')\} + f''\{\Omega''(A'')\}, \qquad (3.2)$$

where the symbols $\Omega, \Omega', \Omega''$ have the same meaning as in § 2, and we have used three distinct functions $f(\,)$, $f'(\,)$, and $f''(\,)$ to allow for the possibility that the functional relation between entropy and Ω may be different for different systems.

To determine these functions, let us consider first the case where $\mathfrak{S}'$ consists of N' particles of one type and $\mathfrak{S}''$ consists of N'' particles of another type. The composition law is then (2.11) which, when substituted into (3.2), gives

$$f(\Omega'\Omega'') = f'(\Omega') + f''(\Omega''), \tag{3.3}$$

where $\Omega' \equiv \Omega'(A')$ and $\Omega'' \equiv \Omega''(A'')$. The natural way to satisfy (3.3) is to use functions $f(\ )$, etc., satisfying the functional equation

$$f(x'x'') = f'(x') + f''(x'') \tag{3.4}$$

for all positive values of the two variables x' and x''. To solve this functional equation we first differentiate both sides partially with respect to x' and set $x'' = 1/x'$. This gives

$$g'(x') = k/x', \tag{3.5}$$

where $g'(x) \equiv df'(x)/dx$ and $k \equiv g(1)$. Integrating (3.5) we obtain

$$f'(x') = k \ln(c'x'), \tag{3.6}$$

where c' is a constant of integration. Similarly, by interchanging the roles of x' and x'' in the derivation of (3.6) we obtain

$$f''(x'') = k \ln(c''x''), \tag{3.7}$$

where $k \equiv g(1)$ as before, and c'' is a new constant of integration. It may be verified by substitution that (3.6) and (3.7) do satisfy the functional equation (3.4) provided

$$f(x) = k \ln(cx) \tag{3.8}$$

where

$$c = c'c''. \tag{3.9}$$

The constant k must be the same for an arbitrary pair of systems $\mathfrak{S}'$ and $\mathfrak{S}''$, each made from a different type of particle, and is therefore a universal constant, determining the units in which entropy is measured. This constant is called Boltzmann's constant. The way in which its numerical value is determined by the practical choice of energy and temperature units will be indicated in § 4; but the only restriction placed on it by purely theoretical considerations is

$$k > 0, \tag{3.10}$$

which is necessary to make $f(\ )$ an increasing rather than a decreasing function.

The other constants c', c'', etc., are different for different systems. Since entropy is unaffected by instantaneous mechanical operations (jumps in the Hamiltonian) these constants depend only on quantities that are unaffected by these jumps, that is, on the number and nature of the particles in the system. We can study the dependence of the constant c' on the number of particles by applying the composition law for a composite system consisting of two subsystems $\mathfrak{S}'$ and $\mathfrak{S}''$ both made from the same type of particle. For classical and Boltzmann statistics, the composition law when all the molecules in $\mathfrak{S}'$ and $\mathfrak{S}''$ are of the same type is (2.13), which when substituted into (3.2) gives

$$f(\Omega'\Omega''N!/N'!N''!) = f'(\Omega') + f''(\Omega''). \tag{3.11}$$

The solution of the resulting functional equation analogous to (3.4) is still given by (3.6), (3.7), and (3.8), but (3.9) is now replaced by

$$cN! = c'c''N'!N''!. \tag{3.12}$$

Since $N = N' + N''$, this condition can be satisfied for all values of N' and N'' by making $\ln(c'N'!)$, $\ln(c''N''!)$, and $\ln(cN!)$ proportional to N', N'', and N respectively. Calling the constant of proportionality $\ln a$, we thus have

$$c'N'! = a^{N'}. \tag{3.13}$$

Substituting into (3.6), dropping the primes, and then substituting the resulting formula into (1.11), we obtain the formula

$$S(A) = k \ln \{a^N\Omega(A)/N!\} \tag{3.14}$$

for the entropy of a system of N identical particles in classical or Boltzmann statistics. The constant a depends only on the particle type; before discussing its value we shall obtain the analogue of (3.14) for Bose and Fermi statistics.

For Bose or Fermi statistics a very similar discussion applies. The expression for Boltzmann entropy now has the form $f\{\Omega_\pm(A)\}$, and the same argument as before shows that the function $f(\ )$ again has the logarithmic form (3.8). The dependence of the constant c on N is simpler than before, because no factorials disfigure the composition law (2.27) applying to two systems of identical bosons or fermions. Consequently the factorials disappear from (3.11), (3.12), and (3.13) too, and the resulting entropy expression is not (3.14) but

$$S(A) = k \ln \{a^N\Omega_\pm(A)\}. \tag{3.15}$$

Once again a is an arbitrary constant, depending only on the particle type.

The constant denoted by a in (3.14) and (3.15) reflects the arbitrariness that remains in the choice of a zero for the measurement of entropy when the additivity condition has been allowed for. By convention this constant is always given the value 1 in Bose and Fermi statistics, and its value for classical or Boltzmann statistics is then determined by the correspondence principle† – that is by making the value of the entropy asymptotically the same whatever type of statistics is used, in the limit (high temperatures and low densities) where the observable properties of the system (such as its equation of state) become independent of the type of statistics used. We shall not go into this calculation here, merely quoting the result, which is

$$a = \left\{ \begin{array}{l} 1 \text{ (Boltzmann, Bose, or Fermi statistics)}, \\ h^{-3} \text{ (classical statistics)}, \end{array} \right\} \tag{3.16}$$

for a three-dimensional system of particles. The formula for entropy that emerges when (3.16) is substituted into (3.14) and (3.15) may be written

$$S(A) = k \ln W(A), \tag{3.17}$$

where $W(A)$ denotes the *weight* of the observational state A, which we define by

$$W(A) \equiv \left\{ \begin{array}{ll} \Omega(A)/h^{3N}N! & \text{for classical statistics}, \\ \Omega(A)/N! & \text{for Boltzmann statistics}, \\ \Omega_+(A) & \text{for Bose statistics}, \\ \Omega_-(A) & \text{for Fermi statistics}, \end{array} \right\} \tag{3.18}$$

for a system of N identical particles. Combining (3.16) with the formula $\pi_A = \Omega(A)/\Omega(Z)$ [eqn. (5.14) of Chap. IV], we see that the weights of the various states in an ergodic set are proportional to their equilibrium probabilities, and hence that $S(A) = \text{const.} + k \ln \pi_A$ within an ergodic set. The connection between entropy and probability was discovered by Boltzmann, although he left it to Planck‡ to express this connection in the elegant general form $S = k \ln W$.

It is interesting to compare the definitions of weight that go with the various types of statistics. From the first two lines of (3.18) we see (remembering that $\Omega(A)$ means a phase-space volume in the first line and a dimension number in the second) that each quantum state carries the same weight as a volume h^{3N} in phase space. The physical interpretation of this fact is that because of the uncertainty principle it is not possible to localize the

† From the purist's point of view this procedure has the disadvantage of bringing Planck's constant into the definition of entropy for a classical system, which gives the false impression that classical statistical mechanics somehow rests on a quantum-mechanical foundation. This disadvantage is outweighed, however, by the practical utility of the correspondence principle.

‡ M. Planck, *Einführung in die Theorie der Wärme* (Hirzel, Leipzig, 1930), pp. 188–9. Boltzmann's result, which applies only to classical gases, is to be found on pp. 55 and 73–5 of *Lectures on Gas Theory*. The formula $S = k \ln (c\Omega)$ with c arbitrary was given by A. Einstein, Theorie der Opaleszenz von homogenen Flüssigkeiten und Flüssigkeitsgemischen in der Nähe des kritischen Zustandes, *Ann. der Phys.* **33**, 1275 (1910).

phase point in a region of phase space of very small volume, and the least possible region of localization has a volume of about

$$\prod_{i=1}^{3N} (\Delta p_i \Delta q_i) \approx h^{3N},$$

and thus corresponds to a single quantum state. Similarly, comparing the second line of (3.18) with the third or fourth, we see that one quantum state in Bose or Fermi statistics carries the same weight as $N!$ quantum states in Boltzmann statistics. The interpretation is that in Boltzmann statistics we count separately the $N!$ dynamical states obtained from a given unsymmetrical dynamical state by permuting the particles, but in Bose or Fermi statistics we count only the single state formed by adding the wave functions of these $N!$ states with appropriate signs; the Boltzmann $\Omega(A)$ is therefore likely to be about $N!$ times as large as $\Omega_+(A)$ or $\Omega_-(A)$.

The definition of entropy provided by (3.17) and (3.18) can be generalized to more complicated cases than a system of N identical particles. The fundamental generalization is to a system composed of a number of different types of particle, say N_1 of type 1, N_2 of type 2, and so on. All the arguments given in § 2 and the foregoing part of § 3 can be extended to cover this case, and the resulting entropy expressions again have the form (3.17), but with (3.18) generalized to

$$W(A) \equiv \begin{cases} \Omega(A)/\prod_i h^{3N_i} N_i! & \text{for classical statistics,} \\ \Omega(A)/\prod_i N_i! & \text{for Boltzmann statistics,} \\ \Omega_*(A) & \text{for Bose and Fermi statistics,} \end{cases} \tag{3.19}$$

where, as in eqn. (4.20) of Chap IV, $\Omega_*(A)$ denotes the dimension number of the part of $\omega(A)$ having the appropriate symmetry (which may be Bose symmetry for some particle types and Fermi for others) under permutations interchanging identical particles.

An interesting feature of the entropy expression implied by (3.19) is the *Gibbs paradox*. Consider a classical mixture consisting of N_1 particles of mass m_1 and N_2 particles of mass m_2, the particles being otherwise identical. If $m_1 \neq m_2$, the entropy of the system for an observational state A is $k \ln W(A)$ with $W(A)$ given by (3.19); if $m_1 = m_2$, however, then the two types of particle are identical, so that $W(A)$ is now given by (3.18), with $N \equiv N_1 + N_2$. Thus, the entropy of an observational state A for $m_1 = m_2$ exceeds the entropy of the corresponding state for $m_1 \neq m_2$ by an amount $k \ln(N!/N_1!N_2!)$. The paradox consists in this apparently unnatural discontinuous dependence of the entropy on the continuous parameter $m_1 - m_2$. Its resolution consists in the realization that the change from a small non-zero value of $m_1 - m_2$ to the exact value zero is by no means as innocent as it looks: although its dynamical influence is small, its influence on the observable properties of the system can be profound. For example, if $m_1 \neq m_2$ then it is possible to observe the concentration N_i/N of each component in the mixture, whereas if $m_1 = m_2$ this is not possible. Likewise, if $m_1 \neq m_2$ then it is possible, at least in principle, to separate the two components of the mixture using a semi-permeable membrane, whereas if $m_1 = m_2$ this is not possible. To allow for these distinctions, it is necessary to use different observational descriptions for the two cases $m_1 \neq m_2$ and $m_1 = m_2$; and since the definition of entropy depends on the choice of observational states,† the definitions of entropy for the two cases are different. Thus the discontinuity noted in the paradox is much more than an "unnatural" mathematical discontinuity in a single function of the continuous variable $m_1 - m_2$; rather it is the natural reflection of a discontinuity in the entire theoretical framework, which arises from the physical distinction between identical and non-identical particles.

If the particles are permanently bound to form molecules, then the approximation of treating each molecule as a rigid body can be used, and instead of (3.19) we obtain

$$W(A) \equiv \begin{cases} \Omega(A)/h^{fN}\sigma^N N! & \text{for classical statistics,} \\ \Omega(A)/\sigma^N N! & \text{for Boltzmann statistics,} \\ \Omega_*(A) & \text{for Bose and Fermi statistics,} \end{cases} \tag{3.20}$$

where N now stands for the number of molecules (assumed for simplicity to be of one type only), σ is the symmetry number and f the number of classical degrees of freedom of a single molecule, and $\Omega_*(A) \equiv tr(J_* J_A)$ with J_* the appropriate symmetrizing and antisymmetrizing operator, written down explicitly in eqn. (4.22) of Chap. IV. As in (3.18) the factors $h^{fN}\sigma^N N!$ and $\sigma^N N!$ come from the correspondence principle; we shall not give the calculation here. These factors can be interpreted physically just like the analogous factors in (3.18), discussed earlier. In quantum mechanics the rigid-molecule approximation can be improved upon by allowing for vibrations of the atoms relative to their rigid-molecule positions and also for electronic excitations; the resulting model is used very successfully to calculate the thermodynamic properties of real gases at low densities.

3.1. Exercises

1. For a one-dimensional harmonic oscillator with natural frequency ν and observational states defined by

$$J_A = \begin{cases} 1 \text{ if } (A-1)\Delta E < H \leqq A\Delta E \\ 0 \text{ if not} \end{cases} \quad (A = 1, 2, \dots)$$

† This fact is stressed by H. Grad, in The Many Faces of Entropy, *Comm. Pure Appl. Math.* **14** 323 (1961).,

with ΔE a positive constant, calculate $\Omega(A)$ (i) for classical mechanics, using (3.14) as the definition of entropy, and (ii) for quantum mechanics. Assuming that the correspondence principle applies to the limit $\Delta E/\hbar\nu \to \infty$, use it to evaluate a for this system, and compare with (3.16).

2. What properties of their Boltzmann entropies characterize the quasi-transient and quasi-equilibrium states defined in Chap. II, § 9? What behaviour of the Boltzmann entropy characterizes a quasi-irreversible transition?

3. In a classical microcanonical ensemble of replicas of $\mathfrak{S}$, show that the ensemble average of any dynamical variable $G'(\alpha')$, whose value depends only on the dynamical state α' of $\mathfrak{S}'$, is $\int G'(\alpha')\Omega''(E - H'(\alpha'))d\alpha'/\Omega(E)$ where $\Omega(E)$ is the volume of the phase-space region where $E - \Delta E < H(\alpha) \leqq E$ for system $\mathfrak{S}$, and $\Omega''(x)$ is the volume where $x - \Delta E < H''(\alpha) \leqq x$ for system $\mathfrak{S}''$. (See Chap. V, eqn. (1.15), for the definitions of H' and H'', and Chap. III, eqn. (3.26), for those of E and ΔE.)

4. Large systems and the connection with thermodynamics

Having arrived at an essentially unique definition of Boltzmann entropy,

$$S(A) = k \ln W(A), \qquad (4.1) = (3.17)$$

exactly satisfying the additivity condition (1.2), we can return to our study of the non-decrease condition (1.1). As indicated in § 1, we can only expect this condition to be perfectly satisfied for deterministic states, which can occur only in the limit of an infinite system, but we can also study the deviations from perfection that are to be expected for non-deterministic states and finite systems.

For deterministic states the shortest way to prove the non-decrease of Boltzmann entropy is to recall that its definition corresponds [see (3.6)] to the choice $k \ln(cx)$ for the function $f(x)$ used in our earlier, more flexible definition (1.11), and that this function is non-decreasing (since $k > 0$) and satisfies the additional condition (1.13). Alternatively we may proceed directly from (4.1), which shows that in the limit of a very large system the change in entropy between two successive instants of observation is

$$\begin{aligned}\lim_{N\to\infty} [S(B, N) - S(A, N)] &= k \lim_{N\to\infty} [W(B, N)/W(A, N)] \\ &= k \lim_{N\to\infty} [\Omega(B, N)/\Omega(A, N)] \\ &\geqq 0 \quad \text{if} \quad w_{BA} = 1, \qquad (4.2)\end{aligned}$$

by (3.18) and (1.8) with the dependence of $S(A)$, $W(A)$, and $\Omega(A)$ on N now shown explicitly. The result (4.2) holds for all types of statistics. It shows that if the initial state of a transition is deterministic then (with probability 1) that transition cannot decrease the entropy, and consequently that if all the states are deterministic, then the entropy is a non-decreasing function of time in conformity with the non-decrease condition (1.1).

In the alternative case where the states are not deterministic, the situation is more complicated. In particular, for persistent states, we can show that there is in general a finite probability for the Boltzmann entropy to decrease. Let A and B be any two persistent states in the same ergodic set; then it follows from eqn. (5.14) of Chap. IV, or one of its analogues, that

$$\begin{aligned} S(B) - S(A) &= k \ln [\Omega(B)/\Omega(A)] \\ &= \ln (\pi_B/\pi_A). \end{aligned} \tag{4.3}$$

In general the equilibrium probabilities π_A, π_B, ... are not all equal and so it is possible to choose A and B to make the right side negative. Since A and B are in the same ergodic set, the probability of a sequence of transitions taking the system from A to B is finite and so the probability for a decrease in entropy is also finite. A neat estimate of this probability can be obtained by combining (4.3) with the inequality (9.32) of Chap. II; after interchanging A and B in the latter we obtain

$$w_{BA}(t) \leqq \exp(-\Delta S/k), \tag{4.4}$$

where $\Delta S \equiv S(A) - S(B)$ is the decrease in entropy. This estimate shows that a transition decreasing the entropy by more than a few k has a very low probability. Even so, the fact remains that the probability for a decrease of Boltzmann entropy is not strictly zero, and we must conclude that Boltzmann's definition of entropy is not, in general, strictly compatible with the non-decrease condition (1.1).

One of the main objectives of our investigation of the concept of entropy is to find the quantity in statistical mechanics whose properties most closely approximate those of the entropy used in thermodynamics. The result of the preceding paragraph shows that Boltzmann's definition (4.1) does not completely achieve this objective, since $S(A)$ can decrease with time whereas the entropy used in thermodynamics does not. The connection between statistical mechanics and thermodynamics is therefore best made through a further development based on Boltzmann's definition, rather than through Boltzmann's definition itself. This development starts from the observation that the entropy decreases allowed by (4.4) are only a vanishingly small fraction of the total entropy: for a non-vanishing probability, ΔS must be finite, but S itself tends to infinity with the size of the system. In fact, the additive property (1.2) indicates that entropy is asymptotically proportional to the size of the system, so that the limit

$$s(A) \equiv \lim_{N\to\infty} \frac{S(A, N)}{N} \tag{4.5}$$

may be expected to exist. We shall call $s(A)$ the entropy per molecule or *molecular entropy*. If the limit (4.5) does exist, then (4.2) and (4.3) imply

$$s(B) \geqq s(A) \quad \text{if} \quad w_{BA} = 1 \tag{4.6}$$

and
$$s(B) = s(A) \quad \text{if} \quad A \text{ and } B \text{ are in the same ergodic set.} \tag{4.7}$$

It follows, both for deterministic and for persistent states, that the probability of a decrease in molecular entropy is zero.

The definition of molecular entropy requires that the limit on the right of (4.5) should exist. In one important case the existence of this limit can be rigorously demonstrated: it is the case of a system of particles with two-body interactions, where A is a persistent state belonging to an ergodic set Z of the form

$$E - \Delta E < H \leqq E \tag{4.8}$$

and the limit $N \to \infty$ is taken in such a way that the energies E and $E - \Delta E$, and also the volume V enclosed by the walls containing the system, are all proportional to N. This way of taking the limit $N \to \infty$ will be called the *thermodynamic limit*; it is like the bulk limit described in Chap. I, § 1, in which $N \to \infty$ and $N/V \to$ const., but more specific in that the ergodic sets and their limiting behaviour are also prescribed. The assumption that the ergodic sets are of the form (4.8) implies that the energy is the only observable constant of the motion; this in turn indicates that the system is a fluid rather than a solid, for in a solid one expects (see Chap. II, § 5) to find further constants of the motion describing the essentially permanent crystalline ordering of the particles. Accordingly, we call a system whose ergodic sets have the form (4.8) a *thermodynamic fluid*. It has been shown by Griffiths† that, with Z defined by (4.8), the limit

$$s(\varepsilon, v) \equiv \lim_{N\to\infty} \frac{k \ln W(Z, N)}{N} \tag{4.9}$$

exists for particles with two-body interactions. As indicated in the notation, s is a (continuous) function of the quantities

$$\varepsilon \equiv \lim_{N\to\infty} \frac{E}{N} \quad \text{and} \quad v \equiv \lim_{N\to\infty} \frac{V}{N}, \tag{4.10}$$

but it does not depend on the limiting value of $\Delta E/N$, provided this last limit is positive. Using (4.5), (4.1), and eqn. (5.14) of Chap. IV in the form $\pi_A = W(A)/W(Z)$, we find that any observational state A in the ergodic

† R. B. Griffiths, Microcanonical Ensemble in Quantum Statistical Mechanics, *J. Math. Phys.* **6**, 1447 (1965). Griffiths also deals with spin systems (having no translational degrees of freedom). His proofs are quantum-mechanical but his method can also be used for classical systems.

set Z defined by (4.8) has the molecular entropy

$$s(A) = \lim_{N\to\infty} \frac{k \ln [\pi_A W(Z, N)]}{N} = s(\varepsilon, v), \tag{4.11}$$

since π_A = const. This completes the proof that $s(A)$ exists for persistent states† of a thermodynamic fluid consisting of particles with two-body interactions.

The definition (4.5) of molecular entropy may be written in the form

$$S(A, N) \sim S_{\text{ther}}(A, N), \tag{4.12}$$

where

$$S_{\text{ther}}(A, N) \equiv Ns(A) \tag{4.13}$$

and the notation $f(N) \sim g(N)$ means, for any functions f and g, that $f(N)/g(N) \to 1$ as $N \to \infty$. The formula (4.12) means that $S_{\text{ther}}(A, N)$, which we call the *thermodynamic entropy*, is a good approximation to the Boltzmann entropy $S(A, N)$, in the sense that when N is large their difference is a small fraction of $S(A, N)$ itself. The thermodynamic entropy so defined is the closest analogue that statistical mechanics can provide for the entropy used in the science of thermodynamics. By virtue of (4.6) and (4.7), the thermodynamic entropy has zero probability of decrease for both the types of transition that our theory can treat rigorously, and it is also asymptotically additive, since (4.12) and the additivity of Boltzmann entropy [eqn. (3.1)] imply

$$S_{\text{ther}}(A'A'', N' + N'') \sim S'_{\text{ther}}(A', N') + S''_{\text{ther}}(A'', N'') \tag{4.14}$$

if the limit $N \to \infty$ is taken in such a way that the ratio of N' to N'' approaches a limiting value. The analogy between our "thermodynamic entropy" S_{ther} and the entropy used in thermodynamics itself is thus almost perfect in the limit of very large systems.

For a thermodynamic fluid the definition (4.13) may be combined with (4.11) to give

$$S_{\text{ther}}(A, N) = Ns(\varepsilon, v). \tag{4.15}$$

This formula shows that the thermodynamic entropy of any persistent observational state A depends only on the three variables N, ε, v, whose values taken together specify the *thermodynamic state* of the fluid. Taken together, the formula (4.15) and the definition (4.9) of $s(\varepsilon, v)$ are of fundamental importance in the applications of statistical mechanics since, (when combined with the definition (4.8) of Z and the relevant definition of $W(Z)$ from § 3, the two formulae provide the basic method for calculating the thermodynamic entropy of a real physical system from the dynamics of its

† Note, however, that if A is transient then $s(A)$ is in general smaller than $s(\varepsilon, v)$. For a discussion of the violations of the non-decrease condition that can arise if we adopt $s(A) = s(\varepsilon, v)$ as a definition of entropy for transient states, see A. B. Pippard's *Classical Thermodynamics* (Cambridge University Press, 1957), p. 98.

constituent molecules.† Once the entropy has been calculated, the other thermodynamic functions, such as the free energy, specific heats, and so on, follow by straightforward application of the methods of thermodynamics.

To illustrate the use of (4.9), we calculate some thermodynamic properties of the classical *ideal gas*, defined as a system of particles confined by hard walls to a region R in physical space, and interacting with each other either not at all or so weakly that their energy of interaction is negligible in comparison with their total kinetic energy. The Hamiltonian of this system is, by Chap. I, eqns. (2.17), (2.22), and (2.23),

$$H \equiv \begin{cases} \sum_{i=1}^{N} p_i^2/2m + H_{int} & \text{if} \quad q_i \in R \quad \text{for} \quad i = 1, \ldots, N \\ 0 & \text{if not} \end{cases} \tag{4.16}$$

where m is the mass of each particle, and H_{int} describes the interactions. Assuming the ergodic sets to have the form (4.8), their weights are given by (3.18) to be

$$W(Z, N) = [\Phi(E, N) - \Phi(E - \Delta E, N)]/h^{3N}N!, \tag{4.17}$$

where

$$\Phi(E, N) \equiv \int\limits_{H(\alpha) \leq E} dp_1 \ldots dp_N \, dq_1 \ldots dq_N \tag{4.18}$$

is the volume of phase space enclosed by the *energy surface* $H(\alpha) = E$. Each of the N integrations over a position vector in (4.18) contributes a factor equal to V, the volume of R. The momentum integrations together contribute a factor which, if H_{int} may be neglected, is equal to the volume of the region in $3N$-dimensional momentum space where $p_1^2 + p_2^2 + \cdots + p_{3N}^2 \leqq 2mE$. For $E < 0$ this region is empty; for $E > 0$ it is a $3N$-dimensional sphere of radius $(2mE)^{\frac{1}{2}}$ and its volume‡ is therefore $(2\pi mE)^{3N/2}/(3N/2)!$. The *factorial function* used here may be defined by

$$n! \equiv \int_0^\infty t^n e^{-t}\, dt \qquad \text{if} \quad n > -1$$

$$= \begin{cases} n(n-1)(n-2)\ldots 3.2.1 \text{ if } n \text{ is a positive integer} \\ n(n-1)(n-2)\ldots \frac{5}{2}\frac{3}{2}\frac{1}{2}\sqrt{\pi} \text{ if } n - \frac{1}{2} \text{ is an integer.} \end{cases} \tag{4.19}$$

† It is more usual to make the connection with thermodynamics through the formula

$$\lim\, [F(T, N, V)/N] = -kT \lim\, [N^{-1} \ln Z(T, N, V)],$$

where F is the Helmholtz free energy and $Z(T, N, V)$ is the *partition function* defined by

$$Z(T, N, V) \equiv \begin{cases} a^N \int \exp\,[-H(\alpha)/kT]d\alpha/N! \text{ (classical statistics),} \\ tr_{\pm}[\exp\,(-H/kT)] \text{ (Bose or Fermi statistics), etc.} \end{cases}$$

The two methods are, however, mathematically equivalent. A proof of this equivalence is given by Griffiths in the paper cited above.

‡ See, for example, D. M. Y. Somerville's *An Introduction to the Geometry of N Dimensions* (Dover, New York, 1958), pp. 135–6.

It is related to the gamma function by $n! = n\Gamma(n) = \Gamma(n+1)$. For large n it may be approximated by *Stirling's formula*†

$$n! = \sqrt{2\pi n}\,(n/e)^n\,[1 + 1/(12n) + \ldots]. \tag{4.20}$$

Combining these various contributions to the integral (4.18) we obtain

$$\Phi(E, N) = V^N(2\pi mE)^{3N/2}/(3N/2)! \quad \text{if} \quad E \geqq 0. \tag{4.21}$$

Substituting this into (4.17) and then (4.9) we obtain

$$s(\varepsilon, v) = \lim_{N\to\infty} \frac{k}{N} \ln\left\{\frac{V^N(2\pi mE)^{3N/2}}{h^{3N}N!(3N/2)!}\left[1 - \left(\frac{E - \Delta E}{E}\right)^{3N/2}\right]\right\}. \tag{4.22}$$

By virtue of the limiting relations (4.10), we may replace V in (4.22) by Nv, E by $N\varepsilon$, and ΔE by $N\delta$ where $0 < \delta < \varepsilon$. Applying Stirling's approximation (4.20) to the factorials in the resulting formula, we obtain

$$s(\varepsilon, v) = k \ln\{v(4\pi m\varepsilon/3)^{3/2} e^{5/2} h^{-3}\} \tag{4.23}$$

provided $\varepsilon \geqq 0$. Equation (4.23) is called the *Sackur–Tetrode* formula for an ideal gas. It can be generalized to any gas whose molecules do not interact, but we shall not do the calculation here, since no essentially new ideas are required.

As an application of (4.23) let us calculate the *adiabatic law* for an ideal gas, that is the law giving the change of energy produced by a reversible and purely mechanical change of volume. Since the process is purely mechanical (no heat entering or leaving) the entropy cannot decrease, and since the process is reversible the entropy cannot increase either (otherwise the reverse process would decrease the entropy). Entropy is therefore constant in time, so that, by (4.23), the adiabatic law is

$$v\varepsilon^{3/2} = \text{const.} \tag{4.24}$$

Using (4.10) and the fact that the process considered does not change the value of N, we can write (4.24) in the alternative form

$$\lim_{N\to\infty} \frac{V_1 E_1^{3/2}}{V_2 E_2^{3/2}} = 1, \tag{4.25}$$

where V_1 and E_1 are the values of V and E at the start of the process, and V_2 and E_2 those at the finish. A purely thermodynamic treatment, ignoring the possibility of observable fluctuations, would have given the simpler

† See R. Courant and D. Hilbert, *Methods of Mathematical Physics* (Interscience, New York, 1953), vol. I, Chap. VII, § 6.1, or E. T. Copson, *Theory of Functions of a Complex Variable* (Oxford University Press, London, 1935), § 9.53.

formulation $V_1E_1^{3/2} = V_2E_2^{3/2}$ for the adiabatic law; but in statistical mechanics the more complicated formulation (4.25) is essential because it is only in the limit $N \to \infty$ that the neglect of fluctuations is strictly justified.

As another simple application of the Sackur–Tetrode formula we may calculate the equation of state of the ideal gas. The pressure p of a thermodynamic fluid may be defined by the condition that the work done by the system in an adiabatic volume change is $p\,dv$ per particle, so that

$$p = -\left(\frac{\partial \varepsilon}{\partial v}\right)_s. \tag{4.26}$$

The absolute temperature, according to (1.6), is

$$T = \left(\frac{\partial \varepsilon}{\partial s}\right)_v. \tag{4.27}$$

These two relations can be combined into the single differential relation

$$d\varepsilon = T\,ds - p\,dv, \tag{4.28}$$

which is the starting point of the derivations of many thermodynamic identities. Equations (4.26) to (4.28) are valid for any thermodynamic fluid. To obtain the equation of state of an ideal gas, we substitute from (4.23) into (4.28), obtaining

$$T = \left(\frac{\partial s}{\partial \varepsilon}\right)_v^{-1} = \frac{2\varepsilon}{3k}, \tag{4.29}$$

$$p = \left(\frac{\partial s}{\partial v}\right)_\varepsilon \Big/ \left(\frac{\partial s}{\partial \varepsilon}\right)_v = \frac{2\varepsilon}{3v}. \tag{4.30}$$

Elimination of ε yields the equation of state in the form

$$pv = kT. \tag{4.31}$$

Comparison with the ideal gas law obtained from experiment shows why we chose the reciprocal function in the definition (1.6) of temperature, and also shows that the value of k is

$$k = R/N_0 = 1{\cdot}38 \times 10^{-16}\ \text{erg/deg C}, \tag{4.32}$$

where R is the ideal gas constant per mole and N_0 is Avogadro's number.

4.1. Exercises

1. Derive the analogue of (4.23), and from it the adiabatic relation between v and ε, for a gas of non-interacting dumb-bell molecules, each with mass m and moment of inertia I.

2. In exercise 3 of Chap. V, § 3.1 show that if $\mathfrak{S}'$ is an ideal gas and the thermodynamic limit is taken for $\mathfrak{S}''$ while the structure of $\mathfrak{S}'$ remains unaltered, then

$$\lim_{N''\to\infty} \langle G'(\alpha')\rangle = C \int G'(\alpha')\, e^{-H'(\alpha')/kT''}\, d\alpha',$$

where C is a normalizing constant. Obtain a simple formula for C. This result shows that a small system in equilibrium with a large ideal gas has a canonical phase-space density. (It is actually true even if $\mathfrak{S}''$ is not an ideal gas.)

3. By applying the result of the preceding exercise in the special case where $\mathfrak{S}'$ consists of a single gas molecule in the same container as $\mathfrak{S}''$ find the average distribution of velocities in a classical ideal gas in equilibrium (*Maxwell's velocity distribution law*).†

5. Equilibrium fluctuations

In the previous section we saw how the Boltzmann entropy satisfies the non-decrease condition in the limit of an infinite system, provided the observational states are chosen in such a way that they are deterministic. One of the consequences of such a choice of observational states is that (assuming there are no periodic ergodic sets) Boltzmann's concept of equilibrium applies perfectly: once the system reaches an equilibrium state it will stay there for as long as the system remains isolated. Molecular motion continues even after the equilibrium state has been reached, but the chosen observational description is too coarse to reveal it.

The purpose of the present section is to apply Boltzmann's definition of entropy to the alternative situation, where the observational description is fine enough to reveal some aspects of the thermal motion. The change from the previous observational description to this new one may be likened to switching from low magnification to high magnification when using a microscope (remembering that even the high magnification is not powerful enough to show the molecular motions in full detail). Each state in the old observational description corresponds to a set of states in the new one, and in particular an equilibrium state in the old corresponds to an ergodic set in the new. According to the new description, a system that is isolated for a long time eventually reaches a persistent state, but after that its observational state, instead of coming to rest, fluctuates irregularly within the ergodic set for as long as the system remains isolated. Consequently the values of observables, in particular the entropy, also fluctuate irregularly.

The statistical properties of these fluctuations can be investigated using the weak ergodic theorem of Chap. II, eqn. (9.30). If X is any observable (for example, the Boltzmann entropy) and the initial state is persistent, lying in an ergodic set Z, then this theorem asserts that the time average of X in a long sequence of observations on an isolated system is likely

† This was Maxwell's own derivation: see *The Scientific Papers of James Clerk Maxwell*, ed. W. D. Niven (Dover reprint, New York), pp. 713–25.

to be close to the mean (or ensemble average)

$$\langle X \rangle = \sum_{A \in Z} X(A)\pi_A \tag{5.1}$$

calculated using the equilibrium probability distribution over Z. This mean value is also called the *equilibrium value* of X. Similarly, the root-mean-square fluctuation of X, defined as the square root of the time average of $[X - \langle X \rangle]^2$, is likely to be close to the equilibrium standard deviation of X, defined in accordance with eqn. (2.16) of Chap. II as

$$\sigma_X \equiv [\text{var}(X)]^{\frac{1}{2}} \equiv [\langle (X - \langle X \rangle)^2 \rangle]^{\frac{1}{2}} \tag{5.2}$$

calculated using the equilibrium probability distribution over Z.

To calculate these statistical properties of the fluctuations we use the basic formula (5.14) of Chap. IV asserting that the equilibrium probabilities of the observational states in an ergodic set Z are proportional to their weights. Using Boltzmann's entropy expression (3.17), this basic formula may be written

$$\pi_A \propto W(A) = \exp[S(A)/k]. \tag{5.3}$$

An alternative form is

$$\pi_A \propto \exp[-\Delta S(A)/k], \tag{5.4}$$

where

$$\Delta S(A) \equiv S(MP) - S(A) \tag{5.5}$$

is the amount by which the value of the entropy for the state A falls short of its value for the *most probable* state in Z, which we denote by MP. By "most probable" state we mean the state having the largest equilibrium probability. Equation (5.4) is *Einstein's fluctuation formula*.†

Unless all the observational states in Z have equal weights, the Boltzmann entropy $S(A)$ will fluctuate just like any other observable. If the system is isolated for a long time, then, by the ergodic theorem, the time average of $\Delta S(A)$ is likely to be close to

$$\begin{aligned}\langle \Delta S(A) \rangle &= \textstyle\sum_A \pi_A [S(MP) - S(A)] \\ &= k \textstyle\sum_A \pi_A \ln(\pi_{MP}/\pi_A).\end{aligned} \tag{5.6}$$

The expression (5.6) must lie in the range

$$0 \leqq \langle \Delta S(A) \rangle \leqq k \ln(M\pi_{MP}), \tag{5.7}$$

where M is the number of states in the ergodic set Z; the left inequality follows from the fact that $\pi_{MP}/\pi_A \geqq 1$ for all A, and the right from the fact that $\ln(1/x)$ is a convex function of x (using eqn. (7.6) of Chap. II with $\varphi(x) = k \ln(1/x)$, $x_A = \pi_{MP}/\pi_A$, $m_A = \pi_A$, $\bar{x} = M\pi_{MP}$). The estimate (5.7)

† A. Einstein, *Ann. der Phys.* **33,** 1275 (1910).

gives quantitative expression to a fact which is already qualitatively evident from (5.4)—that a fluctuation decreasing the entropy by more than a few k below its maximum value is an extremely rare occurrence.

To obtain more specific information about the fluctuations than is provided by the inequalities (5.7), we need more specific information about the values of the equilibrium probabilities. The usual method of proceeding, due to Einstein, is to make the assumption† that there exists an observable Y with the following properties:

(i) Y takes a different value $Y(A)$ in each observational state A of the ergodic set Z.

(ii) Over the range of values of Y that have appreciable probability, the entropy can be approximated by a quadratic function of Y,

$$S(A) = S_0 - \tfrac{1}{2}\alpha[Y(A) - Y_0]^2, \tag{5.8}$$

where S_0, Y_0 and α are constants, and $\alpha > 0$. We shall call S_0 and Y_0 the *modal* values of S and Y. The most probable value of $Y(A)$ is, by (5.3), the allowed value lying closest to Y_0.

(iii) The possible values of Y are equally spaced, and their spacing ΔY is small enough to justify replacing sums such as

$\sum_A \exp\left\{-\dfrac{\alpha}{2k}[Y(A) - Y_0]^2\right\}$ by integrals.‡ That is to say,

$$\alpha(\Delta Y)^2/k \text{ must be less than about } 1. \tag{5.9}$$

Under these assumptions the equilibrium mean and variance of Y, as defined in (5.1) and (5.2), are given approximately by

$$\begin{aligned}\langle Y\rangle &= \frac{\sum_A Y(A)\exp[-\Delta S(A)/k]}{\sum_A \exp[-\Delta S(A)/k]}\\ &= \frac{\int_{-\infty}^{\infty} Y\exp[-\alpha(Y - Y_0)^2/2k]\,dY}{\int_{-\infty}^{\infty}\exp[-\alpha(Y - Y_0)^2/2k]\,dY}\\ &= Y_0\end{aligned} \tag{5.10}$$

† More generally, Y might be a vector with, say, n components $Y_1, \ldots, Y_n$. The expression (5.8) would then be replaced by S_0 minus a quadratic form in $\Delta Y_1, \ldots, \Delta Y_n$.

‡ One estimate of the error in such a replacement is provided by applying Poisson's summation formula (see E. C. Titchmarsh, *Fourier Integrals* (Oxford University Press, 1937), §2.8) to the function $e^{-\frac{1}{2}t^2}$. This gives $\sum_{n=-\infty}^{\infty} e^{-\frac{1}{2}n^2x^2} = (2\pi)^{\frac{1}{2}}x^{-1}\{1+2e^{-2\pi^2/x^2}+\ldots\}$, so that if $x \leqq 1$ the error in replacing this sum by an integral is not more than about $2e^{-2\pi^2} < 10^{-8}$.

and

$$\operatorname{var}(Y) = \frac{\sum_A [Y(A) - \langle Y \rangle]^2 \exp[-\Delta S(A)/k]}{\sum_A \exp[-\Delta S(A)/k]}$$

$$= \frac{\int_{-\infty}^{\infty} (Y - Y_0)^2 \exp[-\alpha(Y - Y_0)^2/2k]\, dY}{\int_{-\infty}^{\infty} \exp[-\alpha(Y - Y_0)^2/2k]\, dY}$$

$$= k/\alpha = -k/(\partial^2 S/\partial Y^2), \tag{5.11}$$

where $\partial^2 S/\partial Y^2$ means the second derivative of the polynomial approximation (5.8) and is written as a partial derivative because the energy (together with any other observables that may be constant over the ergodic set) is held fixed during the differentiation.

Einstein's approximations also lead to a simple formula for the average amount by which $S(A)$ falls short of its maximum value. By (5.5), (5.8), and (5.11), this amount is

$$\langle \Delta S(A) \rangle = \langle S(MP) - S_0 + \tfrac{1}{2}\alpha[Y(A) - Y_0]^2 \rangle$$

$$= S(MP) - S_0 + \tfrac{1}{2}k. \tag{5.12}$$

If Y_0 is one of the allowed values of Y, then (5.8) gives $S(MP) = S_0$. If Y_0 is just halfway between two allowed values, then (5.8) gives $S(MP) = S_0 - \frac{1}{2}\alpha(\frac{1}{2}\Delta Y)^2$. These two cases are the extreme possibilities, and so the fractional error in neglecting the terms $S(MP) - S_0$ from the right-hand side of (5.12) is at most $\alpha(\Delta Y)^2/4k$. Thus we may approximate (5.12) by

$$\langle \Delta S(A) \rangle = \tfrac{1}{2}k \tag{5.13}$$

with an error which, by (5.9), cannot exceed about 25 per cent and is much smaller if $\alpha(\Delta Y)^2/k \ll 1$. This estimate of the average difference between the actual value of the Boltzmann entropy and its most probable value is evidently considerably sharper than our earlier estimate (5.7).

The simplest physical problem to which Einstein's result (5.11) may be applied is to calculate the variance of the fluctuations in the energy of one of two bodies in thermal contact. Here the system, which we denote as usual by $\mathfrak{S}$, consists of two spatially separated interacting subsystems $\mathfrak{S}'$ and $\mathfrak{S}''$; its Hamiltonian has a form discussed earlier [eqn. (1.15)]:

$$H = H' + H'' + H_{\text{int}}, \tag{5.14}$$

where H' and H'' are the Hamiltonians of the two subsystems considered separately, and H_{int} is the interaction. If H_{int} were zero, the subsystems would each be isolated, and their energies would each be constants of the motion; we assume, however, that H_{int} is large enough to transfer appreciable amounts of energy between the subsystems, so that only the total energy is a constant of the motion rather than the individual energies of the subsystems. At the same time we also assume that H_{int} is not large enough to contribute to the value of the energy measured at any particular observation, so that the observed energy E of the total system is related to the observed energies E' and E'' of the two subsystems by

$$E(A'A'') = E'(A') + E''(A''). \tag{5.15}$$

Since the total energy is conserved but not the energies of the subsystems, the ergodic sets of $\mathfrak{S}$ will be sets of observational states $A'A''$ having the property

$$E'(A') + E''(A'') = \text{const.} \tag{5.16}$$

but not $E'(A') = \text{const.}$

To apply Einstein's method, let us take $Y(A'A'') = E'(A')$ and extend the quadratic approximation (5.8) to the entropies of the individual subsystems. Although the total entropy $S'(A') + S''(A'')$ is a maximum for the most probable state, the individual entropies S' and S'' are not; the appropriate quadratic approximation for each therefore includes a term linear in $Y - Y_0$, and we may write

$$S'(A') \;\; = S_0' + \gamma'(Y - Y_0) - \tfrac{1}{2}\alpha'(Y - Y_0)^2, \tag{5.17}$$

$$S''(A'') = S_0'' + \gamma''(Y - Y_0) - \tfrac{1}{2}\alpha''(Y - Y_0)^2, \tag{5.18}$$

where γ', γ'', α', and α'' are constants. Applying the addition law $S(A'A'') = S'(A') + S''(A'')$ and comparing with (5.8), we obtain

$$S_0 = S_0' + S_0'', \tag{5.19}$$

$$0 = \gamma' + \gamma'', \tag{5.20}$$

$$\alpha = \alpha' + \alpha''. \tag{5.21}$$

The first of these relations merely reiterates the additivity of entropy, applied now to the modal values S_0, S'_0, S''_0. The second relates the derivatives of the approximations (5.17) and (5.18) when $Y = Y_0$ and can therefore be interpreted as a statement about temperatures. To formulate this statement, we extend the original thermodynamic definition of temperature [eqn. (1.6)] to the finite systems considered here by defining

$$\left.\begin{aligned} T' &\equiv (dS'/dE')^{-1} = (dS'/dY)^{-1}, \\ T'' &\equiv (dS''/dE'')^{-1} = -(dS''/dY)^{-1}. \end{aligned}\right\} \tag{5.22}$$

The differentiations must be applied to the approximate entropy expressions (5.17) and (5.18) rather than the exact Boltzmann entropies, because the latter are not continuous and therefore cannot be differentiated. Calculating T' and T'' from (5.17) and (5.18), we see that (5.20) asserts that the temperatures of $\mathfrak{S}'$ and $\mathfrak{S}''$ are equal when Y has its modal value.

The third relation in the group (5.19)–(5.21) can also be interpreted in terms of temperature, since it can be written

$$\begin{aligned} \alpha &= -[d^2S'/dY^2 + d^2S''/dY^2] \\ &= -[d(1/T')/dE' + d(1/T'')/dE''] \\ &= (1/C' + 1/C'')/T^2, \end{aligned} \tag{5.23}$$

where

$$C' \equiv dE'/dT', \quad C'' = dE''/dT'' \tag{5.24}$$

are the specific heats of $\mathfrak{S}'$ and $\mathfrak{S}''$ respectively. Substituting (5.23) into (5.11) we finally obtain, for the variance of the fluctuating part of the energy of either system,

$$\text{var}(E') = \text{var}(E'') = \text{var}(Y) = kT^2C'C''/(C' + C''). \tag{5.25}$$

If one of the subsystems, say $\mathfrak{S}''$, has a much larger specific heat than its companion (in which case the larger subsystem is called a *heat bath*) the right side reduces to kT^2C', giving the energy fluctuation formula that is found in most textbook treatments.

An important application of (5.25) is the estimation of the asymptotic behaviour of var(E') in the bulk limit, taken in such a way that the volumes, energies, and particle numbers of $\mathfrak{S}'$ and $\mathfrak{S}''$ are all asymptotically proportional to each other and hence also to N, the total number of particles in $\mathfrak{S}$. By virtue of the definitions (5.24), the specific heats C' and C'' are also extensive (proportional to N in this limit) and it follows by (5.25) that var (E') is extensive. The standard deviation of E' is thus (by (5.2)) asymptotically proportional to $N^{\frac{1}{2}}$ and therefore very much smaller than E' itself, which is proportional to N. Whether or not the energy fluctuations can be neglected depends on the convention that is adopted about what can be regarded as observable—that is, on the choice of observational states. The convention leading to the thermodynamic description of matter is, as indicated in § 4, to assume that the tolerance in any measurement of energy is proportional to the size of the system—in symbols, that $\Delta E \propto N$. With this convention, the energy fluctuations, being proportional only to $N^{\frac{1}{2}}$, are too small to observe when N is large. This neglect of fluctuations is characteristic of the thermodynamic description. An alternative convention, however, would be to assume that the tolerance in energy measurements is proportional to $N^{\frac{1}{2}}$ instead of to N. In this case the energy fluctuations remain observable as the bulk limit is taken. One consequence is that such an observational description cannot be deterministic, in contrast to the thermodynamic description which, in general, is deterministic. This contrast illustrates once again how the choice of observational states is at least as important an ingredient in the statistical mechanics of a given physical system as the choice of dynamical model.

5.1. Exercises

1. If Y is a vector with n components, instead of a scalar, show that (5.11) generalizes to

$$\text{covar}\,(Y_i, Y_j) = -k[\partial^2 S/\partial Y_i\, \partial Y_j]^{-1},$$

where $[\ ]^{-1}$ indicates the inverse of a matrix. Hence evaluate $\langle \Delta S \rangle$ for this case.

2. A box is divided into two compartments of equal volume, and N molecules of an ideal gas, having total energy E, are shared equally between the two compartments. Calculate the variance of the energy and of the number of particles in one of the compartments (i) if energy but not particles can pass from one compartment to the other, (ii) if both energy and particles can pass from one compartment to the other. Make a numerical estimate of the order of magnitude of these fluctuations. (Note that a similar problem is treated in exercise 1 of Chap. II § 2.1.)

6. Equilibrium fluctuations in a classical gas

As a further application of the theory described in the previous section, let us study the fluctuating behaviour of a classical ideal gas in equilibrium. Quite apart from its intrinsic interest, the study of these fluctuations is important because they lie at the root of the historic controversy raised by Boltzmann's work on the kinetic equation for a gas. This controversy will be considered in § 8.

The system we shall consider is, from the dynamical point of view, the same as the ideal gas, considered in § 4: it consists of N not too strongly interacting particles, and its Hamiltonian is given by (4.16). This time, however, we shall work with an observational description more detailed than the one used in § 4, where the energy was taken to be the only observable. To help define the observational states we introduce the 6-dimensional phase space of a single gas molecule; this space is often called *μ-space* (molecule space) to distinguish it from the $6N$-dimensional phase space of the entire gas, which is called *γ-space* (gas space). The basic step in defining the observational states for this system is to partition μ-space into cells of arbitrary shape and size; we denote these cells by $\upsilon_1, \upsilon_2, \ldots$. For example, one way of doing this is to divide the region R (of physical space) enclosed by the containing walls into a number of sub-regions $R_1, R_2, \ldots$, and to say that the dynamical state of a molecule is in the ith cell υ_i of μ-space if and only if the molecule itself is in the ith sub-region R_i of physical space.

For any choice of μ-space cells we define a family of dynamical variables (that is of functions over γ-space) called *occupation numbers*; the occupation number of the cell υ_i, denoted by $N_i(\alpha)$, is defined as the number of molecules whose individual dynamical states are in υ_i when the dynamical state of the entire system is α (a point in γ-space). In the example just considered, where the μ-space cells are defined in terms of sub-regions of the region R available to the system in physical space, $N_i(\alpha)$ is simply the number of molecules inside the sub-region R_i of physical space when the dynamical state of the entire system is α. In general, the occupation numbers are non-negative integers and they satisfy the sum rule

$$\sum_i N_i(\alpha) = N \quad \text{for all} \quad \alpha, \tag{6.1}$$

where the sum goes over the labels of all the μ-space cells.

Our choice for the observational states is equivalent to making the following two assumptions: (i) that all the occupation numbers are observables, and (ii) that all the observables are functions of the occupation numbers. In the example where the cells in μ-space correspond to regions in physical space, the occupation numbers could be measured approximately by measuring the average refractive index in each of the regions. To simplify the calculations, we shall make the (unrealistic) assumption that the occupation numbers can be measured exactly, so that to each observational state there corresponds exactly one set of values for the occupation numbers, and vice versa. We shall write $N_i(A)$ for the value taken by the occupation number of the ith cell when the system is in the observational state A. These observables, which we shall also call occupation numbers, are related to the dynamical variables $N_i(\alpha)$ by the condition

$$N_i(\alpha) = N_i(A) \quad \text{for all } i \quad \text{if and only if } \alpha \text{ is in } \omega(A), \tag{6.2}$$

where $\omega(A)$ denotes, as usual, the dynamical image (in γ-space) of the observational state A. To illustrate how other observables may be expressed in terms of the occupation numbers, let us consider the energy. This can be written

$$E = \sum_i E_i N_i(A), \tag{6.3}$$

where E_i is the observable energy of a single molecule when it is in the μ-space cell υ_i (we treat E_i as a constant even though the single-particle Hamiltonian is not exactly constant over the cell, since our choice of observational states implies that this deviation from exact constancy is too small to measure).

An alternative characterization of the same set of observational states, particularly useful when all the non-vanishing occupation numbers are large, is provided by *Boltzmann's distribution function*. The Boltzmann distribution function corresponding to an observational state A is the function over μ-space defined by

$$f(\xi, A) \equiv N_{i(\xi)}(A)/\Upsilon_{i(\xi)}, \tag{6.4}$$

where $i(\xi)$ is defined as the label of the cell containing the μ-space point ξ, and Υ_i is the μ-space volume of the cell υ_i. This function is thus a constant over each μ-space cell, but it is in general discontinuous at the cell boundaries. Sums over cells involving the occupation numbers can be written as integrals over all μ-space involving the Boltzmann distribution function: for example, the sum rule (6.1) can be written as a normalization condition,

$$N = \int d\xi\, f(\xi, A), \tag{6.5}$$

and the expression (6.3) for energy can be written

$$E = \int d\xi\, f(\xi, A)\, E_{i(\xi)}. \tag{6.6}$$

Since the single-particle Hamiltonian $H_{(1)}(\xi)$ differs from $E_{i(\xi)}$ by an amount too small to measure, we may replace (6.6) by

$$E = \int d\xi\, f(\xi, A)\, H_{(1)}(\xi) \tag{6.7}$$

with no observable loss of accuracy.

In order to apply the theory developed in this book it is necessary to assume that the Markovian postulate applies to the observational states defined here. Since a fairly large number of very accurate simultaneous measurements would be necessary to determine the observational state of a system, this assumption would be extremely difficult to test experimentally. A theoretical test, based on the consistency condition (7.2) of Chap. IV, seems possible in principle but, as indicated at the end of Chap. IV, § 7, the mathematical techniques that would permit such a test have not yet been developed. Whatever justification the assumption has must therefore rest mainly on physical intuition. One necessary condition for its (approximate) validity is that the term H_{int} in the Hamiltonian (4.16) must be very small compared with the total kinetic energy; for if not, eqn. (6.3) would break down and, with the energy no longer an observable, the validity of the observational description in terms of occupation numbers would be very dubious. For any given type of molecule, we can satisfy this condition by working at sufficiently low densities, so that at each instant only a small proportion of the particles are close enough to other particles to interact appreciably with them. Another way of expressing the same requirement is that each particle should move freely most of the time, spending only a small fraction of the time in collision with other particles. A second useful, though perhaps not absolutely necessary, condition is that the time between successive instants of observation be long compared with the duration of a collision; this permits us to treat collisions as impulsive changes of velocity. When these two conditions are satisfied, it is physically reasonable to treat the collisions suffered by each particle between two successive instants of observation as independent random events, whose probabilities depend on the velocity of the particle in question and the distribution of velocities in its neighbourhood, but not directly on the previous history of the particle or its neighbours. Hence, provided the observational description is based on a network of μ-space cells sufficiently fine to give a good estimate of the velocity distribution in all parts of the gas, one would expect the probability of a transition (brought about by collisions) from one observational state to another to depend only on the initial observational state and not on the previous history of the system, so that the Markovian postulate would, indeed, be valid.

The entropies of the observational states defined by (6.2) are easily calculated. For any given observational state A, there are $N!/\Pi_i N_i(A)!$ distinct ways of sharing the particles among the various cells in μ-space, and each of these ways corresponds to a region in γ-space whose volume is $\Pi_i \Upsilon_i^{N_i(A)}$, where Υ_i is the volume of the μ-space cell υ_i. Consequently the total phase-space volume of the dynamical image of A is

$$\Omega(A) = N!\Pi_i[\Upsilon_i^{N_i(A)}/N_i(A)!] \tag{6.8}$$

and hence, by (3.17) and (3.18), the corresponding Boltzmann entropy is

$$S(A) = k\sum_i \ln [\Upsilon_i^{N_i(A)}/N_i(A)!] - Nk \ln h^3. \tag{6.9}$$

Since (6.9) is a sum over cells it may be converted, just like (6.1) or (6.3), into an integral over μ-space involving the Boltzmann distribution function. To obtain this integral in simple form we first apply Stirling's approximation (4.20), in the form

$$\frac{\ln N!}{N} = \ln N - 1 + \frac{\ln(2\pi N)}{2N} + \frac{1}{12N^2} + \cdots, \tag{6.10}$$

to all the terms in the sum (6.9) for which $N_i(A) \neq 0$. Provided these non-vanishing occupation numbers are large enough to justify neglecting the terms which tend to zero for large N in (6.10), this approximation gives

$$\begin{aligned} S(A) &\cong k\textstyle\sum_i' N_i(A)\,[\ln \Upsilon_i - \ln N_i(A) + 1] - Nk \ln h^3 \\ &= -k \textstyle\int d\xi\, f(\xi, A) \ln f(\xi, A) + Nk \ln (e/h^3). \end{aligned} \tag{6.11}$$

The summation is restricted to cells with $N_i(A) \neq 0$, but the integral may be taken over all μ-space if we make the convention

$$f \ln f \equiv 0 \quad \text{when} \quad f = 0. \tag{6.12}$$

The formula (6.11) is due to Boltzmann.† Like the corresponding formula (6.7) for the energy, it can be valid only at low densities; for at higher densi-

† L. Boltzmann, *Lectures on Gas Theory*, pp. 55 and 73–75. The generalization of Boltzmann's entropy formula (6.11) to arbitrary densities depends on the distribution functions for pairs, triplets, etc., as well as for single particles. Relevant papers include J. E. Mayer, Ensembles of Maximum Entropy, *J. Chem. Phys.* **33**, 1484 (1960), and D. Robinson and D. Ruelle, Mean Entropy of States in Classical Statistical Mechanics, *Commun. Math. Phys.* **5**, 288 (1967).

ties the observational description on which it is based ceases to satisfy the Markovian postulate.

The sign of approximate equality, $\cong$, is used in (6.11) to indicate that this result is not a rigorous consequence of the postulates of our theory applied to the observational model, but involves the further approximation that the non-vanishing occupation numbers are all large. To justify (6.11) rigorously would require a limit process in which all the occupation numbers and all the μ-space cell volumes Υ_i tend to infinity in proportion to one another. Once again the appropriate limit process is the bulk limit. We shall not, however, enter into these technicalities here, since even if they were carried out the basic mathematical problem of verifying that the chosen observational states are consistent with the Markov postulate would remain untouched.

To obtain an indication of the behaviour to be expected of a gas that has been isolated for a long time, let us calculate the occupation numbers of the most probable state (i.e. the state of largest equilibrium probability) in an ergodic set, assuming as usual that the energy is the only constant of the motion. This most probable state, MP, may be specified, in accordance with eqns. (5.14) of Chap. IV, (6.1), and (6.3), by the condition that

$$\Omega(A)/\Omega(MP) \leqq 1 \tag{6.13}$$

for all A satisfying

$$\textstyle\sum_i \delta N_i = 0 \quad \text{and} \quad \sum_i E_i\, \delta N_i = 0, \tag{6.14}$$

where

$$\delta N_i \equiv N_i(A) - N_i(MP). \tag{6.15}$$

Using (6.8) with the abbreviated notation N_i for $N_i(MP)$, we may write the left side of (6.13) in the form

$$\begin{aligned}\frac{\Omega(A)}{\Omega(MP)} &= \Pi_i^+ \frac{\Upsilon_i^{\delta N_i}}{(N_i+1)(N_i+2)\cdots(N_i+\delta N_i)} \\ &\quad\times \Pi_i^- \frac{(N_i - |\delta N_i| + 1)(N_i - |\delta N_i| + 2)\cdots N_i}{\Upsilon_i^{|\delta N_i|}},\end{aligned} \tag{6.16}$$

where the first product ranges over those values of i for which $\delta N_i > 0$ and the second over those for which $\delta N_i < 0$. Now, the conditions (6.14) imply that the identity

$$1 = \Pi_i^+ e^{-(\alpha+\beta E_i)\delta N_i} \Pi_i^- e^{(\alpha+\beta E_i)|\delta N_i|}$$

holds for all values of α and β. Multiplying together corresponding sides of (6.16) and this identity, we obtain

$$\begin{aligned}\frac{\Omega(A)}{\Omega(MP)} &= \Pi_i^+ \left(\frac{\Upsilon_i e^{-\alpha-\beta E_i}}{N_i+1}\right)^{\delta N_i} \left(1 + \frac{1}{N_i+1}\right)^{-1} \left(1 + \frac{2}{N_i+1}\right)^{-1} \cdots \\ &\cdots \left(1 + \frac{\delta N_i - 1}{N_i+1}\right) \Pi_i^- \left(\frac{N_i}{\Upsilon_i e^{-\alpha-\beta E_i}}\right)^{|\delta N_i|} \left(1 - \frac{1}{N_i}\right)\left(1 - \frac{2}{N_i}\right)\cdots \\ &\cdots \left(1 - \frac{|\delta N_i| - 1}{N_i}\right).\end{aligned} \tag{6.17}$$

If we now choose α and β so that the set of occupation numbers defined by

$$N_i \leqq \Upsilon_i e^{-\alpha-\beta E_i} < N_i + 1 \quad \text{for all } i \tag{6.18}$$

satisfies the two sum rules (6.1) and (6.3) then for this set of occupation numbers every factor on the right-hand side of (6.17) is manifestly less than or equal to 1. Consequently the set (6.18) must be the actual occupation numbers $N_i(MP)$ of the most probable state.

The formula (6.18) can be written in terms of Boltzmann's distribution function, defined in (6.4). Just as in the case of the entropy, the resulting expression is considerably simplified by the approximation that all the non-vanishing occupation numbers are large; under this approximation it becomes

$$f(\xi, MP) \cong \exp\left[-\alpha - \beta E_{i(\xi)}\right]. \tag{6.19}$$

Using also the approximation that led to (6.7), this can be written

$$f(\xi, MP) \cong \exp\left[-\alpha - \beta H_{(1)}(\xi)\right]. \tag{6.20}$$

This is called the *Maxwell–Boltzmann distribution.* In the case of an ideal gas with no external field of force, (6.20) reduces to

$$f(\xi, MP) \cong \exp\left[-\alpha - \tfrac{1}{2}\beta m v^2\right], \tag{6.21}$$

where $\boldsymbol{v}$ is the velocity vector of a particle whose dynamical state is ξ. The constants α and β in (6.21) can be evaluated by applying the conditions (6.5) and (6.7):

$$N \cong \textstyle\sum_i \Upsilon_i e^{-\alpha-\frac{1}{2}\beta m v_i^2} \cong \int d^3\boldsymbol{x} \int d^3\boldsymbol{v}\, e^{-\alpha-\frac{1}{2}\beta m v^2} \cong V e^{-\alpha}(2\pi/m\beta)^{3/2}, \tag{6.22}$$

$$E \cong \textstyle\sum_i(\frac{1}{2}m v_i^2)\, \Upsilon_i e^{-\alpha-\frac{1}{2}\beta m v_i^2} \cong -\left(\dfrac{\partial N}{\partial \beta}\right)_\alpha \cong \dfrac{3N}{2\beta} \tag{6.23}$$

where x is a position co-ordinate and V is the volume of the gas. Combining (6.23) with (4.29), we obtain

$$\beta \cong \frac{3N}{2E} \cong 1/kT, \tag{6.24}$$

where T is the absolute temperature and k Boltzmann's constant. Using (6.22) and (6.24) we can write (6.21) in the more easily interpreted form

$$f(\xi, MP) \cong \frac{N}{V}\left(\frac{m}{2\pi kT}\right)^{3/2} e^{-mv^2/2kT}. \tag{6.25}$$

This is *Maxwell's velocity distribution law*; it occupies a central position in the kinetic theory of gases.

Besides enabling us to identify the most probable state, (6.17) can also furnish us with information about the equilibrium fluctuations of the occupation numbers in the vicinity of their most probable values. One way of studying these fluctuations is to use the theory based on Einstein's quadratic approximation, described in the preceding section, but the following method is more direct. We rewrite (6.17) in the form

$$\frac{\Omega(A)}{\Omega(MP)} = \Pi_i \frac{\varphi_i[N_i(A)]}{\varphi_i[N_i(MP)]}, \tag{6.26}$$

where $\varphi_i(n) \equiv (\Upsilon_i e^{-\alpha-\beta E_i})^n/n!$ for $n = 0, 1, 2, \ldots.$ (6.27)

We now make the approximation of relaxing the constraints of fixed total energy and particle number [(6.1) and (6.3)], using them only to determine the values of α and β through the condition that the most probable set of occupation numbers, given by (6.18), must satisfy these constraints. The effect of this approximation† is that instead of studying one ergodic set (corresponding to fixed values of E and N) on its own, we mix it with ergodic sets having nearby values of E and N so that the distributions we shall study describe average properties of a class of ergodic sets rather than the individual properties of a single ergodic set.

Applying the principle that equilibrium probabilities are proportional to phase-space volumes [eqn. (5.14) of Chap. IV], we see from (6.26) that our approximation makes the occupation numbers of the various μ-space cells statistically independent at equilibrium, with probability distributions given by

$$Pr(N_i(A) = n) = \varphi_i(n)/\textstyle\sum_r \varphi_i(r) \quad \text{for} \quad n = 0, 1, 2, \ldots. \tag{6.28}$$

From this probability distribution it follows, by (6.27) and the definitions given in Chap. II, § 2, that

$$\langle N_i(A)\rangle = \Upsilon_i e^{-\alpha-\beta E_i}, \tag{6.29}$$

$$\text{var}\,[N_i(A)] = \langle N_i(A)\rangle, \tag{6.30}$$

$$\text{covar}\,[N_i(A), N_j(A)] = 0 \quad \text{if} \quad i \neq j. \tag{6.31}$$

The first of these formulae shows that the difference between the mean and most probable values for $N_i(A)$, the latter being given by (6.18), is at most 1. The second shows that the standard deviation of $N_i(A)$ is $\sqrt{\langle N_i(A)\rangle}$, and is therefore much less than $\langle N_i(A)\rangle$ itself if the μ-space cell is large enough to contain, on the average, many particles.

Once again, as we remarked at the end of § 5, the physical interpretation to be put on results such as (6.30) depends on the choice of observational states. If, as we assumed in the calculations of this section, the occupation

† The relevant equilibrium ensemble is called the *grand canonical ensemble*.

numbers can be measured precisely, then (6.30) implies that their fluctuations are amply large enough to observe, so that the observational behaviour of the system is non-deterministic, and the Boltzmann concept of equilibrium does not apply. On the other hand, if we assume, more realistically, that the tolerance ΔN_i in occupation-number measurements is much greater than $\sqrt{\langle N_i \rangle}$ (e.g. if we take the thermodynamic limit in such a way that N_i/N and $\Delta N_i/N$ tend to finite limits), then (6.30) implies that the fluctuations are too small to observe; in this case the most probable observational state (which corresponds here to a set of tolerance intervals, rather than precise values, for the occupation numbers) has equilibrium probability very close to 1. This state is often called the *overwhelmingly most probable* state; in the language of exercise 1 of Chap. II, § 9.1, it is a *quasi-equilibrium* state.

The result (6.29) can also be written in terms of Boltzmann's distribution function, defined in (6.4); it then takes the form

$$\langle f(\xi, A)\rangle = \exp[-\alpha - \beta E_{i(\xi)}], \tag{6.32}$$

another version of the Maxwell–Boltzmann distribution law. This is a somewhat more accurate statement than the version (6.20) given earlier in this section, since it does not involve the approximation that all occupation numbers are large; on the other hand, the derivation of (6.32) does depend on the approximation of relaxing the constraints of constant energy and particle number. (The most satisfactory derivation of the Maxwell–Boltzmann law is the one indicated for exercise 3 of Chap. V, § 4.1.)

6.1. Exercises

1. Show that the approximate entropy expression (6.11) is maximized, subject to constraints of constant energy and particle number, by the Maxwell–Boltzmann distribution.

2. Show that, in the approximation where all non-vanishing occupation numbers are treated as large, the quadratic approximation for the entropy expression implied by (6.17) is†

$$S(A) \cong S(MP) - k\, \Sigma_i (\delta N_i)^2/2N_i.$$

Use this to estimate the means, variances, and covariances of the occupation numbers, and compare with the expressions (6.29), (6.30), and (6.31).

7. The kinetic equation for a classical gas

The system of observational states defined in terms of occupation numbers, which we used in the last section to discuss equilibrium fluctuations in an ideal classical gas, can also be used to discuss the non-equilibrium behaviour of such a gas. In this and the next section, therefore, we consider an ideal classical gas whose initial observational state has a very small

† See H. Grad, *Comm. Pure Appl. Math.* **14**, 323 (1961), § 3.

equilibrium probability, so that its occupation numbers are initially very different from the Maxwell–Boltzmann distribution (6.20), and we shall study the way these occupation numbers approach the Maxwell–Boltzmann form as time proceeds.

Unlike the equilibrium theory, non-equilibrium theory cannot neglect the interactions between particles, even as a first approximation at very low densities, since it is these very interactions that change the distribution of velocities, enabling it to approach the Maxwellian form (6.25). To simplify the treatment of these interactions, we shall make explicit use of an assumption which, according to the discussion following (6.7), is already implicit in the Markovian postulate for our chosen observational description. This assumption is that the density is so low that each molecule moves independently of the others for most of the time, interrupted only occasionally by a collision with some other molecule. Multiple collisions (involving three or more molecules simultaneously) are assumed to be so rare that they may be ignored. To take advantage of this assumption of rare collisions, we shall suppose the time between successive instants of observation to be not only (as in the discussion following (6.7)) much greater than the duration of a collision, but also much less than the (mean) free time between the successive collisions experienced by any one molecule.† These assumptions imply that during the time interval between any two successive instants of observation most of the molecules move freely (i.e. without any collision), a small but not negligible fraction are free at the beginning and end of the interval but make just one complete binary collision during the interval, and a negligible fraction make more than one binary collision, or make a multiple collision, or are in collision at the beginning or end of the interval. This may be called the *binary collision approximation.*

In order to focus attention on the effect of collisions rather than the free motion between collisions, it is convenient to use μ-space cells chosen so that a particle can get from one μ-space cell to another only by making a collision. In the absence of external forces this is easily done by choosing a network of cells in the *momentum space*‡ of a single particle and defining the μ-space cells by

$$\xi \equiv (\boldsymbol{p}, \boldsymbol{q}) \text{ is in } \upsilon_i \text{ if and only if } \boldsymbol{p} \text{ is in the } i\text{th momentum-space cell.} \quad (7.1)$$

This choice implies that the velocity distribution of the gas molecules can be

† For air at standard temperature and pressure the duration of a collision is roughly 10^{-12} sec and the mean free time is roughly 10^{-9} sec. The geometric mean of these two quantities, about 3×10^{-11} sec, would be a suitable choice for the interval between successive instants of observation. The fact that observations cannot, in practice, be repeated so rapidly does not invalidate our treatment, since there is no obligation actually to make an observation on every occasion when the theory permits it.

‡ It is usual to use velocity space in this connection rather than momentum space, but there is no essential distinction since $\boldsymbol{p} = m\boldsymbol{v}$ for point particles.

observed but not their spatial distribution; such a situation might arise, for example, if the velocity distribution were deduced from the Doppler broadening of a spectral line emitted or scattered by the molecules, the width of the light beams employed being at least as large as the linear dimensions of the specimen of gas under observation.

The choice (7.1) of μ-space cells ensures, by Newton's second law of motion, that a molecule will not change its μ-space cell unless acted upon by a force. This force may be exerted either by other particles in the same system during a collision, or else by other particles outside the system, for example, when the particle in question collides with the wall of the container. Normally the container is much larger than the mean free path, and any specified particle hits the wall much less often than it hits other particles. For the time being, therefore, we shall make the approximation of neglecting all external forces, including collisions with the wall.

Let us define an (ij, kl) collision to be a binary collision transferring one particle from the cell υ_k to the cell υ_i and one from υ_l to υ_j, and $b_{ij,kl}(\xi_r, \xi_s)$ to be a dynamical variable taking the value 1 if the rth and sth particles, having the dynamical states ξ_r and ξ_s at one instant of observation, make an (ij, kl) collision before the next instant of observation. More specifically, we define this dynamical variable to be

$$b_{ij,kl}(\xi_r, \xi_s) \equiv J_i(\xi_r')\, J_j(\xi_s')\, J_k(\xi_r)\, J_l(\xi_s), \tag{7.2}$$

where ξ_r' and ξ_s' are the dynamical states of the two particles at the second instant of observation, and $J_i(\xi)$ is the indicator function of the μ-space cell υ_i, taking the value 1 if ξ is in υ_i and 0 if not. According to the binary collision approximation the interaction of the two particles with the rest of the system can be ignored in calculating ξ_r' and ξ_s'; hence these two final states are unique functions of ξ_r and ξ_s, and may be computed by integrating the equations of motion for the two particles regarded as a self-contained dynamical system.

Using (7.2) and the fact that $\sum_i J_i(\xi) = 1$, we can calculate the contribution of collisions involving the rth and sth particles to the change in the occupation number of some particular cell, say υ_i, between the two successive instants of observation; this contribution is

$$J_i(\xi_r') + J_i(\xi_s') - J_i(\xi_r) - J_i(\xi_s)$$

$$= \sum_j \sum_k \sum_l \{b_{ij,kl}(\xi_r, \xi_s) + b_{ij,kl}(\xi_s, \xi_r) - b_{kl,ij}(\xi_r, \xi_s) - b_{kl,ij}(\xi_s, \xi_r)\}. \tag{7.3}$$

Summing over all pairs of particles we obtain, for the total change ΔN_i in the occupation number of υ_i between the two instants of observation, the formula

$$\Delta N_i = \sum_{r \neq s} \sum_j \sum_k \sum_l \{b_{ij,kl}(\xi_r, \xi_s) - b_{kl,ij}(\xi_r, \xi_s)\}. \tag{7.4}$$

The summation $\sum_{r \neq s}$ ranges over all ordered pairs of unequal positive integers r, s, neither of which is to exceed N: for example, there are separate terms in the sum for $r = 1$, $s = 2$, and for $r = 2$, $s = 1$.

An important statistical property of the random variable ΔN_i defined in (7.4) is its expectation, evaluated in a statistical ensemble whose members are all in the same observational state, say A, at the first of the two instants of observation under consideration. According to the result proved in Chap. IV, § 6, this expectation is the same as the ensemble average calculated in a coarse-grained ensemble whose phase-space density is uniform over the dynamical image (in γ-space) of A. As we indicated in the derivation of (6.8), this image region $\omega(A)$ consists of $N!/\Pi_i N_i(A)!$ congruent components. One of these components is the region $\omega_c(A)$ defined by

$$\left.\begin{array}{l}\alpha \text{ is in } \omega_c(A) \text{ if and only if } \xi_1, \xi_2, \ldots, \xi_{N_1} \text{ are all in } \upsilon_1, \\ \xi_{N_1+1}, \ldots, \xi_{N_1+N_2} \text{ are all in } \upsilon_2, \text{ etc.,}\end{array}\right\} \tag{7.5}$$

where N_1 stands for $N_1(A)$, N_2 for $N_2(A)$, etc. The other components of $\omega(A)$ are obtainable from $\omega_c(A)$ by rearrangements of the suffixes 1, ..., N in (7.5). Since ΔN_i, regarded as a dynamical variable, is a symmetric function of the N μ-space points $\xi_1 \ldots \xi_N$, its average over each component of $\omega(A)$ is the same; consequently we may calculate $\langle \Delta N_i \rangle$ by averaging over the single component $\omega_c(A)$ instead of over the whole of $\omega(A)$. Carrying out this averaging in (7.4), we obtain

$$\langle \Delta N_i \rangle = \sum_j \sum_k \sum_l \Big[\sum_{r \neq s} \langle b_{ij,kl}(\xi_r, \xi_s) \rangle_c - \sum_{r \neq s} \langle b_{kl,ij}(\xi_r, \xi_s) \rangle_c \Big], \tag{7.6}$$

where $\langle \ldots \rangle_c$ indicates an average over the γ-space region $\omega_c(A)$.

Using the definition (7.2), we can evaluate the average of $b_{ij,kl}(\xi_r, \xi_s)$ over $\omega_c(A)$, obtaining

$$\langle b_{ij,kl}(\xi_r, \xi_s) \rangle_c = \frac{\int \ldots \int J_{cA}(\alpha)\, J_i(\xi_r')\, J_j(\xi_s')\, J_k(\xi_r)\, J_l(\xi_s)\, d\alpha}{\int \ldots \int J_{cA}(\alpha)\, d\alpha}, \tag{7.7}$$

where $\alpha \equiv (\xi_1, \ldots, \xi_N)$ and $d\alpha \equiv d\xi_1 \ldots d\xi_N$, the integration ranges over the whole of γ-space, and

$$J_{cA}(\alpha) \equiv J_1(\xi_1)\, J_1(\xi_2) \ldots J_1(\xi_{N_1})\, J_2(\xi_{N_1+1}) \ldots J_2(\xi_{N_1+N_2}) \ldots \tag{7.8}$$

is, in accordance with the definition (7.5), the indicator function of the γ-space region $\omega_c(A)$. The integrations over variables other than ξ_r and ξ_s are easily carried out, and contribute equal factors (the volumes of the appropriate μ-space cells) to numerator and denominator. Removing these cancelling factors, we are left with

$$\langle b_{ij,kl}(\xi_r, \xi_s) \rangle_c = \frac{\iint J_m(\xi_r)\, J_n(\xi_s)\, J_i(\xi_r')\, J_j(\xi_s')\, J_k(\xi_r)\, J_l(\xi_s)\, d\xi_r\, d\xi_s}{\iint J_m(\xi_r)\, J_n(\xi_s)\, d\xi_r\, d\xi_s}, \tag{7.9}$$

where m and n are the suffixes of the rth and sth factors in (7.8), that is, the labels of the cells containing the rth and sth particles when the γ-space point $(\xi_1, \ldots, \xi_N)$ is in $\omega_c(A)$. Unless $m = k$ and $n = l$, the integrand in the numerator of (7.9) vanishes identically; consequently the formula simplifies to

$$\langle b_{ij,kl}(\xi_r, \xi_s)\rangle_c = \begin{cases} \beta_{ij,kl}/\Upsilon_k\Upsilon_l \text{ if } \xi_r \text{ is in } \upsilon_k \text{ and } \xi_s \text{ is in } \upsilon_l \\ 0 \text{ if not,} \end{cases} \quad (7.10)$$

where
$$\beta_{ij,kl} \equiv \iint b_{ij,kl}(\xi_r, \xi_s)\, d\xi_r\, d\xi_s$$
$$= \iint J_i(\xi_1')\, J_j(\xi_2')\, J_k(\xi_1)\, J_l(\xi_2)\, d\xi_1\, d\xi_2, \quad (7.11)$$

the dummy variables of integration ξ_r and ξ_s having been replaced here by ξ_1 and ξ_2 for clarity. The quantities $\beta_{ij,kl}$, which we shall call *collision coefficients*, can be evaluated from the dynamics of two-particle collisions, but the details of this evaluation are unimportant here.

When (7.10) is substituted into (7.6) the summations over r and s can be carried out at once: for example, the first of them contains, if $k \neq l$, just $N_k N_l$ non-vanishing terms, each one equal to $\beta_{ij,kl}/\Upsilon_k\Upsilon_l$; but if $k = l$, the number of terms is only $N_k(N_l - 1)$. In this way (7.6) reduces to

$$\langle \Delta N_i\rangle = \textstyle\sum_j\sum_k\sum_l[\beta_{ij,kl}N_k(N_l - \delta_{kl})/\Upsilon_k\Upsilon_l - \beta_{kl,ij}N_i(N_j - \delta_{ij})/\Upsilon_i\Upsilon_j], \quad (7.12)$$

where $\delta_{kl} \equiv 1$ if $k = l$ and $\equiv 0$ if $k \neq l$.

The averages on the left-hand sides of (7.6) and (7.12) are taken in a special statistical ensemble whose members are all in the same observational state (denoted earlier by A) at some given time, say t. To extend our results to a general statistical ensemble, we multiply both sides of (7.12) by the probability of the state A at time t in this general ensemble and sum over all values of A; recalling the definitions of ΔN_i and N_i [associated with (7.4) and (7.5) respectively], we obtain

$$\langle N_i\rangle_{t+1} - \langle N_i\rangle_t = \textstyle\sum_j\sum_k\sum_l[\beta_{ij,kl}\langle N_k(N_l - \delta_{kl})\rangle_t/\Upsilon_k\Upsilon_l$$
$$- \beta_{kl,ij}\langle N_i(N_j - \delta_{ij})\rangle_t/\Upsilon_i\Upsilon_j], \quad (7.13)$$

where $\langle\rangle_t$ means an average over the general ensemble, taken at any time t when this ensemble is realizable.

In order to compute the temporal evolution of the set of mean occupation numbers $\langle N_i\rangle_t$ from the difference equation (7.13), we need information about the second moments, such as $\langle N_k N_l\rangle$, of these occupation numbers. These second moments are related to the corresponding variances and covariances by the definitions (2.16) and (2.24) of Chap. II, which may be combined to give the identity

$$\langle N_k(N_l - \delta_{kl})\rangle = \langle N_k\rangle\langle N_l\rangle + \text{covar}(N_k, N_l) - \langle N_k\rangle\delta_{kl}, \quad (7.14)$$

where covar (X,X) is merely another way of writing var (X). To estimate the covariance appearing in (7.14), we may refer to (6.30) and (6.31), which indicate that at equilibrium the terms covar (N_k, N_l) and $\langle N_k\rangle\,\delta_{kl}$ cancel, so that (7.14) reduces to

$$\langle N_k(N_l - \delta_{kl})\rangle = \langle N_k\rangle\,\langle N_l\rangle . \tag{7.15}$$

Away from equilibrium, the appropriate covariances and variances are much more difficult to calculate. For a rough estimate, we may make the assumption that the order of magnitude of these variances and covariances is no greater than at equilibrium, so that (7.15) still holds approximately, with error roughly $\langle N_k\rangle\,\delta_{kl}$. The effect of such an error on a sum such as the right-hand side of (7.13) can be estimated by considering its effect on the similar, but simpler, sum $\sum_k\sum_l N_k N_l$. This latter sum has the value N^2, and an error $N_k\delta_{kl}$ in each term would lead to a total error $\sum_k\sum_l N_k\delta_{kl} = N$, which is only $1/N$ times the true value of the sum. Even when the system is not in equilibrium, therefore, we shall assume† that the approximation (7.15) may be used in the right-hand side of (7.13) with an error that is a very small fraction of the whole when N is large.

Substituting the approximation (7.15) into (7.13), we obtain an equation that can be written in the form

$$f_i(t+1) - f_i(t) = \Upsilon_i^{-1}\textstyle\sum_j\sum_k\sum_l[\beta_{ij,kl}f_k(t)f_l(t) - \beta_{kl,ij}f_i(t)f_j(t)] . \tag{7.16}$$

where

$$f_i(t) \equiv \langle N_i\rangle_t/\Upsilon_i \tag{7.17}$$

denotes the statistical expectation, at time t, of the value taken by the Boltzmann distribution function (defined in (6.4)) in the ith μ-space cell. We shall call (7.17) the *kinetic equation* for the gas. By applying it for all values of i, first with $t = 0$ (assuming this to be the initial time, for which the occupation numbers are given), then with $t = 1$, then with $t = 2$, and so on, we can in principle calculate the expectation of the Boltzmann distribution function for any instant of observation after $t = 0$. Equation (7.16) can be converted into an integro-differential equation by replacing the finite difference on its left-hand side by the corresponding differential quotient $\partial f/\partial t$ and expressing the sums on the right-hand side as integrals by the method used for (6.5) and (6.7). The resulting equation, first obtained by Boltzmann‡ and therefore known as *Boltzmann's equation*, is fundamental in the kinetic theory of gases.

Our derivation of the kinetic equation is very similar to Boltzmann's original derivation of his own equation. Boltzmann's derivation, however,

† A rigorous justification of the approximation (7.15) has been given by M. Kac for a simplified model of the collision mechanism in Chap. III, §§ 16 and 17, of his *Probability and Related Topics in Physical Science* (Interscience, London, 1959).

‡ L. Boltzmann, *Lectures on Gas Theory*, §§ 3 and 4.

is based on an *ad hoc* assumption that the gas is not "molecular-ordered". The main virtue of our derivation is to show that this assumption can be dispensed with, since it is already inherent in the Markovian postulate which we use to justify the application of probability methods to the set of observational states defined by occupation numbers. A defect in our derivation is that we have not shown that the consistency condition is satisfied; there is also the drawback that this type of derivation is difficult to generalize to finite densities. A type of derivation that is better adapted to this generalization has been used by Bogolyubov and Cohen† among others. In these derivations the Markovian postulate is replaced by a different *ad hoc* assumption, that at some time in the distant past all the particles were completely uncorrelated – that is, the phase-space density had the form $D(\alpha) = \Pi_i D_{(1)}(\xi_i)$.

The kinetic equation (7.16) is similar in mathematical form to the "master equation" [eqn. (4.21) of Chap. II] in the theory of Markov chains (although it does not share the linearity of the master equation). The two equations also have similar physical meanings, since each of them gives the change with time of a distribution. The kinetic equation, however, refers to a distribution described by occupation numbers rather than by probabilities, a distribution of many interacting molecules (together forming a single dynamical system) over cells in μ-space rather than of many non-interacting systems (together forming an ensemble) over observational states corresponding to cells in γ-space. Despite these differences in meaning between the two equations, their formal similarity suggests that the asymptotic behaviour of the μ-space distribution as $t \to \infty$ can be studied by a method similar to the one used in Chap. II, § 7, for probability distributions. This method will be described in the next section.

8. Boltzmann's *H* theorem

In order to show that a kinetic equation such as (7.16) really does imply that the gas approaches equilibrium, Boltzmann devised a method based on showing that the quantity

$$H(t) \equiv \sum_i \Upsilon_i f_i(t) \ln f_i(t)$$
$$= \sum_i \langle N_i \rangle_t \ln (\langle N_i \rangle_t / \Upsilon_i) \quad (8.1)$$

is a non-decreasing function of the time t. Boltzmann's method is intimately related both to the method used in Chap. II, § 7, to show that the probabilities in a Markov chain approach equilibrium and to the non-decrease property of the entropy. The relation to the method used for Markov chains can be seen by comparing the right-hand side of (8.1) with the expression

$$\sum_A \pi_A \varphi[p_A(t)/\pi_A]$$

† N. N. Bogolyubov, Kinetic Equations, *J. Phys. USSR* **10**, 265 (1946); E. G. D. Cohen, On the Kinetic Theory of Dense Gases, *J. Math. Phys.* **4**, 183 (1963).

which was denoted by $H(t)$ in Chap. II, § 7, and shown there to be a non-decreasing function of time if the function $\varphi(x)$ is convex. The definition (8.1) corresponds to the particular choice $\varphi(x) = x \ln x$ for this function; the choice is much narrower here than in the theory of Markov chains because the kinetic equation (7.16) is non-linear whereas its counterpart in the Markov chain theory, eqn. (4.20) of Chap. II, or the master equation (4.21) of Chap. II, is linear.†

To obtain a relation between $H(t)$ and entropy, let us assume that the formula

$$\langle N_i \rangle = N(MP), \tag{8.2}$$

shown in (6.29) to hold approximately at equilibrium if $\langle N_i \rangle$ is large, also holds approximately away from equilibrium, in the form

$$\langle N_i \rangle_t = N_i[MP(t)], \tag{8.3}$$

where $MP(t)$ denotes the observational state whose probability at the time t is greatest. Substituting (8.3) into (8.1), and using the first line of (6.11), we obtain

$$S[MP(t)] = -kH(t) + Nk \ln(e/h^3) \tag{8.4}$$

provided that all non-vanishing occupation numbers are large. The monotonic decrease of $H(t)$ demonstrated by Boltzmann thus implies that the entropy of the most probable state increases with time.

Boltzmann's proof depends on the symmetry properties of the collision coefficients $\beta_{ij,kl}$ which appear in the kinetic equation (7.16). These symmetries are due to corresponding symmetries in the laws of dynamics which determine the collision coefficients through their definition (7.11). The simplest of these is the symmetry under interchanges of particles, which implies that the integral in (7.11) is invariant under the transformation

$$\left.\begin{aligned} \xi_1 \to \xi_2, \quad \xi_1' \to \xi_2', \\ \xi_2 \to \xi_1, \quad \xi_2' \to \xi_1', \end{aligned}\right\} \tag{8.5}$$

and consequently that the collision coefficient defined by this integral has the symmetry property

$$\beta_{ij,kl} = \beta_{ji,lk}. \tag{8.6}$$

Further symmetry properties of the collision coefficients can be obtained if the Hamiltonian is symmetrical under the operations of *time reversal* (denoted here by τ)

$$\left.\begin{aligned} \mathbf{p}_i \to \tau\mathbf{p}_i \equiv -\mathbf{p}_i, \\ \mathbf{q}_i \to \tau\mathbf{q}_i \equiv \mathbf{q}_i, \end{aligned}\right\} \tag{8.7}$$

† A good discussion of the similarities and differences between the two types of H theorem is given by P. A. P. Moran in Entropy, Markov Processes, and Boltzmann's H Theorem, *Proc. Camb. Phil. Soc.* **57**, 833 (1961).

and *space reversal* (denoted here by σ)

$$\left.\begin{aligned} \boldsymbol{p}_i &\to \sigma\boldsymbol{p}_i \equiv -\boldsymbol{p}_i, \\ \boldsymbol{q}_i &\to \sigma\boldsymbol{q}_i \equiv -\boldsymbol{q}_i. \end{aligned}\right\} \tag{8.8}$$

Writing the Hamiltonian of the gas in the form indicated by eqns. (2.17) and (2.22) of Chap. I for a system of particles in the absence of electromagnetic forces,

$$H = \textstyle\sum_i \boldsymbol{p}_i^2/2m + \sum_i U_{(1)}(\boldsymbol{q}_i) + \sum_{i<j} U_{(2)}(\boldsymbol{q}_i, \boldsymbol{q}_j), \tag{8.9}$$

we see that it possesses both these types of symmetry provided that both the functions $U_{(1)}$ (representing external and wall forces) and $U_{(2)}$ (representing interactions between molecules) are symmetric under space reversal. Normally $U_{(2)}$ is in fact taken to be spherically symmetrical as in the Lennard–Jones potential (2.19) of Chap. I, and is therefore *a fortiori* symmetrical under space reversal; the function $U_{(1)}$ can easily be given the desired symmetry by giving the container a centre of symmetry at the origin.

According to the time-reversal invariance of *H*, if two particles whose initial dynamical states are ξ_1 and ξ_2 finish in the states ξ_1' and ξ_2', then two particles starting in the time-reversed final states $\tau\xi_1'$ and $\tau\xi_2'$ will traverse the same set of configurations in the reversed order and finish in the time-reversed initial states $\tau\xi_1$ and $\tau\xi_2$. The integrand of (7.11) is therefore invariant under the transformation

$$\left.\begin{aligned} \xi_1 &\to \tau\xi_1', \quad \xi_1' \to \tau\xi_1, \\ \xi_2 &\to \tau\xi_2', \quad \xi_2' \to \tau\xi_2, \end{aligned}\right\} \tag{8.10}$$

from which we obtain

$$\beta_{ij,kl} = \iint J_i(\tau\xi_1)\, J_j(\tau\xi_2)\, J_k(\tau\xi_1')\, J_l(\tau\xi_2') \frac{\partial(\tau\xi_1', \tau\xi_2')}{\partial(\xi_1, \xi_2)}\, d\xi_1\, d\xi_2. \tag{8.11}$$

The Jacobian appearing in (8.11) can be written

$$\frac{\partial(\tau\xi_1', \tau\xi_2')}{\partial(\xi_1, \xi_2)} = \frac{\partial(\tau\xi_1', \tau\xi_2')}{\partial(\xi_1', \xi_2')}\, \frac{\partial(\xi_1', \xi_2')}{\partial(\xi_1, \xi_2)} = 1 \times 1. \tag{8.12}$$

The first factor, the Jacobian of the transformation τ applied to ξ_1' and ξ_2', is +1 by (8.10) and (8.7), since an even number of momentum coordinates are reversed. The other factor is also 1 by eqn. (3.19) of Chap. III, since the transformation from (ξ_1, ξ_2) to (ξ_1', ξ_2') is a "Liouville" transformation applied to particles 1 and 2 moving independently of the rest.

To take full advantage of the symmetries in momentum space described by (8.7) and (8.8), the network of cells in momentum space must be chosen to have a centre of symmetry at the origin $\boldsymbol{p} = 0$ of that space. Let us denote by τi the label of the cell obtained from the ith cell by reversing the momentum vector, so that

$$J_i(\tau\xi) = J_{\tau i}(\xi). \tag{8.13}$$

Using (8.12) and (8.13) in (8.11), and then applying the definition (7.11) once again, we obtain

$$\beta_{ij,kl} = \beta_{\tau k \tau l, \tau i \tau j} \tag{8.14}$$

for any system with time-reversal invariance.

The space reversal condition is applied similarly. If two particles with initial states ξ_1 and ξ_2 finish in the states ξ_1' and ξ_2', then particles starting in the spatially opposite states $\sigma\xi_1$ and $\sigma\xi_2$ will traverse the spatially opposite set of configurations in the same order, finishing in the states $\sigma\xi_1'$ and $\sigma\xi_2'$. The integrand of (7.11) is therefore invariant under the transformation

$$\left.\begin{aligned} \xi_1 \to \sigma\xi_1, \quad \xi_1' \to \sigma\xi_1', \\ \xi_2 \to \sigma\xi_2, \quad \xi_2' \to \sigma\xi_2'. \end{aligned}\right\} \tag{8.15}$$

The Jacobian of this transformation, like that of the time-reversal transformation (8.10), is 1. Moreover, since the μ-space cells are defined in terms of $\boldsymbol{p}$-space cells by (7.1), and both τ and σ reverse the momentum vector, these two transformations have the same effect on the μ-space cells, so that

$$J_i(\sigma\xi) = J_i(\tau\xi) = J_{\tau i}(\xi). \tag{8.16}$$

By an argument analogous to that which led to (8.14), we thus obtain

$$\beta_{ij,kl} = \beta_{\tau i \tau j, \tau k \tau l}. \tag{8.17}$$

For a system with both time- and space-reversal invariance, we can combine (8.17) with (8.14), after making the permutation of dummy indices $i \to k$, $j \to l$, $k \to i$, $l \to j$ in (8.14), and we obtain

$$\beta_{ij,kl} = \beta_{kl,ij}. \tag{8.18}$$

In consequence of this symmetry relation, the kinetic equation (7.16) simplifies to

$$f_i(t+1) - f_i(t) = \Upsilon_i^{-1} \textstyle\sum_j \sum_k \sum_l \beta_{ij,kl} [f_k(t) f_l(t) - f_i(t) f_j(t)]. \tag{8.19}$$

The change in $H(t)$ between successive instants of observation can now be calculated. Expanding each term of the last sum in (8.1) in powers of $\langle \Delta N_i \rangle_t \equiv \langle N_i \rangle_{t+1} - \langle N_i \rangle_t$ by means of Taylor's theorem, we may write (omitting the subscripts t where it causes no confusion)

$$\begin{aligned} \Delta H(t) &\equiv H(t+1) - H(t) \\ &= \textstyle\sum_i \langle \Delta N_i \rangle \, [1 + \ln(\langle N_i \rangle / \Upsilon_i) + \langle \Delta N_i \rangle / 2 \langle N_i \rangle + \cdots] \end{aligned} \tag{8.20}$$

The first term in square brackets in (8.20) may be omitted since $\sum_i \langle \Delta N_i \rangle$, the change in the total number of particles, must vanish; the term $\langle \Delta N_i \rangle / 2\langle N_i \rangle$ and the unwritten terms will also be omitted since the binary collision approximation makes them small. Retaining only the logarithm, remembering that (7.17) implies $\langle \Delta N_i \rangle = [f_i(t+1) - f_i(t)]\,\Upsilon_i$, and substituting from (8.19) for $f_i(t+1) - f_i(t)$, we obtain

$$\Delta H(t) = \sum_i \sum_j \sum_k \sum_l \beta_{ij,kl}(f_k f_l - f_i f_j) \ln f_i, \tag{8.21}$$

where

$$f_i \equiv f_i(t). \tag{8.22}$$

By rearranging the dummy suffixes in the sum on the right of (8.21), we may also write this equation in the form

$$\Delta H(t) = \sum\sum\sum\sum \beta_{ji,lk}(f_l f_k - f_j f_i) \ln f_j$$

or

$$\Delta H(t) = \sum\sum\sum\sum \beta_{kl,ij}(f_i f_j - f_k f_l) \ln f_k$$

or

$$\Delta H(t) = \sum\sum\sum\sum \beta_{lk,ji}(f_j f_i - f_l f_k) \ln f_l.$$

Using the symmetry relations (8.6) and (8.18), and averaging the four resulting expressions for $\Delta H(t)$, we obtain

$$\Delta H(t) = \tfrac{1}{4} \sum_i \sum_j \sum_k \sum_l \beta_{ij,kl}(f_k f_l - f_i f_j) \ln (f_i f_j / f_k f_l). \tag{8.23}$$

By (7.11), the collision coefficients $\beta_{ij,kl}$ are non-negative, and the other two factors in the summand must always have opposite signs. Thus (8.23) implies that

$$\Delta H(t) \leqq 0 \tag{8.24}$$

with equality only if

$$f_i f_j - f_k f_l = 0 \quad \text{for all} \quad i, j, k, l \quad \text{such that} \quad \beta_{ij,kl} > 0. \tag{8.25}$$

It follows immediately from (8.24) that $H(t)$ is a non-increasing function of the time t. This result is *Boltzmann's H theorem*.

The condition (8.25) shows that if $H(t)$ is constant in time, then every term in the triple sum in (8.19) must vanish, so that the mean occupation numbers are also constant in time. That is to say, at equilibrium the mean occupation numbers are maintained at constant values, not merely by a net balancing of the total number of particles entering each μ-space cell against the total number leaving it, but by a separate balancing for every different type of collision, the mean number of (ij, kl) collisions being equal to the mean number of the inverse (kl, ij) collisions for every quadruplet $(ijkl)$. This is called the *principle of detailed balancing*.

To apply (8.25) in a further study of the equilibrium distribution, we need to know the conditions under which the collision coefficients vanish identically. The only general principles restricting the types of possible collision are the principles of conservation of energy and momentum. These two principles give rise to two selection rules. First, since the colliding particles are assumed not to interact with other particles during a collision, the total energy of the colliding particles is an invariant of the motion, so that

$$E_i + E_j = E_k + E_l \quad \text{if} \quad \beta_{ij,kl} > 0, \tag{8.26}$$

where, as in (6.3), E_i denotes the energy of a particle in the cell υ_i. Secondly if we neglect external forces and wall forces, the total momentum of the two particles is also conserved, so that

$$\boldsymbol{p}_i + \boldsymbol{p}_j = \boldsymbol{p}_k + \boldsymbol{p}_l \quad \text{if} \quad \beta_{ij,kl} > 0, \tag{8.27}$$

where $\boldsymbol{p}_i$ denotes the momentum of a particle in υ_i. The conservation law (8.27) is only approximately true, however, when wall forces are taken into account.

As in the discussion of Markov chains, we can now deduce the behaviour of the distribution in the limit $t \to \infty$ from the H theorem. Since H is related to S by (8.4), and S is bounded above by the entropy of the most probable observational state, H is bounded below by its value for a Maxwellian distribution. Consequently $H(t)$, being a non-increasing function, must tend to a limit as $t \to \infty$. Taking the limit $t \to \infty$ on both sides of (8.23) we find that (8.25) must be satisfied in this limit. It follows, by the kinetic equation (8.19), that the mean occupation numbers themselves do not change with time in this limit, so that the possibility of periodic behaviour, which had to be allowed for in the discussion of Markov chains (see Chap. II, § 8) does not arise here. Any set of occupation numbers satisfying (8.25) is a possible form for the limiting (equilibrium) distribution. The simplest solution of (8.25) is the Maxwell distribution

$$f_i = \exp(-\alpha - \beta E_i) = \exp(-\alpha - \beta \boldsymbol{p}_i^2/2m), \tag{8.28}$$

where α and β are constants; this satisfies (8.25) by virtue of the conservation of energy at collisions [eqn. (8.26)]. Thus the non-equilibrium theory of this section confirms the result obtained by the simpler methods of equilibrium theory in § 6, namely that the mean occupation numbers at equilibrium are Maxwellian.

Although the Maxwell distribution is a solution of (8.25), it is not the only solution; the distribution

$$f_i = \exp(-\alpha - \beta E_i + \beta \boldsymbol{u} \cdot \boldsymbol{p}_i) = \exp[-\alpha' - \beta(\boldsymbol{p}_i - m\boldsymbol{u})^2/2m], \tag{8.29}$$

where $\boldsymbol{u}$ is a constant vector and $\alpha' \equiv \alpha - \frac{1}{2}\beta m u^2$, also satisfies (8.25), by virtue of the fact [eqn. (8.27)] that momentum as well as energy is conserved at collisions. The distribution (8.29) describes a gas whose distribution would appear to be Maxwellian if viewed by an observer moving with velocity $\boldsymbol{u}$.

Such a distribution therefore describes a gas that is in internal equilibrium but has a bulk motion of velocity $\boldsymbol{u}$; it appears as a possible solution of our equations because we have neglected collisions with the walls. When these wall collisions are allowed for, an extra term of the form $\Upsilon_i^{-1}\sum_j \alpha_{ij}(f_j - f_i)$ appears on the right-hand side of the kinetic equation (8.19), with

$$\alpha_{ij} = \alpha_{ji} \equiv \int J_i\,(U_{1(1)}\xi)\,J_j(\xi)\,d\xi \tag{8.30}$$

where $U_{t(1)}$ is the time evolution operator (see Chap. III, § 3) for a single-particle system and the integration, like the ones in (7.11), is overall μ-space. Since collisions with the wall conserve energy but not momentum, the coefficients α_{ij} satisfy an energy conservation law analogous to (8.26),

$$E_i = E_j \quad \text{if} \quad \alpha_{ij} > 0, \tag{8.31}$$

but no momentum conservation law. The H theorem generalizes in a straightforward way to this case, the contribution of the wall collision term to ΔH being negative unless

$$f_i - f_j = 0 \quad \text{whenever} \quad \alpha_{ij} > 0. \tag{8.32}$$

According to (8.31), the Maxwellian distribution (8.28) still satisfies the equilibrium conditions (8.25) and (8.32) when the walls are allowed for, but the distribution (8.29) describing bulk motion no longer does.

In drawing conclusions from Boltzmann's H theorem it is important to realize that $H(t)$ is defined in (8.1) to depend not on the occupation numbers in a particular replica of the system but on the mean values of these occupation numbers over a statistical ensemble. To see the significance of this fact, let us compare the properties of $H(t)$ with those of the corresponding observable, which is a function of an observational state rather than of the time, and therefore takes a definite value for every individual system in the ensemble at any time t. By analogy with (8.1), this observable is

$$H[A] \equiv \textstyle\sum_i N_i(A) \ln\,[N_i(A)/\Upsilon_i]. \tag{8.33}$$

The relation between $H(t)$ and $H[A]$, according to the approximation (8.3), is

$$H(t) = H[MP(t)]. \tag{8.34}$$

Moreover, our assumption that the random variables N_i have standard deviations with the same order of magnitude as at equilibrium [eqn. (7.15)] implies that, at time t, $N_i(A)$ is not likely to differ greatly from $N_i[MP(t)]$, and hence that the value of $H[A]$ for a given individual system is not likely to differ greatly from its value $H(t)$ in the most probable state at that time.

Boltzmann's result (8.24), when combined with (8.34), implies that $H[MP(t)]$ is a non-increasing function of time:

$$\text{if} \quad 0 \leqq t_1 < t_2, \quad \text{then} \quad H[MP(t_1)] \geqq H[MP(t_2)]. \tag{8.35}$$

To relate this result to the behaviour of individual systems, let us define $A(t)$ to be the observational state of the system at time t: that is $A(0)$, $A(1)$,

$A(2)$, ... are the observational states taken by one and the same system at the successive instants of observation 0, 1, 2, ... Since $H[A(t)]$ is likely to be close to $H[MP(t)]$, it follows from (8.35) that if t_1 and t_2 are instants of observation satisfying $0 \leqq t_1 < t_2$ then the inequality

$$H[A(t_1)] \geqq H[A(t_2)] \tag{8.36}$$

is most unlikely to be violated by a large amount. On the other hand, the possibility of such a violation cannot be completely ignored, and so one cannot deduce from (8.35) that $H[A(t)]$, too, is a non-increasing function of t. The fact that the H theorem is correctly expressed by the probability statements (8.24) or (8.35) which refer, in principle, to ensembles, rather than by the statement (8.36) which refers to a single system only, is known as the *statistical interpretation* of the H theorem.

It is possible to "deduce" (8.36) from the theory if we make the assumption that (7.12) gives not merely the expected but the actual change in N_i. This assumption is known as the *Stosszahlansatz*. The fact that (8.36) is not strictly true thus implies that the *Stosszahlansatz* is not strictly true and in fact must be violated whenever $H[A(t)]$ increases with time.

In the early days of kinetic theory, the distinction between statements such as (8.24) and (8.35), on the one hand, and (8.36), on the other, was not clearly understood. In consequence, objections were raised against Boltzmann's theory on the grounds that (8.36), if true for all t_1, t_2 satisfying $0 \leqq t_1 < t_2$, contradicts the laws of dynamics. These objections are known as the *reversibility paradox* and the *recurrence paradox*. Both depend on constructing artificial motions of the molecules in a gas that are inconsistent with (8.36). In the reversibility paradox,† the artificial motion considered is obtained by taking the molecules of the gas through a sequence of configurations constituting a natural motion, but in the reverse order and with reversed velocities. Since $H(t)$ decreases for the natural motion, it increases for the reversed motion. In the recurrence paradox,‡ the artificial motion is obtained by starting the gas in a state of high H and keeping it isolated for an arbitrarily long time; according to a theorem of Poincaré,§ the dynamical state of

† J. Loschmidt, Über den Zustand des Wärmegleichgewichtes eines Systems von Körpern mit Rücksicht auf die Schwerkraft, *Wien. Ber.* **73**, 128, 366 (1876); **75**, 287; **76**, 209 (1877); L. Boltzmann, *Lectures on Gas Theory*, Chap. I, § 6. The reversibility paradox should not be confused with the paradox of irreversibility, which was discussed in Chap. I, § 6.

‡ E. Zermelo, Über einen Satz der Dynamik und die Mechanische Wärmetheorie, *Ann. Phys.* **57** (3), 485 (1896); L. Boltzmann, *Lectures on Gas Theory*, Chap. II, § 88.

§ H. Poincaré, Sur le Problème de Trois Corps et les Équations de la Dynamique, *Acta math.* **13**, 1 (1890); *Oeuvres* **7**, 262. A nice proof is given by J. W. Gibbs, *Elementary Principles in Statistical Mechanics* (Yale University Press, 1902; Dover reprint, New York, 1960) pp. 139–40. The name *Poincaré period* is often given to the length of time the system takes to return close to its original state, but Poincaré's theorem does not imply that the motion of the molecules is periodic, for the Poincaré period may depend on the initial dynamical state. Boltzmann (*Lectures on Gas Theory*, § 88) estimates the Poincaré periods for 100 cm^3 of gas to be "enormously long compared to $10^{10^{10}}$ years".

the system must (save in highly exceptional cases) eventually return close to its original value, and hence $H(t)$, after the initial decrease predicted by (8.36), must eventually increase again as it returns to its original value. In both paradoxes the artificial motions are inconsistent with (8.36), but this does not imply any inconsistency in the theory as a whole, since the statistical interpretation of the H theorem requires us to admit that it is possible, although highly improbable, for an actual motion of the gas molecules to be inconsistent with (8.36).

8.1. Exercise

1. Show that, for any system (not necessarily a gas) whose observational states are symmetrical under time reversal (i.e. for which $\alpha \in \omega(A)$ implies $\tau\alpha \in \omega(A)$),

$$w_{AB}\pi_B = w_{BA}\pi_A \quad \text{for all} \quad A, B \tag{8.37}$$

(with the convention $\pi_A = 0$ if A is transient) so that for such systems a principle of detailed balancing applies to the master equation (4.21) of Chap. II.†

† The type of time-reversal symmetry that is invoked in this proof is known as the principle of *microscopic reversibility* and was first applied to obtain symmetry relations connecting measurable physical quantities by L. Onsager, Reciprocal Relations and Irreversible Processes, *Phys. Rev.* **38**, 2265 (1931). The applications of Onsager's beautiful theory are treated by S. R. de Groot in *Thermodynamics of Irreversible Processes* (North-Holland, Amsterdam, 1963).

CHAPTER VI

Statistical Entropy

1. The definition of statistical entropy

The discussion in the previous chapter has shown that the Boltzmann entropy $S(A)$, as defined in eqn. (4.1) of Chap. V, does not exactly satisfy the non-decrease condition (1.1) of Chap. V unless the observational states are deterministic; for non-deterministic observational states this entropy can decrease, although the probability for a large decrease is small. In particular, as we showed in Chap. V, § 5, the Boltzmann entropy of a system that has been isolated for a long time fluctuates continually, sometimes increasing and sometimes decreasing, provided only that the ergodic set contains two observational states of different entropies. It is the purpose of the present section to formulate a definition for a new type of entropy which is a rigorously non-decreasing function of time.

The possibility of decreases in the Boltzmann entropy is related to the fact that a non-deterministic Markov chain does not, in general, have equilibrium states (states which cannot be left, once entered). If some of the observational states are equilibrium states (as they are, for example, in the deterministic case), then a system reaching such a state will stay there, so that its entropy reaches a constant final value. If, on the other hand, there are no equilibrium states then the system, however long it has been isolated, will always retain the option of moving to a new observational state; and this move will lead to a decrease of entropy if both states are persistent and the new state has a smaller weight than the old one.

Despite the absence of equilibrium states for a general Markov chain, however, we were able in Chap. II, § 6, to define equilibrium for such chains. This was achieved by regarding equilibrium as a property of a statistical ensemble rather than an individual system: we say that an ensemble is in *statistical equilibrium* if its probabilities $p_t(A)$ are independent of the time t. When an ensemble is in statistical equilibrium, the observational states of its individual systems are continually changing, but these changes do not affect the probabilities $p_t(A)$, which describe only the average properties of the ensemble. Consequently, we can eliminate equilibrium fluctuations of the entropy by making it a function of the probabilities in a statistical ensemble, rather than of the observational state of an individual system. The entropy defined in this new way will be called *statistical entropy*. (Alternatively, it might be called *Gibbs entropy*, since Gibbs was the first to define entropy in terms of ensembles.) To complete the definition, we must specify the functional dependence of the statistical entropy on the probabilities $p_t(A)$.

This functional dependence is to be chosen so that the statistical entropy increases monotonically to a maximum as the ensemble approaches statistical equilibrium.

A plausible choice for this functional dependence is suggested by the calculations of the preceding section: it is to call $S[MP(t)]$ the statistical entropy, where $MP(t)$ is the most probable observational state at the time t. For a dilute gas under the binary collision approximation, $S[MP(t)]$ equals $-kH(t) + \text{const.}$ [eqn. (8.4) of Chap. V], and hence Boltzmann's H theorem implies that $S[MP(t)]$ does have the required non-decrease property. Unfortunately, however, this result of kinetic gas theory cannot be extended to a general non-decrease theorem for $S[MP(t)]$. This can be shown by a counter-example: consider the three-state Markov chain whose transition probability matrix [as defined in eqn. (4.9) of Chap. II] is

$$[w_{BA}] = \begin{bmatrix} \frac{1}{4} & \frac{3}{4} & \frac{3}{4} \\ \frac{1}{2} & 0 & 0 \\ \frac{1}{4} & \frac{1}{4} & \frac{1}{4} \end{bmatrix}. \tag{1.1}$$

Its equilibrium probabilities, determined by solving the equations $\sum_A w_{BA}\pi_A = \pi_B$ [eqn. (6.3) of Chap. II], are

$$\pi_1 = \tfrac{1}{2}, \quad \pi_2 = \pi_3 = \tfrac{1}{4}, \tag{1.2}$$

where we have labelled the states 1, 2, 3.

Suppose now that at the initial instant, $t = 0$, the system is in the state 1, so that its statistical entropy, according to the proposed definition, would be

$$S[MP(0)] = S(1).$$

At the next instant of observation, the probabilities are $p_1(1) = w_{11} = \frac{1}{4}$, $p_1(2) = w_{21} = \frac{1}{2}$ and $p_1(3) = w_{31} = \frac{1}{4}$ by (1.1); consequently the new value of $S[MP(t)]$ is

$$S[MP(1)] = S(2).$$

The change in $S[MP(t)]$ between these two instants is thus

$$\begin{aligned} S(2) - S(1) &= k \ln [\Omega(2)/\Omega(1)] \\ &= k \ln (\pi_2/\pi_1) = -k \ln 2, \end{aligned} \tag{1.3}$$

a decrease. This counter-example shows that no general theorem of non-decrease can be proved for $S[MP(t)]$, and it is therefore not a satisfactory choice for the definition of statistical entropy.

In order to arrive at a satisfactory definition of statistical entropy it is useful to consider the statistical ensemble temporarily as a giant composite dynamical system, whose subsystems are the individual systems of the ensemble. The definition of entropy for this composite system depends, as usual, on the definition chosen for its observational states (which we shall call *composite observational states* to distinguish them from the observational states of a single system). Here there are two different definitions of the composite observational states to be considered. The simpler of these is tantamount to an assumption that each system of the ensemble can be observed individually, so that if the first system is in the state $A^{(1)}$, the second in $A^{(2)}$, and so on, then the ensemble as a whole is in a composite observational state which may be denoted by $(A^{(1)}\, A^{(2)} \ldots)$. Composite observational states defined in this way will be called *complexions*. By the law of addition of entropy, eqn. (3.1) of Chap. V, the entropy of the complexion $(A^{(1)}\, A^{(2)} \ldots)$ is

$$\mathscr{S}(A^{(1)}A^{(2)} \ldots) \equiv \sum_{i=1}^{\mathscr{N}} S(A^{(i)}), \tag{1.4}$$

where $\mathscr{N}$ is the number of systems in the ensemble.

To obtain the average contribution of a single system to the entropy of the complexion of a very large ensemble, we divide both sides of (1.4) by $\mathscr{N}$ and take the limit $\mathscr{N} \to \infty$. The limit of the right-hand side is identical with the statistical expectation [defined in eqn. (2.1) of Chap. II] of the Boltzmann entropy of a single system; accordingly we may write

$$\lim_{\mathscr{N}} \mathscr{S}(A^{(1)}(t)A^{(2)}(t) \ldots) = \langle S(A)\rangle_t \equiv \textstyle\sum_A p_t(A)\, S(A), \tag{1.5}$$

where $\lim_{\mathscr{N}}$ means a limit where $\mathscr{N}$ tends to infinity. We are thus led to consider adopting $\langle S(A)\rangle_t$ as a possible definition of statistical entropy. Like the one we have just rejected, this proposal is at first sight supported by the kinetic theory of dilute gases; for the first line of eqn. (6.11) of Chap. V implies

$$\langle S(A)\rangle_t = -k\textstyle\sum_i \langle N_i \ln (N_i/\Upsilon_i)\rangle + \text{const.} \tag{1.6}$$

when all the occupation numbers are large, and if we also make the approximation of neglecting the standard deviations of these occupation numbers then the right-hand side of (1.6) is the same as $-k\sum\langle N_i\rangle \ln (\langle N_i\rangle/\Upsilon_i)$ + const. and is therefore non-decreasing by virtue of Boltzmann's H theorem. Once again, however, we can show by means of a counter-example that there can be no general law of non-decrease for $\langle S(A)\rangle_t$. The Markov chain with transition probability matrix (1.1) provides such a counter-example: if the initial state is 1, then we have

$$\left.\begin{aligned}\langle S(A)\rangle_0 &= S(1)\\ \langle S(A)\rangle_1 &= \tfrac{1}{4}S(1) + \tfrac{1}{2}S(2) + \tfrac{1}{4}S(3)\\ &= (\tfrac{1}{4} + \tfrac{1}{2} + \tfrac{1}{4})\, S(1) + \tfrac{1}{2}[S(2) - S(1)],\end{aligned}\right\} \tag{1.7}$$

so that

$$\langle S(A)\rangle_1 - \langle S(A)\rangle_0 = -\tfrac{1}{2}k \ln 2 \tag{1.8}$$

by (1.3). Thus $\langle S(A)\rangle_t$, too, can decrease with time, and is therefore no better than $S[MP(t)]$ as a definition of statistical entropy.

This possibility of a decrease in $\langle S(A)\rangle_t$ springs from the fact that composite observational states of the form $(A^{(1)} A^{(2)} \ldots)$ are not in general deterministic (since the observational states of the individual systems are not in general deterministic). Because of this lack of determinism, $\mathscr{S}(A^{(1)} A^{(2)} \ldots)$ can decrease with time, and such a decrease can, by (1.5), lead to a decrease of $\langle S(A)\rangle_t$ with time, such as the one exhibited in (1.8). To obtain a statistical entropy that cannot decrease, therefore, we must choose the composite observational states in a different way, so as to make them deterministic. This can be done if we assume that it is only possible to observe the number of systems in each observational state, not the identities of the individual systems constituting this number. Since the probability of any single-system observational state is proportional (in the limit $\mathscr{N} \to \infty$) to the number of systems of the ensemble that are in that state, our new assumption is tantamount to identifying the composite observational state with the probability distribution over single-system observational states. Composite observational states defined in this new way may therefore be called *distributions*. The Markov equation (4.20) of Chap. II, $p_{t+1}(A) = \sum_B w(AB)\, p_t(B)$, shows that the probability distribution at time $t+1$ is (in the limit $\mathscr{N} \to \infty$) uniquely determined by the probability distribution at time t: consequently the corresponding composite observational states are essentially deterministic, approaching an equilibrium composite observational state (the equilibrium probability distribution) as time progresses, and we can confidently expect the corresponding entropy expression never to decrease with time.

This second method of defining observational states for the entire ensemble is analogous to the method used in Chap. V, §§ 6 and 7, to define them for a classical gas. The systems of the ensemble are analogous to the molecules of the gas: the dynamical space of a single system (γ-space, in classical mechanics) is analogous to the dynamical space of a molecule (μ-space) and the dynamical images of single-system observational states (which, for classical mechanics, are cells in γ-space) are analogous to the cells in μ-space. The observational state of the ensemble is a distribution over the single-system observational states, described mathematically by specifying the set of non-negative integers $\{\mathscr{N}(1), \mathscr{N}(2), \ldots\}$, where $\mathscr{N}(A)$ denotes the number of systems in the single-system observational state A; this description is analogous to the description of observational states of a gas as distributions over μ-space cells, specified by sets of occupation numbers. Pursuing the analogy a little further, it is a simple matter to obtain an expression for the entropy of the compound observational state $\{\mathscr{N}(1), \mathscr{N}(2), \ldots\}$. The number of distinct complexions compatible with this observational state is just $\mathscr{N}!/\Pi_A \mathscr{N}(A)!$. By symmetry, all these complexions correspond to equal weights in dynamical space, and consequently the distribution $\{\mathscr{N}(1),$

$\mathcal{N}(2), \ldots\}$ has a dynamical image whose weight exceeds that of any one of the individual complexions by the factor $\mathcal{N}!/\Pi_A\mathcal{N}(A)!$; the entropy of the distribution (regarded as a single observational state) therefore exceeds the entropy of any one of its complexions by a term

$$k \ln[\mathcal{N}!/\Pi_A\mathcal{N}(A)!].$$

Adding this to the expression (1.4) for the entropy of a single complexion, we obtain the formula

$$\mathcal{S}\{\mathcal{N}(1), \mathcal{N}(2), \ldots\} = \sum_A \mathcal{N}(A)\, S(A) + k \ln [\mathcal{N}!/\prod_A \mathcal{N}(A)!] \quad (1.9)$$

for the entropy of a distribution.

To obtain the average entropy per system we divide both sides of (1.9) by $\mathcal{N}$ and take the limit $\mathcal{N} \to \infty$, just as in the derivation of (1.5). By the definition of probability, eqn. (4.1) of Chap. I, the average entropy per system at time t is therefore

$$\lim_{\mathcal{N}} \mathcal{S}\{\mathcal{N}(1), \mathcal{N}(2), \ldots\}/\mathcal{N}$$
$$= \sum_A p_t(A)\{S(A) + k \lim_{\mathcal{N}} \left[\frac{\ln \mathcal{N}!}{\mathcal{N}} - \frac{\ln \mathcal{N}(A)!}{\mathcal{N}(A)}\right]\},$$

where the sum covers all values of A for which $p_t(A) \neq 0$. Applying Stirling's formula, (6.10) of Chap. V, to the factorials and comparing the resulting expression with (1.5), we obtain

$$\lim_{\mathcal{N}} \mathcal{S}(\text{distribution})/\mathcal{N} = \lim_{\mathcal{N}} \mathcal{S}(\text{complexion})/\mathcal{N} + k \sum_A p_t(A) \ln [1/p_t(A)]. \quad (1.10)$$

Making the usual convention that $x \ln (1/x) = 0$ for $x = 0$, we may extend the sum to all observational states A, and so write (1.10) in the form

$$\lim_{\mathcal{N}} \mathcal{S}(\text{distribution})/\mathcal{N} = \tilde{S}(t) \quad (1.11)$$

where

$$\tilde{S}(t) \equiv \langle S(A)\rangle_t + k\langle \ln [1/p_t(A)]\rangle_t . \quad (1.12)$$

An alternative form for the definition (1.12) can be obtained by substituting from the Boltzmann–Planck formula (3.17) of Chap. V: it is

$$\tilde{S}(t) = k \sum_A p_t(A) \ln [W(A)/p_t(A)], \quad (1.13)$$

where $W(A)$ is the weight of the observational state A, defined in eqn. (3.18) of Chap. V to be $c\Omega(A)$ with c a constant. A third way of writing the definition of $\tilde{S}(t)$ is to use the coarse-grained ensemble density, defined in eqn. (6.27) of Chap. IV by

$$\tilde{D}_t(\alpha) \equiv p_t(A)/\Omega(A) \quad \text{when} \quad \alpha \in \omega(A) \quad (1.14)$$

for classical mechanics and by an analogous expression, eqn. (6.30) of Chap. IV, for quantum mechanics. Using (1.14) in (1.13) we obtain

$$\tilde{S}(t) = -k \int d\alpha \tilde{D}_t(\alpha) \ln \tilde{D}_t(\alpha) + k \ln c. \tag{1.15}$$

Comparing this with the formula (6.11) of Chap. V giving the Boltzmann entropy of a gas in terms of its distribution function, we see that the analogy used in obtaining (1.9) persists, $\tilde{D}(\alpha)$ for the ensemble being the analogue of $f(\xi)$ for the gas. The quantum analogue of (1.15) is

$$\tilde{S}(t) = -k\, tr(\tilde{D}_t \ln \tilde{D}_t) + k \ln c. \tag{1.16}$$

Expressions similar to the right-hand side of (1.15) were first studied by Gibbs in his theory of ensembles,† and he showed that such expressions had properties analogous to those of thermodynamic entropy. Accordingly we shall adopt $\tilde{S}(t)$ as our definition of *statistical entropy*.

To show that this definition gives a statistical entropy that is rigorously non-decreasing, we must consider both deterministic and persistent states, and also the possibility of mechanical operations on the system, just as we did in Chap. V, §§ 1 and 4, when we considered the change of $S(A)$ with time. The case where the single-system observational states are deterministic is easily dealt with, since there all probabilities are either 0 or 1, so that (1.12) makes the statistical entropy identical with the Boltzmann entropy, which we have already shown in Chap. V, § 4, to have a vanishing probability of decrease in this case. For the case of persistent states (the only non-deterministic states we can treat rigorously: see Chap. IV, § 8), we apply the theorem of Chap. II, § 7, which states that any quantity of the form

$$-\textstyle\sum_A \pi_A\, \varphi\{p_t(A)/\pi_A\} \tag{1.17}$$

is a non-decreasing function of time if $\varphi(x)$ is convex for $x \geqq 0$, the sum goes over the ergodic set Z containing the initial state K, and π_A is the equilibrium probability of the state A. If we take $\varphi(x)$ to be

$$\varphi(x) = \begin{cases} -kx \ln [W(Z)/x] & \text{if } x > 0, \\ 0 & \text{if } x = 0, \end{cases} \tag{1.18}$$

† J. W. Gibbs, *Elementary Principles in Statistical Mechanics*, pp. 44–45, and also p. 168. Gibbs takes $\ln c = 0$ and uses the symbol η in place of $\ln \tilde{D}$. His notation does not distinguish between $\tilde{D}$ and the fine-grained phase-space density D defined in Chapter III. The distinction is important because the expression obtained by replacing $\tilde{D}$ by D on the right-hand side of (1.15) or (1.16) is not suitable as a definition of entropy for non-equilibrium situations, for this expression (the fine-grained entropy) can be shown, using Liouville's theorem, to be independent of the time. See exercises 4 of Chap. III, § 3.1, and 3 of Chap. III, § 5.1.

then the expression (1.17) becomes

$$k\sum_A p_t(A) \ln [W(Z)\,\pi_A/p_t(A)]. \tag{1.19}$$

The summation in (1.19) may be extended to cover all observational states, since $p_t(A) = 0$ when A is outside Z. By virtue of the basic formula (5.14) of Chap. IV—or, in quantum mechanics, eqn. (5.25) of Chap. IV—relating equilibrium probabilities to weights, the expression (1.19) is equal to $\tilde{S}(t)$ as defined in (1.13); it follows, by the theorem of Chap. II, § 7, that $\tilde{S}(t)$ cannot decrease with time if the initial state is persistent and the system isolated. Mechanical operations, too, cannot decrease $\tilde{S}(t)$: according to the idealization explained in Chap. I, § 3, any such operations are represented here as a succession of impulsive changes in the Hamiltonian carried out at instants of observation. As we showed in Chap. V, § 1, such changes do not affect the weights $W(A)$, nor do they immediately affect the observational state of any system of the ensemble; consequently they also do not immediately affect the probability distribution, so that the monotonic increase of $\tilde{S}(t)$ is uninterrupted by these impulsive changes of the Hamiltonian. This completes the proof of the non-decrease law for statistical entropy. The law may be written as a formula,

$$\tilde{S}(t_1) \leqq \tilde{S}(t_2) \quad \text{if} \quad t_1 \leqq t_2. \tag{1.20}$$

Unlike Boltzmann's H theorem and the non-decrease law for Boltzmann entropy, the theorem (1.20) requires no special assumptions, approximations or limit operations: it is a direct consequence of the basic postulates of our theory.

Equation (1.12) shows that the statistical entropy exceeds the expectation of the Boltzmann entropy by an amount $k\sum_A p_t(A) \ln[1/p_t(A)]$. This excess is called† the *entropy of the probability distribution* $p_t(\ )$. It represents the part of the statistical entropy arising from the fact that we use a relatively coarse observational description of the ensemble, expressible in terms of probabilities and ensemble averages, rather than the finer description in terms of complexions that would be possible if we followed the behaviour of individual systems. This contribution to the statistical entropy lies in the range

$$0 \leqq k\sum_A p_t(A) \ln [1/p_t(A)] \leqq k \ln M, \tag{1.21}$$

where M is the number of observational states of non-vanishing probability. The left-hand inequality holds because all probabilities are non-negative; it becomes a nequality only if all the probabilities are 0 or 1 — that is, at the initial time, or for deterministic observational states. The right-hand inequality is obtained by applying the fundamental property of convex functions (the inequality (7.6) of Chap. II) to the function $\ln(1/x)$, with $m_A =$

† C. E. Shannon and W. Weaver, *The Mathematical Theory of Communication*, p. 20.

$p_t(A)$ and $x_A = 1/p_t(A)$. The two inequalities together show that the order of magnitude of $\sum_A p_t(A) \ln[1/p_t(A)]$ is a few k, so that its value is comparable with the likely decreases or fluctuations of Boltzmann entropy for individual systems of the ensemble, but (for any macroscopic system) very small compared with the entropy values themselves. That is to say, the term $k\langle \ln(1/p) \rangle$ in (1.12) is a small correction, just large enough to cancel out any possible decrease in $\langle S(A) \rangle_t$, such as the one described by (1.8).

Because of the probabilities $p_t(A)$ appearing in the definition (1.12) or (1.13) of statistical entropy, this concept requires a more sophisticated physical interpretation than Boltzmann entropy. The Boltzmann entropy of a particular system at a particular time t depends only on its observational state at that time; its statistical entropy, on the other hand, depends on what ensemble the system is to be regarded as belonging to, that is, on the experimental procedure to which the system was subjected before time t and which, if replicated many times, would generate this ensemble. In the language of Chap. I, § 5, the Boltzmann entropy depends on the result of the observational stage of a trial, whereas the statistical entropy depends on the instructions specifying the preparation and development stages. Unless the observational states are deterministic, a specification of this experimental procedure, while sufficient to determine the statistical entropy, would not, in general, be sufficient to determine the observational state and hence would not be sufficient to determine the Boltzmann entropy $S(A)$; on the other hand, a specification of the current observational state, although sufficient to determine the current Boltzmann entropy, would not be sufficient to determine the statistical entropy. Thus the two definitions of entropy refer to different physical situations and are, in a sense, complementary.

1.1. Exercises

1. Show that if all the observational states of non-vanishing probability are confined to a single ergodic set Z, then $\tilde{S} \leqq k \ln W(Z)$. Under what conditions does the equality sign hold?

2. The *fine-grained entropy* of a classical Gibbs ensemble (see exercise 3 of Chap. II, § 7.1) is defined by

$$S_{\text{fine}} \equiv k \int D(\alpha) \ln [c/D(\alpha)] \, d\alpha,$$

where $c \equiv W(A)/\Omega(A)$ is the constant of proportionality defined in eqn. (3.18) of Chap. V. Show that $S_{\text{fine}} \leqq \tilde{S}$.

For what type of ensemble does the equality sign hold? If Z is an ergodic set and $D(\alpha)$ vanishes unless $\alpha \in \omega(Z)$, what can we deduce from the equation $S_{\text{fine}} = k \ln W(Z)$?

3. Generalize the result proved in (1.8) by showing that, for any ensemble whose initial state is the most probable state MP in an ergodic set that contains at least one state with equilibrium probability less than that of MP, values of t exist for which $\langle S(A) \rangle_t < \langle S(A) \rangle_0$.

4. An *assembly* is defined as a large collection of identical systems weakly interacting with one another. Taking over the definitions and notation used in the foregoing for ensembles, show that the most probable distribution compatible with a given total energy $\mathscr{E}$ (that is, with a given mean energy $\langle E\rangle \equiv \mathscr{E}/\mathscr{N}$ per system) has the form

$$\mathscr{N}(A)/\mathscr{N} \propto W(A)\, e^{-\beta E(A)} \text{ for large } \mathscr{N},$$

where β is determined by the condition

$$\frac{\sum_A E(A) W(A)\, e^{-\beta E(A)}}{\sum_A W(A)\, e^{-\beta E(A)}} = \langle E\rangle.$$

By taking the limit $\mathscr{N} \to \infty$ and then allowing the weights of the dynamical images $\omega(A)$ to become small, we can derive from this result the canonical distribution for a system in equilibrium with a very large heat bath;† but the derivation indicated in exercise 2 of Chap. V, § 4.1, is much more straightforward.

2. Additivity properties of statistical entropy

In Chap. V, § 1, two conditions were specified which an ideal definition of entropy would satisfy: the non-decrease condition (1.1) and the additivity condition (1.2). Having shown in the preceding section that statistical entropy satisfies the non-decrease condition, we now consider the extent to which it satisfies the additivity condition.

As in the discussion of additivity for Boltzmann entropy (Chap. V, §§ 2 and 3) we consider a composite system $\mathfrak{S}$ consisting of two subsystems $\mathfrak{S}'$ and $\mathfrak{S}''$. For this composite system the Boltzmann entropies exactly satisfy an addition law, given in eqn. (3.1) of Chap. V. Averaging both sides of this equation over an ensemble of such systems, prepared in some composite state $K'K''$ at time 0, we obtain

$$\langle S(A'A'')\rangle_t = \langle S'(A')\rangle_t + \langle S''(A'')\rangle_t \tag{2.1}$$

using the notation of Chap. V, § 2. Combining (2.1) with the definition (1.12), we obtain

$$\tilde{S}(t) = \tilde{S}'(t) + \tilde{S}''(t) - C(t), \tag{2.2}$$

where

$$C(t) \equiv k\langle \ln [p_t(A'A'')/p_t(A')\, p_t(A'')]\rangle_t. \tag{2.3}$$

Thus statistical entropies are not exactly additive, unless the probability distribution is such that the expression defining $C(t)$ vanishes.

The quantity $C(t)$, which measures the deviation from perfect additivity of statistical entropies, lies in the range

$$0 \leqq C(t) \leqq k \ln \min(M', M''), \tag{2.4}$$

where M' is the number of observational states of non-vanishing probability for the subsystem $\mathfrak{S}'$, and M'' the corresponding number for $\mathfrak{S}''$. To

† See E. Schrödinger, *Statistical Thermodynamics* (Cambridge University Press, 1946), Chap. II. The "heat bath" here consists of a large number of weakly interacting replicas of the system itself.

prove the left inequality of (2.4) we combine the definition (2.3) with the elementary inequality $\ln x \leqq x - 1$, obtaining†

$$\begin{aligned} C(t) &= -k \sum_{A'} \sum_{A''} p_t(A'A'') \ln [p_t(A')\, p_t(A'')/p_t(A'A'')] \\ &\geqq -k \sum_{A'} \sum_{A''} p_t(A'A'') \left[\frac{p_t(A')\, p_t(A'')}{p_t(A'A'')} - 1 \right] \\ &= 0 \end{aligned} \tag{2.5}$$

since the probability distributions $p_t(A'A'')$, $p_t(A')$, and $p_t(A'')$ all satisfy the normalization condition (1.5) of Chap. II. To prove the right inequality of (2.4) we combine (2.3) with the inequality $p_t(A'A'') \leqq p_t(A'')$, a consequence of eqn. (1.15) of Chap. II; thus we find that $C(t)$ cannot exceed $k \langle \ln[1/p_t(A')] \rangle$, which in turn cannot exceed $k \ln M'$ because of (1.21). In a similar way we can show that $C(t)$ cannot exceed $k \ln M''$, and the right inequality of (2.4) follows at once. The bounds (2.4), taken together with (2.2), show that for large systems the deviation from perfect additivity of statistical entropies is at most a very small fraction of the total entropy.

Although $C(t)$ is negligible in comparison with thermodynamic entropies, it is a quantity well worth studying in its own right, because it provides a useful measure of the amount of correlation between the observational states of $\mathfrak{S}'$ and $\mathfrak{S}''$. To justify this interpretation of $C(t)$ we note that, by (2.3), $C(t)$ vanishes if

$$p_t(A'A'') = p_t(A')\, p_t(A'') \quad \text{for all} \quad A', A''. \tag{2.6}$$

Moreover, the converse is also true, for if $C(t) = 0$ then the two double sums in (2.5) must be equal term by term and this implies that (2.6) holds. Thus $C(t)$ vanishes if and only if (2.6) holds, or, in the language of Chap. II, § 3, if and only if the observational states of $\mathfrak{S}'$ and $\mathfrak{S}''$ are statistically independent. The magnitude of $C(t)$ therefore provides a measure of the deviation from perfect statistical independence of the observational states of $\mathfrak{S}'$ and $\mathfrak{S}''$; we shall call $C(t)$ the *logarithmic correlation*‡ between $\mathfrak{S}'$ and $\mathfrak{S}''$.

Using the basic results of statistical mechanics we can prove some simple theorems concerning the numerical value of $C(t)$ in various special situ-

† The idea of this proof is due to J. W. Gibbs, *Elementary Principles in Statistical Mechanics*, pp. 133–5.

‡ The importance of logarithmic correlation and some of its basic properties were brought to my attention by I. C. Percival (unpublished manuscript "Entropy and Logarithmic Correlation", 1954). See also L. Berger, Quantité d'Information et Systèmes Physiques, *Helv. Phys. Acta* **31**, 159 (1958).

The logarithmic correlation should not be confused with the *coefficient of correlation* used in statistics. The coefficient of correlation between two random variables X and Y is defined to be covar $(X, Y)/[\text{var}(X)\ \text{var}(Y)]^{\frac{1}{2}}$, in the notation of eqns. (2.16) and (2.24) of Chap. II. Unlike the logarithmic correlation, it can take negative values, and its vanishing does not imply that X and Y are statistically independent.

ations. The simplest of these theorems is that $C(t)$ vanishes if all the probabilities at time t are 0 or 1. This follows from (2.6), since $Pr(A'A'') = 1$ if and only if $Pr(A') = 1$ and $Pr(A'') = 1$. Consequently $C(t)$ must vanish at the initial instant $t = 0$; moreover, if the Markov chain of observational states for the composite system is deterministic, then $C(t)$ vanishes for all t. That is, the type of correlation measured by $C(t)$ can be observed only in a genuinely statistical situation.

A less trivial, though still intuitively obvious, theorem is that there can be no correlation without past interaction: if $\mathfrak{S}'$ and $\mathfrak{S}''$ are both isolated between the initial time and the time t, then $C(t)$ vanishes. Another way of stating the theorem is that if a compound trial $\mathfrak{T}$ is conducted by carrying out two simultaneous sub-trials $\mathfrak{T}'$ and $\mathfrak{T}''$ on two different isolated systems at the same time, then $\mathfrak{T}'$ and $\mathfrak{T}''$ are statistically independent. This theorem is a companion to the result of Chap. II, § 3, that sub-trials conducted with the same system at different times are statistically independent. Normally this is taken for granted – for example if the trial is to toss two coins simultaneously we would take it for granted that the results of the two tosses are statistically independent – but we shall show here that nothing need be taken for granted here beyond the original postulates of the theory, in particular the Markovian postulate.

To prove the theorem, we start once again from the basic law (6.13) of Chap. IV giving probabilities in terms of weights. Summing both sides of (6.13) of Chap. IV over $B_1, C_2, \ldots$, and F_{n-1}, and changing to the notation of the present section, we obtain

$$\begin{aligned} p_t(A'A'') &= Pr(A'_tA''_t|K'K'') \\ &= \Omega(A'_tA''_tK'K'')/\Omega(K'K''). \end{aligned} \tag{2.7}$$

We must show that this can be reduced to the form (2.6) if $\mathfrak{S}'$ and $\mathfrak{S}''$ are isolated. The reduction is simplest when $\mathfrak{S}'$ and $\mathfrak{S}''$ have no particle type in common and obey classical mechanics. For this case, the denominator in (2.7) can be factorized, using eqn. (2.11) of Chap. V, into the form $\Omega'(K')$ $\Omega''(K'')$. To obtain the corresponding factorization for the numerator, we use eqns. (6.14) and (1.2) of Chap. IV to put this numerator into the form

$$\Omega(A'_tA''_tK'K'') = \int d\boldsymbol{P} \int d\boldsymbol{Q}\, J_{A'A''}(\boldsymbol{P}_t, \boldsymbol{Q}_t)\, J_{K'K''}(\boldsymbol{P}, \boldsymbol{Q}), \tag{2.8}$$

where the integration covers all points $(\boldsymbol{P}, \boldsymbol{Q})$ in the phase space of $\mathfrak{S}$, and $(\boldsymbol{P}_t, \boldsymbol{Q}_t) \equiv U_t(\boldsymbol{P}, \boldsymbol{Q})$ with U_t the evolution operator of $\mathfrak{S}$, defined in Chap. III, § 3. The right-hand side of (2.8) can be broken down by a method analogous to the one applied to the expression (2.10) of Chap. V for $\Omega(A'A'')$. The result, analogous to the first line of eqn. (2.11) of Chap. V, is

$$\begin{aligned} \Omega(A'_tA''_tK'K'') = \int d\boldsymbol{P}' \int d\boldsymbol{P}'' \int_{C'} d\boldsymbol{Q}' \int_{C''} d\boldsymbol{Q}''\, J_{A'}(\boldsymbol{P}'_t, \boldsymbol{Q}'_t)\, J_{A''}(\boldsymbol{P}''_t, \boldsymbol{Q}''_t) \\ \times J_{K'}(\boldsymbol{P}', \boldsymbol{Q}')\, J_{K''}(\boldsymbol{P}'', \boldsymbol{Q}''), \end{aligned} \tag{2.9}$$

where C' and C'' are the configuration spaces of $\mathfrak{S}'$ and $\mathfrak{S}''$ respectively, $(\boldsymbol{P}', \boldsymbol{Q}')$ and $(\boldsymbol{P}'', \boldsymbol{Q}'')$ are the dynamical states of $\mathfrak{S}'$ and $\mathfrak{S}''$ when $\mathfrak{S}$ has the dynamical state $(\boldsymbol{P}, \boldsymbol{Q})$, and $(\boldsymbol{P}'_t, \boldsymbol{Q}'_t)$ and $(\boldsymbol{P}''_t, \boldsymbol{Q}''_t)$ are analogously related to $(\boldsymbol{P}_t, \boldsymbol{Q}_t)$. Since $\mathfrak{S}'$ and $\mathfrak{S}''$ are isolated sub-systems, the dynamical state $(\boldsymbol{P}'_t, \boldsymbol{Q}'_t)$ of $\mathfrak{S}'$ at time t is uniquely determined by its dynamical state $(\boldsymbol{P}', \boldsymbol{Q}')$ at time 0, and is the same function of $(\boldsymbol{P}', \boldsymbol{Q}')$ as if $\mathfrak{S}''$ were absent. Consequently the integral in (2.9) can be factorized to give

$$\Omega(A'_tA''_tK'K'') = \int d\boldsymbol{P}' \int_{C'} d\boldsymbol{Q}' J_{A'}(\boldsymbol{P}'_t, \boldsymbol{Q}'_t)\, J_{K'}(\boldsymbol{P}', \boldsymbol{Q}')$$

$$\times \int d\boldsymbol{P}'' \int_{C''} d\boldsymbol{Q}'' J_{A''}(\boldsymbol{P}''_t, \boldsymbol{Q}''_t)\, J_{K''}(\boldsymbol{P}'', \boldsymbol{Q}'')$$

$$= \Omega'(A'_tK')\,\Omega''(A''_tK'') \tag{2.10}$$

by eqns. (1.2) and (6.14) of Chap. IV. Substituting (2.10) into (2.7), and also using the corresponding factorization, (2.11) of Chap. V, for the denominator of (2.7), we obtain

$$Pr(A'_tA''_t|K'K'') = \frac{\Omega'(A'_tK')\,\Omega''(A''_tK'')}{\Omega'(K')\,\Omega''(K'')} = Pr(A'_t|K')\,Pr(A''_t|K''). \tag{2.11}$$

Allowing for the difference in notation, the result (2.11) is the same as (2.6), and therefore completes the proof that there can be no correlation without interaction. This multiplication law can also be proved, by analogous methods, for the cases where $\mathfrak{S}'$ and $\mathfrak{S}''$ do have particle types in common, and where they obey quantum rather than classical mechanics.

This result is actually a special case of a stronger theorem, that the logarithmic correlation between $\mathfrak{S}'$ and $\mathfrak{S}''$ cannot increase during any time interval throughout which both subsystems are isolated. To prove this stronger theorem it is sufficient to show that $C(t+1) - C(t) \leqq 0$ if $\mathfrak{S}'$ and $\mathfrak{S}''$ are isolated between the instants t and $t+1$. By the definition (2.3), this increase in correlation is given by

$$C(t+1) - C(t) = k \sum_{B'B''} p_{t+1}(B'B'') \ln \frac{p_{t+1}(B'B'')}{p_{t+1}(B')\,p_{t+1}(B'')}$$

$$- k \sum_{A'A''} p_t(A'A'') \ln \frac{p_t(A'A'')}{p_t(A')\,p_t(A'')}. \tag{2.12}$$

According to the Markovian equation (4.20) of Chap. II, the probabilities at time $t+1$ are given by

$$p_{t+1}(B'B'') = \sum_{A'A''} w(B'B'', A'A'')\,p_t(A'A''), \tag{2.13}$$

where, by analogy with eqn. (4.9) of Chap. II, we have defined the transition probabilities

$$w(B'B'', A'A'') \equiv Pr(B'_{t+1}B''_{t+1}|A'_tA''_t). \tag{2.14}$$

Transforming the first sum in (2.12) by means of (2.13) and the second by means of the normalization condition for the transition probabilities in (2.14), we obtain

$$C(t+1) - C(t) = k \sum_{A'A''} \sum_{B'B''} w(B'B'', A'A'')\, p_t(A'A'') \times \left[\ln \frac{p_{t+1}(B'B'')}{p_{t+1}(B')\, p_{t+1}(B'')} - \ln \frac{p_t(A'A'')}{p_t(A')\, p_t(A'')} \right]. \tag{2.15}$$

Using the elementary inequality $\ln x - \ln y = \ln (x/y) \leqq (x/y) - 1$, we deduce from (2.13) the upper bound

$$C(t+1) - C(t) \leqq k \sum_{A'A''} \sum_{B'B''} w(B'B'', A'A'') \times \left[\frac{p_{t+1}(B'B'')\, p_t(A')\, p_t(A'')}{p_{t+1}(B')\, p_{t+1}(B'')} - p_t(A'A'') \right]. \tag{2.16}$$

Provided $\mathfrak{S}'$ and $\mathfrak{S}''$ do not interact between the times t and $t+1$, the right-hand side of (2.16) can be simplified by applying the product law (2.11), with initial time t rather than 0 and final time $t+1$ rather than t, to the right-hand side of (2.14). This gives

$$\begin{aligned} w(B'B'', A'A'') &= Pr(B'_{t+1}|A'_t)\, Pr(B''_{t+1}|A''_t) \\ &= w'(B', A')\, w''(B'', A''), \end{aligned} \tag{2.17}$$

where $w'(B', A')$ is the transition probability for $\mathfrak{S}'$ considered on its own. By combining (2.17) with the Markovian law (4.20) of Chap. II applied to $\mathfrak{S}'$ and $\mathfrak{S}''$ separately, we obtain the sum rule

$$\sum_{A'A''} w(B'B'', A'A'')\, p_t(A')\, p_t(A'') = p_{t+1}(B')\, p_{t+1}(B''). \tag{2.18}$$

Using (2.18) to simplify the contribution from the first term in square brackets to the sum in (2.16), and (2.13) to simplify that from the second, we find that the right-hand side of (2.16) vanishes, and hence that

$$C(t+1) \leqq C(t) \tag{2.19}$$

if $\mathfrak{S}'$ and $\mathfrak{S}''$ are isolated between the instants t and $t+1$. Thus, the only times when $C(t)$ can increase are the times when $\mathfrak{S}'$ and $\mathfrak{S}''$ interact.

2.1. Exercises

1. When does $C(t) = k \ln \min (M', M'')$?

2. Show that if the composite system $\mathfrak{S}' + \mathfrak{S}''$ is in equilibrium the nthe logarithmic correlation between its two subsystems is zero.

3. Show that if a classical composite system comprises two non-interacting subsystems both of which obey the consistency condition analogous to eqn. (7.2) of Chap. IV, then the composite system also obeys the corresponding consistency condition.

3. Perpetual motion

The non-decrease law for statistical entropy, proved in § 1, is the closest analogue in statistical mechanics for the thermodynamic law of non-decreasing entropy, usually called the second law of thermodynamics. The second law is often formulated as a principle asserting that it is impossible to build a perpetual motion machine of the second kind – one that extracts heat from its surroundings and converts it all into mechanical work. It is the purpose of the present section to provide an analogous formulation for the law of non-decrease for statistical entropy, with a view to providing a concrete physical interpretation for this law and for the term $k\langle \ln 1/p(A)\rangle$ in the definition of statistical entropy.

Following Planck,† we may think of a perpetual motion machine of the second kind as one that could be attached to a system containing a heat reservoir and a weight and whose cycle of operations would have the net effect of extracting heat from the reservoir and raising the weight. In other words, at the time (call it u) when the cycle of operations ended the machine (call it $\mathfrak{M}$) would be in the same observational state as it was at the time (say 0) when the cycle began, and the attached system (call it $\mathfrak{S}'$) would be in an observational state of lower entropy than at time 0. The composite system $\mathfrak{S} \equiv \mathfrak{S}' + \mathfrak{M}$ would thus have the property that its entropy could decrease with time. In thermodynamics, where all thermal fluctuations are treated as unobservable, the assertion of the second law that no perpetual motion machine of the second kind can exist is thus equivalent to an assertion that the entropy of an isolated system cannot decrease.

In statistical mechanics, on the other hand, fluctuations cannot be ignored and, as we saw in Chap. V, § 4, the possibility of a decrease in the (Boltzmann) entropy does exist. As a specific example, suppose that $\mathfrak{S}'$ consists of a Brownian particle in suspension in a fluid. Even without the intervention of any hypothetical perpetual motion machine, there is a good probability that the second of two successive observations of the Brownian particle will reveal it to be higher than it was at the first; if this happens then the observations are open to the interpretation that some heat has been taken from a reservoir (the fluid) and has been used to raise a weight (the Brownian particle) . This interpretation may seem perverse, but it cannot be denied that the Boltzmann entropy of the system decreases whenever the Brownian particle is observed to move upwards. The amount of the decrease is $mg\delta h/T$, where m is the mass of the particle, δh its upward displacement, and T the temperature of the fluid, defined in eqn. (1.6) of Chap. V. Although this decrease is very unlikely to be more than a few k, the fact that it can occur at all compels us to consider the possibility, not recognized in thermodynamics, of a perpetual motion machine whose principle of

† M. Planck, *Thermodynamics* (Dover reprint, New York, 1945; translated from the 7th German edition of 1922 by A. Ogg), § 116.

operation depends on some fluctuation phenomenon such as Brownian motion.

In a literal sense, of course, Brownian motion itself is a kind of perpetual motion, but it lacks a feature that would be essential in the operation of any useful perpetual motion machine: this is the feature of reliability. Brownian motion is completely capricious and unreliable, the particle moving sometimes upwards, sometimes downwards. Thus, although there is a probability of roughly half that the first observed displacement of the Brownian particle will be upwards, indicating that a little heat has been converted into work, the probability that during a succession of observations the upward displacements will exceed the downward by a margin large enough to amount to a useful quantity of mechanical work is much less than the probability of breaking the bank at Monte Carlo.

This property of reliability which would distinguish a useful perpetual motion machine from the capricious "perpetual motion" of a Brownian particle may be idealized by requiring that the operation of the perpetual motion machine be *cyclically deterministic*. By this we mean that the (observational) state of the composite system $\mathfrak{S}' + \mathfrak{M}$ at the end of each cycle of operations must be uniquely determined by its state at the beginning, even though the intermediate states need not be. If K is the state at the beginning of a cycle (time 0) and B the state at the end (time u), with B uniquely determined by K, then we have

$$Pr(B_u|K) = 1. \tag{3.1}$$

We have seen earlier, however, [eqn. (4.4) of Chap V] that probabilities such as the one in (3.1) have upper bounds of the form

$$Pr(B_u|K) \leqq \exp{[S(B) - S(K)]/k}. \tag{3.2}$$

Equations (3.1) and (3.2) together imply that $S(B) \geqq S(K)$ so that the pertpeual motion machine cannot do what is required of it, namely to reduce het entropy. Accordingly, the following revision of Planck's form of the second law makes it compatible with the possibility of observable fluctuations: there can be no cyclically deterministic perpetual motion machine of the second kind.

Although this statement of the second law refers directly only to deterministic situations, for which the Boltzmann definition of entropy is adequate, we can also make use of it in discussing the non-deterministic situations for which the statistical definition is more suitable, and so obtain a physical interpretation of the law of non-decreasing statistical entropy. To do this, we take advantage of the fact that the intermediate stages in the cycle of operations need not be deterministic. Let t be any intermediate time during the cycle of operations to which (3.1) refers, so that $0 \leqq t \leqq u$, and let $p_t(A)$ be the probability distribution at time t. The non-decrease law (1.20), when combined with the definition (1.12) of statistical entropy, gives

$$S(K) - \langle S(A)\rangle_t \leqq k\langle \ln[1/p_t(A)]\rangle_t, \tag{3.3}$$

$$S(B) - \langle S(A)\rangle_t \geqq k\langle \ln[1/p_t(A)]\rangle_t. \tag{3.4}$$

Equation (3.3) shows that if the Boltzmann entropy is decreased, on the average, by the operations in the part of the cycle between times 0 and t, then this decrease is paid for by an equal or greater increase in the entropy of the probability distribution. (That such decreases of Boltzmann entropy are possible is shown by the artificial example to which eqns. (1.7) and (1.8) refer. A less artificial example is provided by the case of Brownian motion if K is a state for which the particle is on the bottom of the container, so that it is bound to move upwards at first).

According to the inequality (1.21), the right-hand side of (3.3) cannot exceed $k \ln M$, where M is the number of observational states of the composite system $\mathfrak{S}' + \mathfrak{M}$ compatible with its initial energy. The values of M that would be necessary to achieve a significant increase in the Boltzmann entropy of $\mathfrak{S}'$ with no compensating increase in that of $\mathfrak{M}$ are fantastically large; for example, the entropy decrease on cooling 1 g of water from 300° to 299°K is (1/300) cal/deg $= 10^{21}\,k$, corresponding to a minimum value for M of $\exp(10^{21})$, a number whose decimal representation requires more than 10^{20} digits. If $\mathfrak{S}'$ is a normal physical system (such as 1 g of water), its number of observational states is utterly negligible on this scale, and so the only way for $\mathfrak{S}' + \mathfrak{M}$ to have so many observational states is for $\mathfrak{M}$ itself to be complicated enough to have about $\exp(10^{21})$ observational states. For example, if $\mathfrak{M}$ were a machine whose moving parts had an average of e different observational states each and occupied a volume of 1 mm^3 each, then 10^{21} of these moving parts would be required, and the smallest cube that could accommodate such a machine would have a side of 10 km.

Even supposing that such a machine could be built and operated without any significant entropy increases due to its own internal friction, there would still be no way of utilizing any entropy decreases it might be able to produce in $\mathfrak{S}'$ for perpetual motion purposes; for to complete the cycle of operations the machine must return to its original state, and (3.4) shows that on the average this "resetting" operation must be accompanied by an increase of Boltzmann entropy at least as great as the previous decrease. This application of (3.4) suggests a physical interpretation for $k\langle \ln [1/p_t(A)]\rangle_t$, the entropy of the probability distribution $p_t(\ )$: if the systems comprising an ensemble have the probability distribution $p_t(\ )$ at an instant t and are afterwards all brought to a single observational state by an operation that is the same for all the systems, then this operation must be accompanied by an increase of Boltzmann entropy averaging at least $k\langle\ln[1/p_t(A)]\rangle_t$ per system. Thus the term $k\langle\ln[1/p_t(A)]\rangle_t$ in the definition (1.12) of statistical entropy can be thought of as a latent contribution to the mean Boltzmann entropy, which becomes manifest only when we perform a *setting operation* that brings all members of the ensemble to the same observational state.

As an illustration, consider a classical system consisting of a single particle in a box that is separated by a partition into two compartments of volumes V_R and V_L (Fig. 12). We treat the total energy (of particle and box),

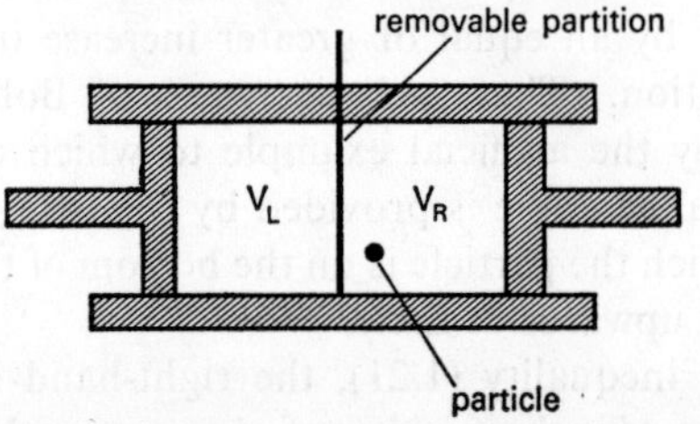

FIG. 12. Example illustrating the increase of Boltzmann entropy in a setting process. The basic observables are the total energy and an observable capable of just two "values": *L* if the particle is in the left compartment and *R* if it is in the right. The setting process is to remove the partition, move the right-hand piston half-way to the left, replace the partition, and withdraw the piston. The left-hand piston is not used in this particular process.

and the label (L or R) of the compartment occupied by the particle, as the only observables. Consider an ensemble of such systems, all with the same energy E, in which a fraction $p(L)$ of the replica systems have the particle in the left compartment and a fraction $p(R) = 1 - p(L)$ have it in the right compartment. Since the phase-space volumes of states having the same energy but the particle in different compartments are proportional to the volumes of the compartments, the mean Boltzmann entropy in this ensemble has the form

$$\langle S(A)\rangle = k\,\{p(L)\ln V_L + p(R)\ln V_R + f(E)\}, \tag{3.5}$$

where $f(\ )$ is independent of $p(L)$ and $p(R)$. Suppose now that all the replica systems in this ensemble are set to a state with the particle in the left compartment: this can be done by removing the partition, sweeping the particle into the left compartment with the piston shown on the right in Fig.12, replacing the partition, and finally withdrawing the piston again. This operation alters the mean Boltzmann entropy in two ways: firstly, the system is slightly heated because of the work the piston must do to compress the one-particle "gas" in the double compartment, and secondly (unless $V_L = V_R$), the mean value of $\ln V$ is altered. The ensemble average of the work done by the piston is

$$-\int P\,dV = -\int (kT/V)\,dV = kT\ln[(V_L + V_R)/V_L],$$

where P and V are the pressure and volume of the one-particle gas; since $dS/dE = 1/T$, this work increases the term $kf(E)$ in the expression (3.5) by an amount $k \ln[(V_L + V_R)/V_L]$. Adding to this increase the change in the average value of $k \ln V$, we obtain

$$\Delta\langle S(A)\rangle = k\{[1 - p(L)] \ln V_L - p(R) \ln V_R + \ln [(V_L + V_R)/V_L]\}$$

$$= k\{\ln(V_L + V_R) - p(L) \ln V_L - p(R) \ln V_R\}.$$

A little elementary calculus shows that this expression (regarded as a function of V_L and V_R) has the minimum value

$$k\{p(L) \ln [1/p(L)] + p(R) \ln [1/p(R)]\},$$

which is achieved if $V_L/(V_L + V_R) = p(L)$, $V_R/(V_L + V_R) = p(R)$. Thus we verify, for this particular model, our general interpretation of $k\langle \ln[1/p(A)]\rangle$ as the least possible increase of mean Boltzmann entropy in a process that brings all the systems in an ensemble with probability distribution $p(\ )$ to a single observational state.

An animate variant of the perpetual motion machine of the second kind is *Maxwell's Demon*; this was defined by Maxwell himself in the following passage.†

> If we conceive a being whose faculties are so sharpened that he can follow every molecule in its course, such a being, whose attributes are still as essentially finite as our own, would be able to do what is at present impossible to us.... Now let us suppose that... a vessel [full of air at uniform temperature] is divided into two portions A and B, by a division in which there is a small hole, and that a being, who can see the individual molecules, opens and closes this hole, so as to allow only the swifter molecules to pass from A to B, and only the slower ones to pass from B to A. He will thus, without expenditure of work, raise the temperature of B and lower that of A, in contradiction to the second law of thermodynamics.

If we adopt the mechanistic standpoint in relation to the biology of this hypothetical demon, there is no essential distinction between his operation and that of a perpetual motion machine of the second kind. Once again, if we combine the demon (denoted by $\mathfrak{D}$) and the system he acts on (denoted by $\mathfrak{S}'$) into a single composite system $\mathfrak{D} + \mathfrak{S}'$, then any decrease in the Boltzmann entropy of this composite system must, on the average, be compensated by an equal increase in the entropy of the probability distribution over observational states of $\mathfrak{D} + \mathfrak{S}'$. The large number of distinct observational states that the Maxwell demon must have in order to make significant entropy reductions possible may be thought of as a large memory capacity in which the demon stores the information about the system which he acquires as he works reducing its entropy. As soon as the

† J. C. Maxwell, *Theory of Heat* (Longmans, London, 1871), p. 328. For a review of the work inspired by Maxwell's idea, see W. Ehrenburg, Maxwell's Demon (*Scientific American*, November 1967).

demon's memory is completely filled,† however, (that is, as soon as $k\langle \ln[p(A)]\rangle$ has reached its maximum value $k \ln M$, where M is the number of observational states of $\mathfrak{D} + \mathfrak{S}'$ compatible with its initial energy) he can achieve no further reduction of the Boltzmann entropy. He gains nothing, for example, by deliberately forgetting or erasing some of his stored information in order to make more memory capacity available; for the erasure, being a setting process, itself increases the entropy by an amount at least as great as the entropy decrease made possible by the newly available memory capacity.

3.1. *Exercise*

1. A Brownian particle is at the bottom end of a weightless thread, the top end of which is controlled by an observer. The observer adjusts the length of the thread to prevent the particle from moving downwards, but without doing any work on the particle—that is, every time he observes the particle to have moved upwards he shortens the thread just enough to prevent the particle from moving down again. In the long run, since the particle cannot move downwards, it will rise, the work to raise it being provided by the fluid it is suspended in. Explain why this "ratchet process" does not violate the law of non-decreasing entropy, (i) if the observer is included in the system under consideration, and (ii) if he is not.

4. Entropy and information

The connection between the memory capacity of a Maxwell demon and the magnitude of the entropy decreases he can achieve, noted at the end of the previous section, is a particular case of a more general connection between entropy and quantity of information, which was discovered by Szilard.‡ This connection, which is the basis of the mathematical theory of communication,§ will be studied in this section and the next. In the theory of communication, a message is regarded as a random sequence of symbols

† " 'You see', he [Sherlock Holmes] exclaimed, 'I consider that a man's brain originally is like a little empty attic, and you have to stock it with such furniture as you choose. ... It is a mistake to think that that little room has elastic walls and can distend to any extent. Depend upon it, there comes a time when for every addition of knowledge you forget something that you knew before. It is of the highest importance, therefore, not to have useless facts elbowing out the useful ones.' " (A. Conan Doyle, *A Study in Scarlet*.)

If Sherlock Holmes's brain had contained the normal 10^{10} neurons, and each of them had had just two observational states, then his whole brain would have had $2^{10^{10}}$ distinct observational states. A Maxwell demon with this amount of memory capacity could bring about entropy decreases not exceeding $10^{10}\,k \ln 2$, which is roughly 10^{-11} of the entropy decrease when 1 g of water is cooled from 300° to 299°K.

‡ L. Szilard, Über die Entropieverminderung in einem thermodynamischen System bei Eingriffen intelligenter Wesen (On the Decrease of Entropy in a Thermodynamic System in case of Intervention of Intelligent Beings), *Z. Phys.* **53**, 840 (1929).

§ C. E. Shannon and W. Weaver, *The Mathematical Theory of Communication*.

chosen from a given "alphabet" in accordance with a given probability law. The alphabet will include not only letters but also other symbols such as punctuation marks, spaces, and so on. If the probability of the nth symbol in the message being the A^nth in the alphabet, conditional on the first $n-1$ symbols of the message being the A^1th, ..., A^{n-1}th of the alphabet, is $p_n \equiv Pr(A^n|A^{n-1} \ldots A^1)$ then this nth symbol of the message may be said to convey a *quantity of information*†

$$I_n \equiv k' \ln [1/Pr(A^n|A^{n-1} \ldots A^1)]. \tag{4.1}$$

The constant k', which fixes the unit of information, is usually given the value $1/\ln 2$ (the corresponding unit of information being the *binary digit* or *bit*) but here we shall take $k' \equiv k$ in order to simplify the connection with entropy. The reason for the choice of the logarithmic function can be seen by considering the total quantity of information conveyed by a succession of symbols, say the mth to nth symbols of the message. By the multiplication law (1.20) of Chap. II for conditional probabilities, this total is

$$\begin{aligned} I_m + I_{m+1} + \cdots + I_n &= k \ln [1/Pr(A^n|A^{n-1} \ldots A^1)\, Pr(A^{n-1}|A^{n-2} \ldots A^1) \\ &\qquad \ldots Pr(A^m|A^{m-1} \ldots A^1)] \\ &= k \ln [1/Pr(A^n \ldots A^m|A^{m-1} \ldots A^1)]. \end{aligned} \tag{4.2}$$

By a natural extension of (4.1), this last expression may also be regarded as the quantity information conveyed by the mth to nth symbols in the message considered together instead of separately. Thus the quantity of information conveyed by a word as a whole is equal to the sum of the quantities of information conveyed by its letters individually. Note also that the most unlikely words and messages convey the greatest quantity of information: a common word such as "of" or "the" conveys much less information than an uncommon one such as "pink".

In order to relate these ideas to statistical mechanics, let us consider a trial $\mathfrak{T}$ which consists of preparing a system in a state K at time 0 and then observing it at predetermined instants of observation t^1, t^2, etc., satisfying

$$0 < t^1 < t^2 < \cdots. \tag{4.3}$$

The observational states of the system at the times t^1, t^2, etc., which we shall denote‡ by A^1, A^2, etc., may be regarded as the successive symbols of a "message" transmitted from system to observer. Each of these symbols is,

† More commonly, the expection of I_n is called the quantity of information, but our usage agrees with that of I. J. Good in *Probability and the Weighing of Evidence* (Griffin, London, 1950), § 6.9, and of G. A. Barnard, The Theory of Information, *J. Roy. Statist. Soc. B* **13**, 46 (1951).

‡ In the notation of eqn. (8.36) of Chap. V A^1, A^2, etc., would be written $A(t^1)$, $A(t^2)$, etc.

of course, a random variable. According to (4.1), the information conveyed at the first observation is

$$I_1 = k \ln [1/Pr(A^1_{t^1})], \tag{4.4}$$

where $Pr(A^1_{t^1})$ denotes the probability of the state A^1 at the time t^1. The average of the random variable I_1 over the ensemble $\mathscr{E}$ obtained by replicating the trial $\mathfrak{T}$ many times is

$$\langle I_1 \rangle = k \langle \ln [1/Pr(A^1_{t^1})] \rangle \tag{4.5}$$

which, allowing for differences of notation, is identical with the expression used in § 1 for the entropy of the probability distribution $p_{t^1}(\ )$ in the ensemble $\mathscr{E}$ at time t^1. Thus the entropy of a probability distribution at any instant measures the average quantity of information that would be obtained about a system in the corresponding statistical ensemble by observing it at that instant.

The result (4.5) can be generalized to the later observations in the sequence. The quantity of information conveyed by the nth observation, whose result we are denoting by A^n, is determined by the conditional probability of this result given that the state of the system was A^{n-1} at time t^{n-1}, A^{n-2} at time t^{n-2}, and so on. This quantity of information is therefore

$$I_n = k \ln [1/Pr(A^n_{t^n} | A^{n-1}_{t^{n-1}} \dots A^1_{t^1})] \tag{4.6}$$

which simplifies, by the Markovian postulate, to

$$I_n = k \ln [1/Pr(A^n_{t^n} | A^{n-1}_{t^{n-1}})]. \tag{4.7}$$

Taking the statistical expectation on both sides, we obtain

$$\langle I_n \rangle = k \textstyle\sum_{A^{n-1}} \{ \sum_{A^n} \ln [1/Pr(A^n | A^{n-1})] \, Pr(A^n | A^{n-1}) \} \, Pr(A^{n-1}), \tag{4.8}$$

where we have used the definition of conditional probability, (1.13) of Chap. II, suppressing the suffixes t^n and t^{n-1} for brevity. Comparing the expression in braces with the expression [see eqn. (1.21)] for the entropy of a probability distribution, we see that the former is just the entropy of the conditional probability distribution $Pr(A^n | A^{n-1})$.

This conditional probability distribution can be interpreted in two distinct ways. According to the physical interpretation of probability upon which this book is based, it is to be interpreted as the distribution of relative frequencies in a sub-ensemble of the original ensemble $\mathscr{E}$; this sub-ensemble, which we shall call $\mathscr{E}(A^{n-1}_{t^{n-1}})$, or $\mathscr{E}^{n-1}$ for short, comprises all the members of $\mathscr{E}$ that are in the state A^{n-1} at the time t^{n-1}. Alternatively, it may be interpreted subjectively, as the subjective probability distribution that would be used by a person who knew the result of the $(n-1)$th observation and was awaiting that of the nth. We have avoided the subjective interpretation up to now, but here it is useful because the notion of information itself has a strongly subjective flavour.

To make this subjective aspect of the theory more complete let us define the *observer probability distribution* $q_t(A|B)$ to be the subjective probability distribution over observational states A that would be held at time t by an observer who knew the result B of the most recent observation but nothing about the subsequent behaviour of the system. The observer probability distribution is given by

$$q_t(A|B) \equiv Pr(A_t|B_{t^m}), \tag{4.9}$$

where m denotes the serial number of the latest observation before time t, so that

$$t^m < t \leqq t^{m+1}. \tag{4.10}$$

We may also define an *observer statistical entropy*, by analogy with the standard definition (1.13) of statistical entropy, to be

$$\tilde{S}_{\text{obs}}(t|B) \equiv k\textstyle\sum_A q_t(A|B) \ln [W(A)/q_t(A|B)]. \tag{4.11}$$

To obtain a non-decrease law for observer statistical entropy, analogous to the non-decrease law (1.20) for the ordinary statistical entropy, it is convenient to return to the interpretation of probabilities in terms of ensembles. According to the definitions (4.9) and (4.11), we may interpret $q_t(A|B)$ and $\tilde{S}_{\text{obs}}(t|B)$ as the probability distribution and statistical entropy in a sub-ensemble $\mathscr{E}(B_{t^m})$ that comprises all the members of $\mathscr{E}$ whose state is B at time t^m. Throughout any time interval in which no observation is made, the value of m stays constant (by (4.10)) and consequently the sub-ensemble $\mathscr{E}(B_{t^m})$ comprises the same individual systems throughout the interval. Applying the non-decrease law (1.20) to this sub-ensemble, and noting also that the statistical entropy of this sub-ensemble at time t^m is $k \ln W(B)$, we find that

$$k \ln W(B) \leqq \tilde{S}_{\text{obs}}(t_1|B) \leqq \tilde{S}_{\text{obs}}(t_2|B) \tag{4.12}$$

for any t_1, t_2 satisfying

$$t^m < t_1 < t_2 \leqq t^{m+1}. \tag{4.13}$$

To obtain the corresponding result for the case where the time interval (t_1, t_2) does contain an observation, say the nth, we must allow for the fact that m increases by 1 during the interval, so that the composition of the sub-ensemble $\mathscr{E}(B_{t^m})$ changes and the simple non-decrease law (1.20) no longer applies. In fact, $\tilde{S}_{\text{obs}}$ does show a tendency to decrease at the moment of an observation. To estimate the decrease, we calculate $\langle\tilde{S}_{\text{obs}}(t^n)\rangle$, the ensemble average of $\tilde{S}_{\text{obs}}(t|B)$ at the time t^n when the nth observation is made. Since the fraction of the members of $\mathscr{E}$ that belong to $\mathscr{E}(B_{t^{n-1}})$ is $Pr(B_{t^{n-1}})$, we obtain

$$\begin{aligned}\langle\tilde{S}_{\text{obs}}(t^n)\rangle &= \textstyle\sum_B \tilde{S}_{\text{obs}}(t^n|B)\, Pr(B_{t^{n-1}}) \\ &= k\textstyle\sum_B\sum_A \ln [W(A)/Pr(A_{t^n}|B_{t^{n-1}})]\, Pr(A_{t^n}B_{t^{n-1}})\end{aligned} \tag{4.14}$$

by (4.11) and (4.9). Using (4.8) we can simplify this to

$$\langle \tilde{S}_{\text{obs}}(t^n) \rangle = k \sum_A \ln W(A)\, Pr(A_{t^n}) + \langle I_n \rangle. \tag{4.15}$$

Combining (4.15) with (4.12) we obtain the inequality

$$\begin{aligned} \langle \tilde{S}_{\text{obs}}(t_1) \rangle \leqq \langle \tilde{S}_{\text{obs}}(t^n) \rangle &= \langle k \ln W(A) + I_n \rangle \\ &\leqq \langle \tilde{S}_{\text{obs}}(t_2) + I_n \rangle \end{aligned} \tag{4.16}$$

for any t_1, t_2 satisfying

$$t^{n-1} < t_1 \leqq t^n < t_2 \leqq t^{n+1}. \tag{4.17}$$

Thus it is possible for the observer statistical entropy to decrease at the moment when an observation is made, but the decrease cannot, on the average, exceed the quantity of information conveyed by that observation.

The two inequalities (4.12) and (4.17) can be combined if we first define

$$I(t) \equiv I_1 + I_2 + \cdots + I_{m(t)} \tag{4.18}$$

which, by (4.10) and the addition law (4.2), is the total quantity of information conveyed by all the observations made before the time t. Then (4.12) and (4.17) together give the new non-decrease law†

$$\langle \tilde{S}_{\text{obs}}(t_1) + I(t_1) \rangle \leqq \langle \tilde{S}_{\text{obs}}(t_2) + I(t_2) \rangle \quad \text{if} \quad t_1 \leqq t_2. \tag{4.19}$$

Like the simpler law of non-decrease for ordinary statistical entropy, the law (4.19) can be interpreted in terms of perpetual motion attempts and setting processes, along the lines laid down in § 3. Since $\tilde{S}_{\text{obs}}(t|B)$ is the statistical entropy of the sub-ensemble $\mathscr{E}(B_{t^m})$, it may be interpreted [see eqn. (3.4)] as a lower bound on the Boltzmann entropy of a state to which all the members of $\mathscr{E}(B_{t^m})$ can be brought by a suitable setting operation. Its ensemble average is thus a lower bound on the average Boltzmann entropy of the ensemble $\mathscr{E}$ after a setting process which, unlike the ones considered in § 3, may use different operations to set the different sub-ensembles. Such a *discriminative* setting process can indeed be applied, in principle, by a person (or machine) outside the ensemble who knows the results of the mth measurement and therefore also knows which system is in which sub-ensemble.

With this physical interpretation of $\tilde{S}_{\text{obs}}(t|B)$, it is easy to interpret the non-decrease law (4.19): the interpretation is that the more information an observer acquires about the ensemble, the more diverse are the possibilities available to him for discriminating between different systems in the setting process, and therefore the lower is the final Boltzmann entropy he can hope to achieve by means of such a process. In other words, he can use his information about the ensemble to reduce its statistical entropy.

† This is L. Brillouin's "negentropy principle of information" (*Science and Information Theory*; Academic Press, New York, 1962; p. 153).

As an illustration of a discriminative setting process, consider again the one-particle model discussed in the preceding section (see Fig. 12). In the symmetrical case where $V_L = V_R$ and $p(L) = p(R) = \frac{1}{2}$, the non-discriminative setting process considered previously increases the mean Boltzmann entropy by at least $k \ln 2$. This increase can be completely eliminated by using instead a discriminative setting process that treats the system differently when the particle is on the right and when it is on the left. This process is to do nothing at all if the particle is on the left, and to turn the entire system round if the particle is (initially) on the right. Since this discriminative setting process does not use the pistons, it does not alter the mean Boltzmann entropy; moreover, it reduces the quantity $k \langle \ln (1/p) \rangle$ from $k \ln 2$ to zero, and thus reduces the total statistical entropy by $\langle I \rangle = k \ln 2$.

This possibility of reducing the statistical entropy of a system does not violate the second law because the principle that entropy cannot decrease applies only to a thermally isolated system—one whose Hamiltonian is either invariant or else changes with time in a predetermined way (as in the type of setting operation considered in § 3). As soon as we permit an outside observer to interfere with the system in a way that is not predetermined but chosen by him during the course of the trial (as in the type of setting process considered in this section), the system ceases to satisfy the definition of thermal isolation, and there is no longer any reason why its entropy should not decrease. In order to keep track of all the entropy changes in this case we must consider not only the entropy of the system but that of the observer. This will be done in the next section.

5. Entropy changes in the observer

The fact demonstrated in the previous section, that an observer who has obtained a quantity I of information by observing a system $\mathfrak{S}'$ can use it to reduce the statistical entropy of $\mathfrak{S}'$ by an amount that does not, on the average, exceed I, led Szilard† to reason that the entropy of the observer must be increased, on the average, by I when he acquires this information. In this way, Szilard argued, the law of non-decreasing entropy would retain its validity when applied to the composite system consisting of both $\mathfrak{S}'$ and the observer, even though it does not apply to the system $\mathfrak{S}'$ taken by itself. In the present section, adopting Szilard's point of view, we analyse the changes of entropy and of quantity of information that can occur in such a composite system.

In order to apply the theory of statistical entropy in composite systems, we imagine the observations on the system $\mathfrak{S}'$ to be carried out not by a human observer but by an automaton,‡ which we denote by $\mathfrak{S}''$. The auto-

† L. Szilard, *Z. Phys.* **53**, 840 (1929).

‡ A similar device is used in the mathematical theory of communication, where $\mathfrak{S}'$ is called the transmitter and $\mathfrak{S}''$ the receiver.

maton might, for example, be a motion-picture camera, or it might be one of the very complicated hypothetical "perpetual motion" machines discussed in § 3. The composite system $\mathfrak{S}' + \mathfrak{S}''$ is denoted by $\mathfrak{S}$ and assumed, as in § 2, to be thermally isolated.

The first step in the argument is to obtain a measure of the quantity of information about the subsystem $\mathfrak{S}'$ that is in the possession of the automaton $\mathfrak{S}''$. To do this, we note that a complete observation of the composite system $\mathfrak{S}$ (made not by the automaton $\mathfrak{S}''$ but by our usual idealized observer outside $\mathfrak{S}$) consists of two parts, an observation of $\mathfrak{S}'$ and an observation of $\mathfrak{S}''$. In principle, these may be carried out either simultaneously or in rapid succession. If they are carried out simultaneously at time t the quantity of information obtained by the observer has the statistical expectation

$$k\langle \ln [1/Pr(A_t'A_t'')]\rangle . \tag{5.1}$$

If $\mathfrak{S}'$ is observed first, say at time $t-0$, and then $\mathfrak{S}''$ is observed immediately afterwards, say at time $t+0$, the same total quantity of information is obtained, but in two parts; the statistical expectations of these two parts are

$$\left.\begin{aligned} &k\langle \ln [1/Pr(A_t')]\rangle \text{ for the observation of } \mathfrak{S}' \\ \text{and} \quad & \\ &k\langle \ln [1/Pr(A_t''|A_t')]\rangle \text{ for that of } \mathfrak{S}''. \end{aligned}\right\} \tag{5.2}$$

On the other hand, if $\mathfrak{S}''$ is observed first and $\mathfrak{S}'$ immediately afterwards the corresponding expectations are

$$\left.\begin{aligned} &k\langle \ln [1/Pr(A_t'')]\rangle \text{ for the observation of } \mathfrak{S}'' \\ \text{and} \quad & \\ &k\langle \ln [1/Pr(A_t'|A_t'')]\rangle \text{ for that of } \mathfrak{S}'. \end{aligned}\right\} \tag{5.3}$$

The expectation of the quantity of information obtained by observing $\mathfrak{S}'$ is greater if it is observed before $\mathfrak{S}''$ than after; according to (5.2) and (5.3), the excess is

$$k\langle \ln [1/Pr(A_t')]\rangle - k\langle \ln [1/Pr(A_t'|A_t'')]\rangle = C(t) \tag{5.4}$$

by the definitions (1.13) of Chap. II for conditional probability and (2.3) for $C(t)$. This quantity, the logarithmic correlation between the two systems, is positive or 0, by (2.4).

The result (5.4) shows that, on the average, making an observation of $\mathfrak{S}''$ brings about a reduction $C(t)$ in the quantity of information obtainable by observing $\mathfrak{S}'$. The natural interpretation is that if $\mathfrak{S}''$ is observed first then a part $C(t)$ of the quantity of information obtained at this observation is also pertinent to the current observational state of $\mathfrak{S}'$. This suggests that if $\mathfrak{S}''$ is an automaton then $C(t)$ should be interpreted as the quantity of information, relevant to the current state of its companion subsystem $\mathfrak{S}'$, that $\mathfrak{S}''$ has obtained by making measurements on $\mathfrak{S}'$ prior to the instant t.

This interpretation is supported by the theorem, proved in § 2, that $C(t)$ cannot increase if $\mathfrak{S}'$ and $\mathfrak{S}''$ are isolated, for if $\mathfrak{S}'$ and $\mathfrak{S}''$ are both isolated then $\mathfrak{S}''$ cannot make any measurements and so can obtain no new information about $\mathfrak{S}'$. Further support for the interpretation can be obtained by considering the case where the automaton $\mathfrak{S}''$ is an *ideal measuring instrument*, that is, where the observational state of $\mathfrak{S}'$ can be deduced from that of $\mathfrak{S}''$; for then $Pr(A'_t|A''_t)$ takes the values 0 and 1 only, so that $\langle \ln [1/Pr(A'_t|A''_t)] \rangle = 0$ and (5.4) reduces to

$$C(t) = k\langle \ln [1/Pr(A'_t)] \rangle . \tag{5.5}$$

That is, if $\mathfrak{S}''$ is an ideal measuring instrument, then the quantity $C(t)$ of information it possesses about the state of $\mathfrak{S}'$ is identical with the expression (4.4) for the quantity of information about $\mathfrak{S}'$ that an outside observer can obtain directly.

It should be noted that we are interpreting $C(t)$ not as the total quantity of information about the object subsystem $\mathfrak{S}'$ held by the automaton $\mathfrak{S}''$ at the time t, but as that part of it which is pertinent to the current observational state of $\mathfrak{S}'$. The total quantity of information that has been conveyed by measurements from $\mathfrak{S}'$ to $\mathfrak{S}''$ may be much greater than $C(t)$, but if so the remaining part of it has a purely historical character and is not pertinent to the present or future behaviour of $\mathfrak{S}''$. We may call the two parts the *current* and the *historical* part of the information about $\mathfrak{S}'$ held by $\mathfrak{S}''$. For example, the automaton might make a permanent record of the observational states taken by its companion subsystem at all the times $t^1, t^2, \ldots$, etc.; but (because of the Markovian postulate) only the most recent entry in this record would be pertinent to the current observational state of the object system. In this case the most recent entry in the record would constitute the current information, and the earlier ones would constitute the historical information.†

In § 2 we proved the theorem that $C(t)$ can increase only when the subsystems $\mathfrak{S}'$ and $\mathfrak{S}''$ are not both isolated. If $C(t)$ is interpreted as the quantity of current information about $\mathfrak{S}'$ held by $\mathfrak{S}''$, the theorem has the interpretation that this quantity of information can increase only as the result of some interaction, such as a measurement. During any time interval when there is no interaction, there can be no increase in the quantity of current information, and in general it will decrease, either by becoming out of date and converted into historical information like yesterday's newspaper, or by being obliterated as a footprint in sand is obliterated by the elements.

† The need to make such a distinction when linking information to entropy is pointed out by L. Brillouin in *Science and Information Theory*, Chap. 12. He uses the words "bound" and "free" rather than "current" and "historical".

During any time interval, say $t_1 \leqq t \leqq t_2$, when the automaton and the object subsystem do interact, the value of $C(t)$ may increase. The magnitude of the increase is related, through the addition law obtained in § 2,

$$\tilde{S}(t) = \tilde{S}'(t) + \tilde{S}''(t) - C(t), \qquad (5.6) = (2.2)$$

to the changes of statistical entropy in $\mathfrak{S}'$ and $\mathfrak{S}''$: by combining (5.6) with the law of non-decrease (1.20) for the statistical entropy of the composite system, we obtain the inequality

$$\Delta C \leqq \Delta\tilde{S}' + \Delta\tilde{S}'', \qquad (5.7)$$

where $\Delta C \equiv C(t_2) - C(t_1)$, etc. With our interpretation of $C(t)$ as a quantity of information, this inequality provides a general formulation of Szilard's principle, cited at the beginning of this section, that the information obtained by $\mathfrak{S}''$ in any measurement of $\mathfrak{S}'$ is paid for by entropy increases in $\mathfrak{S}'$ and $\mathfrak{S}''$ totalling at least as much as the quantity of information gained.

Using (5.7), we can now elucidate the changes of entropy and of current information that take place during a process of the type discussed by Szilard in the original derivation of his principle. This process consists of two stages: in the first the automaton observes its companion subsystem $\mathfrak{S}'$ and in the second it uses the information so obtained to reduce the entropy of $\mathfrak{S}'$. We suppose for simplicity that during the first stage the automaton does not influence $\mathfrak{S}'$, so that $\Delta_1\tilde{S}'$, the change in $\tilde{S}'(t)$ during this first stage, is non-negative. Likewise, we suppose that during the second stage the automaton does not allow $\mathfrak{S}'$ to influence it, so that $\Delta_2\tilde{S}''$ is non-negative. Assuming also the most favourable case, where all processes are reversible and all the information obtained by $\mathfrak{S}''$ during the first stage is used up during the second (so that $\Delta_1 C + \Delta_2 C = 0$), we find by applying (5.7) that

$$\left.\begin{aligned} \Delta\tilde{S}'' &\equiv (\Delta_1 + \Delta_2)\tilde{S}'' = \Delta_1 C + 0, \\ \Delta\tilde{S}' &= 0 + \Delta_2 C = -\Delta_1 C. \end{aligned}\right\} \qquad (5.8)$$

That is, the net result of the process is to transfer an amount of statistical entropy, just equal to the amount $\Delta_1 C$ of information obtained during the measurement, from $\mathfrak{S}'$ to $\mathfrak{S}''$.

As a very simple (and very artificial) example, suppose that $\mathfrak{S}'$ and $\mathfrak{S}''$ have two observational states each, which we label 1′, 2′ and 1″, 2″, and that the four observational states of $\mathfrak{S}$, which are 1′1″, 1′2″, 2′1″ and 2′2″, each have weight 1. The probability and entropy relationships for this model are shown in the first panel of Table 1. Suppose that before the measurement, say at time t_1, the object subsystem $\mathfrak{S}'$ is in equilibrium, so that $Pr(1') = Pr(2') = \frac{1}{2}$, and the "automaton" $\mathfrak{S}''$ is in the state 1″. The probabilities and entropies for this time are shown in the second panel of Table 1.

We represent the measurement stage of the process by allowing $\mathfrak{S}'$ and $\mathfrak{S}''$ to interact in accordance with the "law of motion" that the observational

TABLE 1. Probability and Entropy Relationships for a Model Illustrating Szilard's Principle

General Scheme and Formulae

$Pr(1'1'')$	$Pr(1'2'')$	$Pr(1')$
$Pr(2'1'')$	$Pr(2'2'')$	$Pr(2')$
$Pr(1'')$	$Pr(2'')$	

$$S' = k\,\Sigma_A\, Pr(A')\ln\frac{1}{Pr(A')}$$

$$\tilde{S} = k\,\Sigma_A\,\Sigma_B\, Pr(A'B'')\ln\frac{1}{Pr(A'B'')}$$

$$\tilde{S}'' = k\,\Sigma_B\, Pr(B'')\ln\frac{1}{Pr(B'')} \qquad C = \tilde{S}' + \tilde{S}'' - \tilde{S}$$

Values at time t_1

$\frac{1}{2}$	0	$\frac{1}{2}$
$\frac{1}{2}$	0	$\frac{1}{2}$
1	0	

$\tilde{S}' = k\ln 2$

$\tilde{S} = k\ln 2$

$\tilde{S}'' = 0 \qquad C = 0$

Values at time t_2

$\frac{1}{2}$	0	$\frac{1}{2}$
0	$\frac{1}{2}$	$\frac{1}{2}$
$\frac{1}{2}$	$\frac{1}{2}$	

$\tilde{S}' = k\ln 2$

$\tilde{S} = k\ln 2$

$\tilde{S}'' = k\ln 2 \qquad C = k\ln 2$

Values at time t_3

$\frac{1}{2}$	$\frac{1}{2}$	1
0	0	0
$\frac{1}{2}$	$\frac{1}{2}$	

$\tilde{S}' = 0$

$\tilde{S} = k\ln 2$

$\tilde{S}'' = k\ln 2 \qquad C = 0$

state of $\mathfrak{S}'$ does not change and the observational state of $\mathfrak{S}''$ changes if and only if $\mathfrak{S}'$ is in the state $2'$. The transitions that this operation can bring about are

$$\left.\begin{aligned} 1'1'' \to 1'1'',\quad 1'2'' \to 1'2'',\\ 2'1'' \to 2'2'',\quad 2'2'' \to 2'1''.\end{aligned}\right\} \tag{5.9}$$

Since $\mathfrak{S}''$ is started in the state $1''$, its state after this stage of the process is the same as the state of $\mathfrak{S}'$, so that the operation may truly be said to constitute a measurement of the state of $\mathfrak{S}'$ carried out by $\mathfrak{S}''$. The probabilities immediately after the measurement (say at time t_2) can be calculated from those at time t_1 by applying the transformation (5.9); they are shown, together with the corresponding entropies, in the third panel of Table 1. It will be seen that C has increased from 0 to $k \ln 2$, showing that $\mathfrak{S}''$ now holds a quantity $k \ln 2$ (i.e. one bit) of current information about $\mathfrak{S}'$, and that $\tilde{S}''$ has also increased by the same amount in accordance with Szilard's principle.

The second stage of the process, in which the information obtained in the measurement is used to diminish the entropy of $\mathfrak{S}'$, can be achieved by an operation whose law of motion is that the state of $\mathfrak{S}''$ does not change and the state of $\mathfrak{S}'$ changes only if $\mathfrak{S}''$ is in the state $2''$. The transitions that are possible in this stage of the process (which ends, say, at time t_3) are

$$\left.\begin{array}{ll} 1'1'' \to 1'1'', & 1'2'' \to 2'2'', \\ 2'1'' \to 2'1'', & 2'2'' \to 1'2''. \end{array}\right\} \qquad (5.10)$$

From the dynamical point of view this transformation is a mirror image of the one described by (5.9), but its effects are quite different because the initial probability distribution is different. These effects are shown in the last panel of Table 1. It will be seen that the net result of the transaction has been to transfer from $\mathfrak{S}'$ to $\mathfrak{S}''$ an amount of entropy that is equal to the quantity $k \ln 2$ of information that was obtained in the measurement. Note also that $C(t)$ returns to zero at the end, so that the automaton $\mathfrak{S}''$ no longer has any current information about $\mathfrak{S}'$. It is true that the final state of $\mathfrak{S}''$ is the same as the original state of $\mathfrak{S}'$, so that $\mathfrak{S}'$ may still be said to hold information about $\mathfrak{S}''$; but this information is merely historical, with no relevance to the current (i.e. final) observational state of $\mathfrak{S}'$.

An interesting feature of this model is that the entire process can be reversed, by applying the same two operations again in the reverse order. Thus the increase in $\tilde{S}' + \tilde{S}''$ that is demanded by Szilard's principle (5.7) every time a measurement takes place does not necessarily correspond to any irreversibility in the measurement process: although $\tilde{S}' + \tilde{S}''$ must increase when C increases, it may also decrease if C decreases. If the measurement increased $\tilde{S}$ as well as $\tilde{S}' + \tilde{S}''$, then by the non-decrease law (1.20), it would, indeed, be irreversible, but the example shows that measurements not increasing $\tilde{S}$ are perfectly consistent with our theory. It is true that the real measurement processes used in physics laboratories are irreversible, normally producing increases in $\tilde{S}$ that are much larger than the quantity of information they yield, but this fact does not seem to have any fundamental significance,† at least, not in the idealized theory this book is about.

† For a contrary opinion see Chap. 13 of L. Brillouin's *Science and Information Theory*, where references to other work on this topic are also given.

The discussion of entropy changes given in this section hinges on the fact that the quantities $\tilde{S}$ and $\tilde{S}' + \tilde{S}''$ are not identical. As (5.6) shows, the entropy of the whole is less than the sum of the entropies of its parts. The physical distinction between $\tilde{S}$ and $\tilde{S}' + \tilde{S}''$ can be seen by using the interpretation of statistical entropies proposed in § 3. According to this interpretation, we are to regard $\tilde{S}$ as a lower bound on the Boltzmann entropy of the final state in a setting operation applied to the composite system $\mathfrak{S}$, and $\tilde{S}' + \tilde{S}''$ as a lower bound on the Boltzmann entropy obtainable when setting operations are applied independently to the two subsystems comprising $\mathfrak{S}$. The performance of independent setting operations on the two subsystems is equivalent, however, to the performance of a single setting operation on the composite system in which its two subsystems do not interact. Since the minimum final entropy obtainable without interaction cannot be less than the minimum obtainable if interaction is permitted, we have $\tilde{S}' + \tilde{S}'' \geqq \tilde{S}$. The amount of the excess is the greatest decrease in $\tilde{S}' + \tilde{S}''$ that can be achieved in a process where $\mathfrak{S}'$ and $\mathfrak{S}''$ do interact; according to (5.7), this greatest decrease is precisely C, so that we recover (5.6).

It remains to find the connection between the results of this section and the non-decrease law (4.19) for the quantity $\langle \tilde{S}_{\text{obs}}(t) + I(t) \rangle$, obtained in the preceding section. There we defined $\tilde{S}_{\text{obs}}(t)$ in terms of $q_t(A|A^m)$, the subjective probability for the observed system to be in a state A at time t, held by an observer whose mental state at this time is sufficiently specified by the statement that he knows the result A^m of the most recent observation, which was made at time t^m. In the present section, we have replaced the observer by an automaton $\mathfrak{S}''$, and so the analogue of the observer's mental state is the observational state of the automaton, which we denote by B''. Consequently the analogue of the observer probability distribution $q_t(A|A^m)$ is $Pr(A'_t|B'')$, and we may write

$$q_t(A|A^m) \to p_t(A'|B'') \equiv p_t(A'B'')/p_t(B''). \tag{5.11}$$

Making this replacement in the definition (4.11) of $\tilde{S}_{\text{obs}}(t|A^m)$ and taking the ensemble average over B'', as in (4.14), we obtain

$$\begin{aligned} \langle \tilde{S}_{\text{obs}}(t) \rangle &\to k \textstyle\sum_B \sum_A p_t(A'B'') \ln [W'(A')\, p_t(B'')/p_t(A'B'')] \\ &= \tilde{S}'(t) - C(t) \end{aligned} \tag{5.12}$$

by the definitions (1.13) and (2.3). To obtain the analogue of $\langle I(t) \rangle$, we start from the definitions (4.18) and (4.6), which combine to give

$$I(t) = k \ln [1/Pr(A^m_{t^m} A^{m-1}_{t^{m-1}} \ldots A^1_{t^1})]. \tag{5.13}$$

Let us assume that all the observational states of the automaton have equal weights, say W'', and that they are in one-to-one correspondence with the possible observational histories $A^m A^{m-1} \dots A^1$ of the observed system. Then the analogue of (5.13), averaged over all observational histories, is

$$\begin{aligned}\langle I(t)\rangle &\to k \textstyle\sum_B p_t(B) \ln [1/p_t(B)] \\ &= \tilde{S}''(t) - k \ln W''. \qquad (5.14)\end{aligned}$$

Combining (5.12) with (5.14) we obtain

$$\begin{aligned}\langle \tilde{S}_{\text{obs}}(t) + I(t)\rangle &\to \tilde{S}'(t) + \tilde{S}''(t) - C(t) - k \ln W'' \\ &= \tilde{S}(t) - k \ln W'' \qquad (5.15)\end{aligned}$$

so that the non-decrease property proved in § 4 is no more than the special case of (5.7) that applies to an automaton satisfying the assumptions stated just after (5.13). These assumptions may be thought of as a description of an ideal observing automaton: one that makes the most accurate observations permitted by the theory, faithfully records all their results, and works reversibly in the sense that its own Boltzmann entropy never increases.

5.1. Exercises

1. Show that the mean quantity of historical information at time t furnished by a sequence of observations made on a system at times $t^1, t^2, \dots$ is, in the notation of § 4,

$$k\langle \ln [1/\Pr(A^1{}_{t^1}, A^2{}_{t^2} \dots A^m{}_{t^m} | A_t)]\rangle,$$

where m is the serial number of the most recent observation and A is the current observational state.

2. Generalize (5.15) to the case where the observational states of the automaton do not all have equal weight.

Solutions to Exercises

Chapter I, § 2.1

1. $\dot{q}_1 = p_1/m, \dot{q}_2 = p_2/m, \dot{q}_3 = p_3/m, \dot{q}_4 = p_4/I, \dot{q}_5 = p_5/I \sin^2 q_4, \dot{p}_1 = \dot{p}_2 = \dot{p}_3 = \dot{p}_5 = 0, \dot{p}_4 = p_5{}^2 \cos q_4/I \sin^3 q_4$.

2. The Hamiltonian (2.11) acquires a new term $-E\mu \cos q_4$, where μ is the dipole moment of the dumb-bell. Only the equation of motion for p_4 is altered, acquiring a term $-E\mu \sin q_4$ on the right. The generalized force belonging to E is $\partial H/\partial E = -\mu \cos q_4$, the component of the dipole moment in a direction opposite to that of E.

3. Taking $q_1 = r, q_2 = \theta$, we have $H = (p_1{}^2 + p_2{}^2/q_1{}^2)/2m$, $\dot{q}_1 = p_1/m$, $\dot{q}_2 = p_2/mq_1{}^2$, $\dot{p}_1 = p_2{}^2/mq_1{}^3$, $\dot{p}_2 = 0$. For constant velocity the motion is $q_1 = (r_0{}^2 + v^2t^2)^{\frac{1}{2}}$, $q_2 = \theta_0 + \arctan(vt/r_0)$, where r_0, θ_0, and v are constants.

Chapter I. § 3.1

1. Any 2 × 2 matrix with unit diagonal sum and unit determinant.

2. $(J_1J_2)^2 = J_1J_2J_1J_2 = J_1{}^2J_2{}^2 = J_1J_2$; $(J_1J_2)^\dagger = J_2{}^\dagger J_1{}^\dagger = J_2J_1 = J_1J_2$. If $J_1J_2|\alpha\rangle = |\alpha\rangle$ then $J_1|\alpha\rangle = J_1(J_1J_2)|\alpha\rangle = J_1J_2|\alpha\rangle = |\alpha\rangle$, and likewise $J_2|\alpha\rangle = |\alpha\rangle$. Conversely, if $J_1|\alpha\rangle = J_2|\alpha\rangle = |\alpha\rangle$, then $J_1J_2|\alpha\rangle = J_1|\alpha\rangle = |\alpha\rangle$.

Chapter I, § 5.1

1. By conservation of energy, this system (if isolated) never changes its observational state; the observational states are therefore deterministic and hence Markovian.

2. The Markovian postulate is not satisfied, because the probability of state A at time t is not determined by the state at time $t-1$: it depends also on the state (say B) at time $t-2$, and is in fact 1 if $B \neq A$ and 0 if $B = A$. If, however, we use compound observational states consisting of pairs of successive simple observational states, the observational description becomes deterministic and therefore Markovian.

Chapter II, § 1.1

1. By (1.10), (1.21), and (1.9) we have, for example, $Pr(E_1')\,Pr(E_2) = Pr(E_2) - Pr(E_1) \times Pr(E_2) = Pr(E_2) - Pr(E_1E_2) = Pr(E_1'E_2)$.

2. Let p be the required probability; then $1 - p$ = probability that all birthdays are

$$\text{different} = \frac{364}{365}\cdot\frac{363}{365}\cdots\frac{365-(r-1)}{365} < \exp\left(-\frac{1}{365} - \frac{2}{365} - \cdots - \frac{r-1}{365}\right)$$

$$\leq \exp\left(-\frac{253}{365}\right) = 0.499998.$$

Chapter II, § 2.1

1. Let $J_i \equiv 1$ if the ith molecule is in the lower half and $\equiv 0$ if not; then by (2.8) and (2.31) we have

$$\langle J_iJ_k\rangle - \langle J_i\rangle\langle J_k\rangle = \tfrac{1}{2} - \tfrac{1}{4} \text{ if } i = k, \text{ and } = 0 \text{ if not.}$$

The definition (2.16) of variance now gives var $(\Sigma_i J_i) = (\frac{1}{2} - \frac{1}{4}) \times 10^{22}$, and so (2.17) becomes

$$Pr\{|\Sigma_i J_i - \tfrac{1}{2} \times 10^{22}| \geqq 10^{-6} \times \tfrac{1}{2} \times 10^{22}\} \leqq \tfrac{1}{4} \times 10^{22}/(\tfrac{1}{2} \times 10^{16})^2 = 10^{-10}.$$

Chapter II, § 3.1

1. By repeating the proof of (3.5) with the following modifications:

$$\mathfrak{T}_2 \to \mathfrak{T}_3,\ \mathfrak{T}_1 \to \mathfrak{T}_2\mathfrak{T}_1,\ B_2 \to C_3,\ A_1 \to B_2A_1,$$

we obtain $Pr(C_3B_2A_1) = Pr(C_3)Pr(B_2A_1)$, which combines with the original (3.5) to give the desired result.

2. The proof follows that of (3.13) with the modifications

$$J_{A,n} \to X_s,\ Pr(A|\mathfrak{T}) \to \langle X \rangle,\ \bar{J}_{A,n} \to \bar{X}_s,\ 1/4n \to [\text{var}(X)]/s,\ n \to s.$$

Chapter II, § 4.1

1. $w_{BA} = cA$ if $B = A - 1$, $c(N - A)$ if $B = A + 1$, $1 - cN$ if $B = A$, 0 otherwise.

2. By (2.25), (4.8), and (4.16) we have, for any a chosen from the set $\{0, 1, \ldots, N\}$,

$$\mathscr{E}[A(t+1)|A(t) = a] = (a-1)\,a/N + (a+1)(1 - a/N) = \tfrac{1}{2}N + (1 - 2/N)(a - N/2),$$

which is the first result asked for. Applying the formula (2.29) we obtain

$$\mathscr{E}[A(t+1)] = \tfrac{1}{2}N + (1 - 2/N)\{\mathscr{E}[A(t)] - N/2\}$$

from which it follows by induction that $\mathscr{E}[A(t)] = \frac{1}{2}N[1 - (1 - 2/N)^t]$.

3. Using the label 0 for the lumped state we have

$$Pr(1_3|0_1 0_0) = Pr(1_3 0_1|0_0)/Pr(0_1|0_0) = 0/Pr(0_1|0_0) = 0, \text{ but}$$

$$Pr(1_3|0_1 1_0) = Pr(1_3 0_1|1_0)/Pr(0_1|1_0) = 1/1 = 1,$$

since the states at times 0 and 3 must be the same; but if the states 0 and 1 were Markovian these two conditional probabilities would be equal by (4.5).

4. Writing (BA) for the compound observational state in which the simple observational states A and B occur in immediate succession, and using the Markovian character of the simple observational states, we can verify (4.6) for the compound states: for example,

$$Pr[(FE)_{t+1}(DC)_t|(BA)_{t-1}] = Pr(F_{t+1}E_tD_tC_{t-1}|B_{t-1}A_{t-2}) =$$

$$Pr(F_{t+1}E_t|D_tC_{t-1})\,Pr(D_tC_{t-1}\,|\,B_{t-1}A_{t-2}) = Pr[(FE)_{t+1}|(DC)_t]Pr[(DC)_t|(BA)_{t-1}].$$

Chapter II, § 5.1

1. $0 \rightleftarrows 1 \rightleftarrows 2 \rightleftarrows 3 \rightleftarrows 4 \rightleftarrows 5$.

2. Even and odd-numbered states alternate. There is one ergodic set.

Chapter II, § 6.1

1. Within each ergodic set the equilibrium probabilities are all equal.

2. From (4.16) and (6.3) we have $\pi_B = [(B+1)/N]\,\pi_{B+1} + [1 - (B-1)/N]\,\pi_{B-1}$ for $B = 0, 1, \ldots, N$, with the convention $\pi_{-1} = \pi_{N+1} = 0$. Applying the equation successively with $B = 0, 1, \ldots$, we find that $\pi_B = \pi_0 N!/B!(N-B)!$, and it follows by (6.4) that $\pi_B = 2^{-N}N!/B!\,(N-B)!$.

3. If the possible states are numbered $0, 1, \ldots, N$, the transition probabilities are given by $w_{BA} = \frac{1}{2}$ if $|A - B| = 1$, $w_{00} = \frac{1}{2}$, $w_{NN} = \frac{1}{2}$, and $w_{BA} = 0$ in all other cases. The equilibrium probabilities are $\pi_0 = \pi_1 = \cdots = \pi_N = 1/(N+1)$.

Chapter II, § 7.1

1. By (5.10), the probability of eventually reaching a persistent state is 1. Denoting by q_Z the probability that the persistent state eventually reached is in the ergodic set Z, we have

$$\lim_{t\to\infty} p_A(t) = \begin{cases} 0 \text{ if } A \text{ is transient,} \\ q_{Z(A)}\pi_A \text{ if } A \text{ is persistent,} \end{cases}$$

where $Z(A)$ denotes the ergodic set containing A.

2. $(d^2/dx^2)\,[x\varphi(1/x)] = x^{-3}\,\varphi''(1/x) \geqq 0.$

3. Since $-x \ln (c/x)$ is a convex function of x, the integral analogue of (7.8) gives

$$\int_{\omega_t} \bar{D}(\alpha) \ln [c/\bar{D}(\alpha)]\, d\alpha \geqq S_{\text{fine}},$$

where

$$\bar{D}(\alpha) \equiv \Omega_t^{-1} \int_{\omega_t} D(\alpha) d\alpha = \Omega_t^{-1} \text{ for all } \alpha \text{ in } \omega_t, \text{ and } \Omega_t \equiv \int_{\omega_t} d\alpha.$$

Maximizing S_{fine} is thus equivalent to choosing $D(\alpha) = \bar{D}(\alpha)$, which leads at once to eqn. (6.5) of Chap. I.

Chapter II, § 8.1

1. The new ergodic sets are the original sets $Z_1, \ldots, Z_q$: by (8.16), (8.15) and (8.9) we have $v_{BA} = q$ if and only if A and B are both in the same set Z_r, and it follows by (8.5) and (8.10) that each Z_r is an aperiodic ergodic set under the new method of observation. The new equilibrium probabilities $\pi_A{}'$ are given in terms of the old ones by $\pi_A{}' = q\pi_A$ [see (8.17) and (8.21)].

2. The quantities $\Sigma_{A\in Z_r}\, p_A(t^*)$ $(r = 1, 2, \ldots, q)$ must all be equal.

Chapter II, § 9.1

1. By (9.32) and (6.4) we have $w_{AB}(t) \leqq \pi_A/\pi_B \leqq (1-\pi_B)/\pi_B \cong 1 - \pi_B$.

2. Over a long time interval of duration τ, the state A occurs about $\pi_A\tau$ times [by (9.11)], and it occurs twice in succession about $w_{AA}\pi_A\tau$ times [by (9.20)]. The number of occurrences of A that are not immediately preceded by occurrences of A is therefore about $(1 - w_{AA})\pi_A\tau$, and the desired result follows.

3. Let $\bar{J}_{A,s,r} \equiv s^{-1} \sum_{n=1}^{s} J_{A,r+nq}$, where q is the period, and let $E_{A,s,r}$ denote the event $|\bar{J}_{A,s,r} - \pi_A| > \delta$ for some $\delta > 0$. Applying the law of large numbers (9.11) to the trial described in exercise 1 of Chap. II, § 8.1, in which observations are made at every qth opportunity, we have $Pr(E_{A,s,r}) \to 0$ as $s \to \infty$, for any r and any $\delta > 0$. From the definitions we have $\bar{J}_{A,qs} - \pi_A = q^{-1} \sum_{r=1}^{q} (\bar{J}_{A,s,r} - \pi_A)$, so that if the event $E_{A,qs}$ (defined as $|\bar{J}_{A,qs} - \pi_A| > \delta$) occurs then at least one of the events $E_{A,s,r}$ $(r = 1, 2, \ldots, q)$ must occur. It follows by (1.12) that $Pr(E_{A,qs}) \leqq Pr(E_{A,s,1} \oplus \ldots \oplus E_{A,s,q}) \leqq Pr(E_{A,s,1}) + \cdots + Pr(E_{A,s,q})$, and since every term on the right tends to zero as $s \to \infty$, the result follows.

4. Applied to the compound observational states defined in exercise 4 of Chap. II, § 4.1, the law of large numbers (9.11) gives $\lim_{s\to\infty} Pr(|\bar{J}_{AB,s+1} - \pi_{BA}| > \delta) = 0$, where π_{BA} is the equilibrium probability of the compound observational state (BA); but it follows from (9.15) that this equilibrium probability is equal to $w_{BA}\pi_A$.

Chapter III, § 2.1

1.
$$\frac{\partial(x, y, p_x, p_y)}{\partial(r, \theta, p_r, p_\theta)} = \begin{vmatrix} \cos\theta & -r\sin\theta & 0 & 0 \\ \sin\theta & r\cos\theta & 0 & 0 \\ \text{immaterial} & \text{immaterial} & \cos\theta & -r^{-1}\sin\theta \\ \text{immaterial} & \text{immaterial} & \sin\theta & r^{-1}\cos\theta \end{vmatrix}.$$

Chapter III, § 3.1

1. $\langle (H - \langle H \rangle)^2 \rangle = \partial^2(\ln \Phi)/\partial\beta^2$.

2. Since α is in R if and only if $\alpha_t \equiv U_t\alpha$ is in R_t, we have $\int_R d\alpha = \int_{R_t} d(\alpha_t) = \int_{R_t} [\partial(\alpha_t)/\partial(\alpha)]\, d\alpha$; but the Jacobian $\partial(\alpha_t)/\partial(\alpha)$ is 1 by (3.19).

3. Verify that $p_t \equiv U_t p$ and $q_t \equiv U_t q$, with p and q held fixed, satisfy the Hamiltonian equations of motion $dp_t/dt = -\omega q_t$, etc. The trajectories are circles centred at the origin, and $D_t(p, q) = 4/\pi$ if $p < 0$, $q > 0$, and $p^2 + q^2 < 1$.

4. Writing $\alpha_0 \equiv U_t^{-1}\alpha$ and using (3.20) and (3.19), we have $S_{\text{fine}}(t) = k \int D_t(\alpha) \ln [c/D_t(\alpha)]\, d\alpha = k \int D_0(\alpha_0) \ln [c/D_0(\alpha)_0]\, [\partial(\alpha)/\partial(\alpha_0)]\, d\alpha_0 = S_{\text{fine}}(0)$.

Chapter III, § 4.1

1. For (i), (ii), and (iv), use (4.16). For (iii), the definition (4.16) implies $(tr)\mathsf{F}^\dagger = [tr(\mathsf{F})]^*$ where the dagger denotes a Hermitian conjugate and the asterisk a complex conjugate. Taking $\mathsf{F} = \mathsf{F}_1\mathsf{F}_2$ and using (4.19) and the Hermitian character of F_1 and F_2, we obtain $[tr(\mathsf{F}_1\mathsf{F}_2)]^* = tr(\mathsf{F}_2{}^\dagger\mathsf{F}_1{}^\dagger) = tr(\mathsf{F}_2\mathsf{F}_1) = tr(\mathsf{F}_1\mathsf{F}_2)$.

2. In (4.16) use the eigenstates of D for $|\beta_1\rangle, |\beta_2\rangle, \ldots$; then the terms of the sum are its eigenvalues, and they are non-negative by (4.25) and sum to 1 by (4.23). The eigenvalues must therefore lie between 0 and 1, and the result (ii) follows at once.

3. For the first result, we note that

$$\int dp\, h^{-1} \int_{-\infty}^{\infty} D(q + \tfrac{1}{2}u, q - \tfrac{1}{2}u)\, e^{-ipu/\hbar}\, du = D(q, q)$$

by the inversion formula for Fourier transforms. For the second, we use the transformation theory for matrix elements,† obtaining

$$\begin{aligned} tr(\varphi(p)\mathsf{D}) &= \int dp\varphi(p) \langle p|\mathsf{D}|p\rangle \\ &= \int dp\varphi(p) \int dq' \int dq''\, e^{-ipq'/\hbar} \langle q'|\mathsf{D}|q''\rangle\, e^{ipq''/\hbar}/h \end{aligned}$$

which takes the desired form if we put $q = \frac{1}{2}(q' + q'')$, $u = q' - q''$.

Chapter III, § 5.1

1. Use (5.8) and the energy levels $E_n = (n + \frac{1}{2})\hbar\omega$.

2. The wave functions of the stationary states have the form

$$\psi_n(x, y, z) = L^{-3/2} \exp [2\pi i(n_1 x + n_2 y + n_3 z)/L],$$

where n_1, n_2, n_3, are integers (of either sign) or zero. Their energies are $E_n = (\hbar^2/2m)(2\pi/L)^2 (n_1{}^2 + n_2{}^2 + n_3{}^2)$. By (4.12), the canonical density matrix is therefore given by

$$D(x, y, z; x', y', z') \propto \Sigma_{n_1} \Sigma_{n_2} \Sigma_{n_3} \exp \{2\pi i[n_1(x - x') + n_2(y - y') + n_3(z - z')]/L - (\beta\hbar^2/2m)(2\pi/L)^2(n_1{}^2 + n_2{}^2 + n_3{}^2)\}.$$

The formula given in the question is obtained by approximating the sums by integrals.

3. $S_{\text{fine}}(t) = -k\, tr[\mathsf{D}_t \ln (c^{-1}\mathsf{D}_t)] = -k\, tr[\mathsf{U}_t\mathsf{D}_0 \ln (c^{-1}\mathsf{D}_0)\mathsf{U}_t{}^\dagger] = S_{\text{fine}}(0)$. The second equality comes from (5.6) and the identity $f(\mathsf{UDU}^\dagger) = \mathsf{U}f(\mathsf{D})\mathsf{U}^\dagger$, where U is any unitary operator and D any Hermitian operator; the last equality comes from (4.20).

Chapter IV, § 1.1

1. Since the operators $\mathsf{J}_{A,t}$ all commute, they can be simultaneously diagonalized‡. Let $|\beta_1\rangle, |\beta_2\rangle, \ldots$ be the basic vectors of a representation that accomplishes this; then we have [by (1.8)] $\mathsf{U}_t{}^\dagger \mathsf{J}_A \mathsf{U}_t |\beta_n\rangle = |\beta_n\rangle$ or 0 for all A, t, n, and it follows that $\mathsf{J}_A\mathsf{U}_t|\beta_n\rangle = \mathsf{U}_t|\beta_n\rangle$ or 0 for all A, t, n. The argument is reversible.

† P. A. M. Dirac, *The Principles of Quantum Mechanics*, § 23.
‡ *Ibid.*, § 13.

Chapter IV, § 2.1

1. Let the binary representations be $p = 0.p_1p_2p_3 \ldots$ (that is, $p = \Sigma_n p_n 2^{-n}$ with $p_n = 0$ or 1 for each positive integer n) and $q = 0.q_1q_2q_3 \ldots$. Then $U_t p = 0.p_{t+1}p_{t+2}p_{t+3}\ldots$ and $U_t q = 0.p_t p_{t-1} \ldots p_2 p_1 q_1 q_2 q_3 \ldots$.

Chapter IV, § 3.1

1. Let E in (3.13) be the complement of F, and use eqn. (1.19) of Chap. II.

2. Let Z_t denote the event that the state of the system at time t is in the ergodic set Z, so that $U_t\omega(Z) = \omega(Z_t)$. By the definition of an ergodic set we have $Pr(Z_t'|K) = 0$ for any K in Z, where Z_t' is the event complementary to Z_t. It follows by (3.10) that $\Omega(Z_t'K) = 0$ and hence that $\Omega(Z_t'Z) = 0$. By the principle of conservation of phase-space volume (exercise 2 of Chap. III, § 3.1) we have $\Omega(Z) = \Omega(Z_t)$; subtracting $\Omega(Z_tZ)$ from both sides we obtain $\Omega(Z_t'Z) = \Omega(Z'Z_t)$, and the required result $\Omega(Z'Z_t) = 0$ follows.

Chapter IV, § 4.1

1. Tetrahedron 12, cube 24, octahedron 24, dodecahedron 60, icosahedron 60, benzene (plane hexagon) 12.

2. Let $|\beta_1\rangle, \beta_2\rangle, \ldots$ be any complete orthonormal set; then by eqns. (4.16), and (4.16) of Chap. III, we have $tr\_(F) = \Sigma_{i,j}\langle\beta_j|\alpha_i\rangle\langle\alpha_i|F|\beta_j\rangle = \Sigma \langle\alpha_i|F|\beta_j\rangle\langle\beta_j|\alpha_i\rangle = \Sigma_i\langle\alpha_i|F\ \alpha_i\rangle$.

Chapter IV, § 5.1

1. By Liouville's theorem (3.20) of Chap. III, we have $D_t(U_t\alpha) = D_Z(\alpha)$, and hence, by (5.16), $D_t(\alpha') = 1/\Omega(Z)$ if α' is in $U_t\omega(Z)$ and $= 0$ if not. It follows by the result of exercise 2 of Chap. IV, § 3.1, that the set of points on which the two functions $D_0(\alpha)$ and $D_t(\alpha)$ differ is the union of $\omega(Z_t'Z)$ and $\omega(Z'Z_t)$, and has zero phase-space volume.

2. By eqn. (8.17) of Chap. II, the generalization is $\lim_{t\to\infty} \int J_A(\alpha)\,[D_t(\alpha) - D_{Z_t}(\alpha)]\,d\alpha = 0$, where D_{Z_t} is defined in analogy with D_Z in (5.16), and Z_t is the Z_r of eqn. (8.16) of Chap. II for which $t - r$ is a multiple of the period q.

Chapter IV, § 6.1

1. By (1.13) of Chap. II and (6.13), we obtain $Pr(G_n|F_{n-1} \ldots B_1A) = Pr(G_nF_{n-1} \ldots |A)/Pr(F_{n-1} \ldots |A) = \Omega(G_n \ldots A)/\Omega(F_{n-1} \ldots A)$.

2. According to (6.13) [or (6.16)], the left-hand side of (3.13)(i) is precisely $Pr(E|K)$.

3. Provided the system is isolated throughout the relevant time interval [0 to 1 in (6.10), 0 to n in (6.13)] the results still hold if the system is not isolated outside this time interval. According to the model used in this book, the Hamiltonian can be altered only at instants of observation (see Chap. I, § 3); consequently the system is isolated throughout the interval between any pair of successive instants of observation t and $t + 1$, and the generalization of (6.11) to arbitrary initial instants, $Pr(B_{t+1}|A_t) = \Omega(B_{t+1}A_t)/\Omega(A_t)$, holds for any t.

Chapter IV, § 7.1

1. All states are persistent, forming a single aperiodic ergodic set. All equilibrium probabilities are $\frac{1}{4}$.

2. The given statement, applied to the event $F_n \ldots B_1$, gives [by (6.26)] $\int J_{F,n} \ldots J_{B,1} D_A\, d\alpha = \int J_{F,n} \ldots J_{B,1} \Sigma_B D_B w_{BA} d\alpha$ which reduces, by (6.19) and (6.14), to

$$\Omega(F_n \ldots B_1 A)/\Omega(A) = \Omega(F_n \ldots B_1)\,\Omega(B_1A)/\Omega(B)\Omega(A).$$

By time-shift invariance, we also have D_B at time 1 observationally equivalent to $\Sigma_C D_C w_{CB}$ at time 2, whence

$$\Omega(F_n \dots B_1)/\Omega(B) = \Omega(F_n \dots C_2)\,\Omega(C_2B_1)/\Omega(C)\Omega(B).$$

Continuing in this way, and combining all the equations so obtained, we arrive at (7.2). Moreover, the argument is reversible, since observational equivalence with respect to all events of the form $F_n \dots B_1$ implies complete observational equivalence for times $\geqq 1$, etc.

3. The argument based on (2.9) shows that two ensembles for which the functions $\varphi_t(p)$ are the same for some particular value of t are observationally equivalent for all times $\geqq t$. If the phase-space density at time 0 is D_A, then $\varphi_1(p)$ is $1 + (-1)^A$ for $0 \leqq p < \frac{1}{2}$, and $1 - (-1)^A$ for $\frac{1}{2} \leqq p < 1$; if the phase-space density at time 1 is $\Sigma_B D_B w_{BA}$, with w_{BA} given by (7.7), then $\varphi_1(p)$ is the same. The proof of consistency follows as in exercise 2.

4. Compatibility follows from the result of exercise 1 of Chap. IV, §1.1, the complete set of quantum states being here the set of states $|a_1a_2 \dots a_N\rangle$. Let us label the observational states so that $J_A|a_1a_2\dots\rangle = |a_1a_2\dots\rangle$ if $A = 2a_1 + a_2$ and $=0$ if not ($A = 0, 1, 2, 3$). Then we have $\Omega(A) = 2^{N-2}$, and $\Omega(B_1A) = 2^{N-3}$ if $N \geqq 3$ and one of the two top lines on the right of (7.6) is true, but $=0$ if the bottom line is true; it follows that $\Omega(B_1A) = w_{BA}\Omega(A)$, with w_{BA} given by (7.7), if $N \geqq 3$. In similar fashion we find that $\Omega(G_n \dots B_1A) = w_{GF} \dots w_{BA}\,\Omega(A) = 2^{N-2-n}$ or 0, provided that $N \geqq n+2$; consequently the consistency condition (7.2) is satisfied if (and only if) $n \leqq N-2$, and (7.19) is therefore satisfied. For this model the recurrence time is N.

Chapter IV, § 8.1

1. Except in trivial periodic cases, deterministic states can arise only for the limiting case of an infinite system; accordingly the form (7.19), rather than (7.2), should be used. Equation (7.19) may, however, be weakened to

$$\lim_{N\to\infty} \Omega(G_n \dots A)/\Omega(A) = \lim_{N\to\infty} [\Omega(G_1F)/\Omega(F)]\,[\Omega(F_{n-1} \dots A)/\Omega(A)]$$
$$\text{if } \lim_{N\to\infty} \Omega(F_{n-1} \dots A)/\Omega(A) = 1,$$

since in the contrary case where $\lim_{N\to\infty}\Omega(F_{n-1} \dots A)/\Omega(A) = 0$ the expression $\lim_{N\to\infty}$ $\Omega(G_nF_{n-1} \dots A)/\Omega(A)$ vanishes *a fortiori*.

Chapter V, § 2.1

1. The combinatorial factor in (2.13) generalizes to $(N_1'+N_1'')!\,(N_2'+N_2'')!/N_1'!\,N_1''!\,N_2'!\,N_2''!$, reducing to (2.11) when $N_1'' = N_2' = 0$ and to (2.13) when $N_2' = N_2'' = 0$.

Chapter V, § 3.1

1. (i) $S(A) = k \ln (a\Delta E/\nu)$; (ii) $S(A) = k \ln \{W(A\Delta E) - W[(A-1)\Delta E]\}$, where $W(E)$ is the integer nearest to $E/h\nu$. The correspondence principle gives $W(E) \approx E/h\nu$ and hence $a = h^{-1}$.

2. The entropy of a quasi-transient state is several k less than that of the most probable state in its ergodic set; that of a quasi-equilibrium state is several k greater than that of any other state in its ergodic set. In a quasi-irreversible transition the entropy increases by several k.

3. The microcanonical phase-space density, defined in eqn. (3.26) of Chap. III, may be written $D(\alpha) = \Delta[E - H(\alpha)]/\Omega(E)$, where $\Delta(x) \equiv 1$ if $0 \leqq x < \Delta E$, and $\equiv 0$ if not. This gives

$$\langle G'(\alpha')\rangle = \iint G'(\alpha')\Delta[E - H'(\alpha') - H''(\alpha'')]\, d\alpha' d\alpha''/\Omega(E),$$

which reduces to the given expression, since

$$\Omega''(x) = \int \Delta[x - H''(\alpha'')]\, d\alpha''.$$

Chapter V, § 4.1

1. In (4.16) replace $p_i^2/2m$ by the single-molecule Hamiltonian defined in eqn. (2.11) of Chap. I. In (4.18) carry out the momentum integrations first (the region of integration is now a $5N$-dimensional ellipsoid). This gives

$$\Phi(E, N) = (4\pi V)^N (2\pi m E)^{3N/2} (2\pi I E)^N / (5N/2)!$$

and $s(\varepsilon, v) = k \ln \{8\,\pi^2 I\,(2\pi m)^{3/2}\, v(2\varepsilon/5)^{5/2}\, e^{7/2}/h^5\sigma\}$ [see (3.20)]. The adiabatic law is $v\varepsilon^{5/2} = \text{const.}$

2. Equations (4.18) and (4.21) give

$$\frac{\Omega''(E-x)}{\Phi(E,N'')} = \left(\frac{E-x}{E}\right)^{3N''/2} - \left(\frac{E-\Delta E - x}{E}\right)^{3N''/2}$$

$$= \left(1 - \frac{x}{E}\right)^{3N''/2}\left[1 - \left(1 - \frac{\Delta E}{E-x}\right)^{3N''/2}\right]$$

and the right-hand side tends to $e^{-3x/2\varepsilon''}$ in the thermodynamic limit $N'' \to \infty$, $E/N'' \to \varepsilon''$, $\Delta E/N'' \to \delta > 0$. Using (4.29) and substituting into the result of exercise 3 of Chap. V, § 3.1, we obtain the given formula if $C = \lim \Phi(E, N'')/\Omega(E)$. A simple formula for C, obtained by setting $G = 1$ in the given formula for $\langle G \rangle$, is $C = 1/\int e^{-H'(\alpha')/kT''} d\alpha'$.

3. Let R be any region in velocity space and let $J_R(p) \equiv 1$ if the velocity p/m lies in R and $\equiv 0$ if not. Since all N particles are identical, the average number of particles with velocities in R is given by

$$\langle N_R \rangle \equiv \sum_{i=1}^{N} \langle J_R(p_i) \rangle = N \langle J_R(p_1) \rangle = N \int_R \exp(-p_1^2/2mkT)\, d^3p_1 / V(2\pi mkT)^{3/2},$$

by the result of exercise 2 with $H'(\alpha')$ taken as $p_1^2/2m$. Maxwell's velocity distribution law, that if R is infinitesimal and has volume d^3v then $\langle N_R \rangle = (N/V)(m/2\pi kT)^{3/2}\, e^{-\frac{1}{2}mv^2/kT}\, d^3v$, follows immediately.

Chapter V, § 5.1

1. Equation (5.8) generalizes to $\Delta S(A) = \frac{1}{2}\Sigma_i \Sigma_j \Delta Y_i(A) \alpha_{ij} \Delta Y_j(A)$. By the definition [(2.24) of Chap. II] of covariance, we obtain

$$\text{covar}\,(Y_i, Y_j) = -2(\partial/\partial\alpha_{ij}) \ln \int \ldots \int \exp\,[-\Delta S(A)/k]\, dY_1 \ldots dY_n.$$

The integral can be evaluated by transforming to the principal axes of the quadratic form for $\Delta S(A)$, and is equal to $(2\pi k)^{n/2}/\{\det\,[\alpha_{ij}]\}^{\frac{1}{2}}$.

2. By (4.15) and (4.23) we have

$$S_{\text{ther}} = Nk \ln\,[(V/N)\,(4\pi m E/3N)^{3/2}\, e^{5/2}\, h^{-3}].$$

Using this formula to approximate the entropies of the two parts of the gas, one of which has energy E' and particle number N' and the other $E'' \equiv E - E'$ and $N'' \equiv N - N'$, we obtain

$$\frac{\partial^2(S' + S'')}{\partial E'^2} = -\frac{3k}{2}\left(\frac{N'}{E'^2} + \frac{N''}{E''^2}\right), \quad \frac{\partial^2(S' + S'')}{\partial N' \partial E'} = \frac{3k}{2}\left(\frac{1}{E'} + \frac{1}{E''}\right),$$

$$\frac{\partial^2(S' + S'')}{\partial N'^2} = -\frac{5k}{2}\left(\frac{1}{N'} + \frac{1}{N''}\right).$$

In case (i), $N' = \frac{1}{2}N \equiv$ const., and by (5.11) with $Y \equiv E'$ we obtain

$$\text{var}\,(E') = \tfrac{2}{3}\,[N'/E'^2 + N''/E''^2]^{-1} = E^2/6N, \quad \text{var}\,(N') = 0.$$

In case (ii) we have, by the answer to exercise 1,

$$\begin{bmatrix} \text{var}(E') & \text{covar}(E', N') \\ \text{covar}(E', N') & \text{var}(N') \end{bmatrix} = \begin{bmatrix} 6N/E^2 & -6/E \\ -6/E & 10/N \end{bmatrix}^{-1} = \begin{bmatrix} 5E^2/12N & E/4 \\ E/4 & N/4 \end{bmatrix}.$$

Under the conditions of exercise 1 of Chap. II, § 2.1, the probability of E' deviating from its mean value $\frac{1}{2}E$ by more than one part in 10^6 is less than var $(E')/(10^{-6}.\frac{1}{2}E)^2 = 5\times 10^{12}/3N = (5/3)10^{-10}$; the corresponding probability for N' is treated in the answer to that exercise.

Chapter V, § 6.1

1. The maximizing function must have the property $\delta S + \lambda_1\, \delta E + \lambda_2\, \delta N = 0$ for arbitrary infinitesimal variations in the function $f(\xi)$; here λ_1 and λ_2 are Lagrange multipliers. Writing $\delta f(\xi)$ for the variation in $f(\xi)$, and using (6.5) and (6.7), the required property becomes $\int d\xi[-k \ln f(\xi) + \lambda_1 H_{(1)}(\xi) + \lambda_2]\, \delta f(\xi) = 0$, and so $f(\xi)$ must have the form (6.20) with $\alpha = -\lambda_2/k$, $\beta = -\lambda_1/k$. That this is a maximum can be shown by considering the second variation, which is $-\frac{1}{2}k \int d\xi\, (\delta f)^2/f$ and therefore negative.

2. Substituting (6.17) into (4.3), with $B = MP$, and using the approximation $\ln(1 + x/N_i) \cong \ln[1 + x/(N_i + 1)] \cong x/N_i$, we obtain

$$S(A) = k \sum_i{}^{+} \left[\delta N_i \frac{N_i{}^{0} - N_i - 1}{N_i} - \frac{\delta N_i(\delta N_i - 1)}{2N_i}\right]$$

$$+ k \sum_i{}^{-} \left[|\delta N_i| \frac{N_i - N_i{}^{0}}{N_i} - \frac{|\delta N_i|\,(|\delta N_i| - 1)}{2N_i}\right] + S(MP),$$

where $N_i{}^{0} \equiv \Upsilon_i \exp(-\alpha - \beta E_i)$, and $N_i = N_i(MP)$. By (6.18), the total contribution to the sums from terms linear in the δN_is cannot exceed $\frac{1}{2}\Sigma_i |\delta N_i|/N_i$; neglecting this contribution we obtain the quadratic approximation given in the question. By (5.10), (5.11), and exercise 1 of Chap. V, § 5.1, we obtain $\langle N_i(A)\rangle \cong N_i(MP)$, var $[N_i(A)] \cong N_i(MP)$, covar $[N_i(A), N_j(A)] \cong 0$ if $i \neq j$.

Chapter V, § 8.1

1. By eqn. (6.9) of Chap. IV it is sufficient to show that $\Omega(B_1A) = \Omega(A_1B)$. By eqn. (3.8) of Chap. IV, $\Omega(B_1A)$ is the volume of a region $\omega(B_1A)$ comprising all states α such that $\alpha \in \omega(A)$ and $U_1\alpha \in \omega(B)$. For every such α there is a phase-space trajectory taking the system from α at time 0 to $U_1\alpha$ at time 1. By the symmetry of the laws of motion, there is also a trajectory (traversing the same configurations in the reverse order with reversed velocities) that takes the system from $\tau U_1\alpha$ at time 0 to $\tau\alpha$ at time 1. By the assumed symmetry of the observational states, the initial point $\alpha' \equiv \tau U_1\alpha$ of this trajectory is in $\omega(B)$ and the final point $U_1\alpha' = \tau\alpha$ is in $\omega(A)$, so that α' is in $\omega(A_1B)$. Thus we obtain

$$\Omega(B_1A) = \int_{\omega(B_1A)} d\alpha = \int_{\omega(A_1B)} d\alpha' \frac{\partial(\alpha)}{\partial(\alpha')} = \int_{\omega(A_1B)} d\alpha' = \Omega(A_1B)$$

since the Jacobian of the transformation $\alpha \rightarrow \alpha' \equiv \tau U_1\alpha$ is 1 by eqn. (3.19) of Chap. III and (8.7). A similar proof exists in quantum mechanics, where the effect of τ is to replace the wave function by its complex conjugate.

Chapter VI, § 1.1

1. Use eqn. (7.6) of Chap. II with m_t replaced by π_A, x_t replaced by $p_t(A)/\pi_A$, and $\varphi(x)$ having the form (1.18). Equality holds for the equilibrium probability distribution.

2. Writing $S_{\text{fine}} = \Sigma_A k \int_{\omega(A)} D(\alpha) \ln[c/D(\alpha)]\, d\alpha$ and applying to each term in the sum an argument similar to the one used for exercise 3 of Chap. II, § 7.1, we find that S_{fine}

cannot exceed the right-hand side of (1.15), with equality whenever $D(\alpha) = \tilde{D}(\alpha)$. If $S_{\text{fine}} = k \ln W(Z)$, then $D(\alpha) = \tilde{D}(\alpha) = D_Z(\alpha)$, defined in eqn. (5.16) of Chap. IV.

3. By eqn. (5.3) of Chap. V we have $\langle S(A)\rangle_0 - \langle S(A)\rangle_t = \sum_A p_t(A) \ln (\pi_{MP}/\pi_A)$ which is positive for any t such that some state with equilibrium probability less than π_{MP} is reachable from MP in t steps.

4. The proof is parallel to that of eqn. (6.18) of Chap. V, with the following correspondences: system→ ensemble, molecule→ system, $\Upsilon_i \to W(A)$, $E_i \to E(A)$, $N_i \to \mathcal{N}(A)$, $E \to \mathscr{E}$, $N \to \mathcal{N}$, etc.

Chapter VI, § 2.1

1. One of the following two conditions must be satisfied: (a) The state of $\mathfrak{S}'$ is uniquely determined by the state of $\mathfrak{S}''$, and all the possible states of $\mathfrak{S}'$ have equal probabilities; (b) the same with primes and double primes interchanged.

2. By eqn. (5.14) of Chap. IV and eqn. (2.11) of Chap. V, the condition (2.6) is satisfied at equilibrium.

3. Generalizing the derivation of (2.10) we can show that

$$\Omega(G_n F_{n-1} \ldots B_1 A) = \Omega'(G_n' \ldots A')\, \Omega''(G_n'' \ldots A'')$$

so that each side of eqn. (7.2) of Chap. IV equals the product of an expression with primes on all letters and a similar one with double primes on all letters. If eqn. (7.2) of Chap. IV holds for the primed and doubly primed subsystems, it therefore holds also for the composite (unprimed) system.

Chapter VI, § 3.1

1. The "observer" in this question is in a situation very similar to that of Maxwell's hypothetical demon, and the same discussion applies. In case (i) any decrease in the entropy of the fluid must be compensated either by an increase in the entropy of the probability distribution over observational states of the entire system, or (if the observer does not have enough states to make this possible) by an increase in the entropy of the observer due to his forgetting some of the past history of the particle's motion. In case (ii) there is no violation of the law because the system is not thermally isolated: the observer alters its Hamiltonian H every time he shortens the thread, and so H is not a predetermined function of time. In both cases it makes no difference if the shortening of the thread is accomplished not by an observer but by a mechanical device, such as a spool with a ratchet to prevent the string from unwinding and a (very weak) spring to wind up any slack.

Chapter VI, § 5.1

1. Let the observations be made and recorded by an automaton whose state at time t we identify with the string of symbols $A^m{}_{tm} \ldots A^1{}_{t1}$ comprising the observations made by it prior to t. Adapting (2.3) to the present notation, we have, for the quantity of current information held at time t by the automaton about the observed system, the formula

$$I_{\text{current}}(t) = k \ln \left[\frac{Pr(A_t A^m{}_{tm} \ldots A^1{}_{t1})}{Pr(A_t)\, Pr(A^m{}_{tm} \ldots A^1{}_{t1})} \right].$$

The total quantity of information is, by (4.18) and (4.2), $I(t) = k \ln [1/Pr(A^m{}_{tm} \ldots A^1{}_{t_1})]$. The quantity of historical information is $I(t) - I_{\text{current}}(t)$, which reduces to the given formula by virtue of the definition [(1.13) of Chap. II] of conditional probability.

2. $\langle \tilde{S}_{\text{obs}}(t) + I(t) + S''[B(t)]\rangle = \tilde{S}(t)$ is nondecreasing.

Index

Index

Index

Index

Index

OTHER TITLES IN THE SERIES IN NATURAL PHILOSOPHY

Vol. 1. DAVYDOV—Quantum Mechanics
Vol. 2. FOKKER—Time and Space, Weight and Inertia
Vol. 3. KAPLAN—Interstellar Gas Dynamics
Vol. 4. ABRIKOSOV, GOR'KOV and DZYALOSHINSKII—Quantum Field Theoretical Methods in Statistical Physics
Vol. 5. OKUN'—Weak Interaction of Elementary Particles
Vol. 6. SHKLOVSKII—Physics of the Solar Corona
Vol. 7. AKHIEZER *et al.*—Collective Oscillations in a Plasma
Vol. 8. KIRZHNITS—Field Theoretical Methods in Many-body Systems
Vol. 9. KLIMONTOVICH—The Statistical Theory of Non-equilibrium Processes in a Plasma
Vol. 10. KURTH—Introduction to Stellar Statistics
Vol. 11. CHALMERS—Atmospheric Electricity, 2nd edition
Vol. 12. RENNER—Current Algebras and their Applications
Vol. 13. FAIN and KHANIN—Quantum Electronics Vol. 1—Basic Theory
Vol. 14. FAIN and KHANIN—Quantum Electronics Vol. 2—Maser Amplifiers and Oscillators
Vol. 15. MARCH—Liquid Metals
Vol. 16. HORI—Spectral Properties of Disordered Chains and Lattices
Vol. 17. SAINT JAMES, SARMA and THOMAS—Type II Superconductivity
Vol. 18. MARGENAU and KESTNER—Theory of Intermolecular Forces
Vol. 19. JANCEL—Foundations of Classical and Quantum Statistical Mechanics
Vol. 20. TAKAHASHI—An Introduction to Field Quantization
Vol. 21. YVON—Correlations and Entropy in Classical Statistical Mechanics